Reviews for *War Without Oversight*

'Paddy Walker's War Without Oversight *is a must-read for military professionals and defence analysts. It dives into the ethical, operational and technological challenges of autonomous weapon systems (AWS). Walker emphasizes the need for human oversight to ensure compliance with international laws and ethical standards. Despite advancements in AWS, the human element remains crucial on the modern battlefield.*'

        Major General Simon Bernard, Deputy Commander.
        Canadian Joint Operations Command

'*An important examination of the myriad complexities and challenges involved in fielding practical lethal autonomous weapons systems that are so often glossed over or ignored by both novel technology evangelists and those campaigning to regulate and restrict lethal autonomy.*'

        Justin Bronk, SRF for Airpower and Technology at RUSI

'*This important book shows precisely why the principles and practices of international arms control and non-proliferation – too often considered to be mere relics of the Cold War – need urgently to be revisited and modernised. And with a view to the arrival on the battlefield of robotics, artificial intelligence and autonomous weapons systems,* War Without Oversight *makes a more fundamental point. If the organised use of violent force is to be ethically and legally constrained (and who would argue the opposite?) then humans must have oversight and must be responsible and accountable for decisions and actions taken in war.*'

        Professor Paul Cornish, Professor of Strategic Studies,
        and Director, University of Exeter

'*Autonomous weapon systems are highly likely to proliferate in the near future, introducing new types of weapon and ways of fighting onto the battlefield as they do. The narrative that is told on the way is already shaping the way they are perceived and assessed. Few if any indicate that they will be short of game changing. But, as Walker shows in* War Without Oversight *this outcome is far from certain. The book provides a sound technical grounding in the things that make autonomous weapons work, as well as the legal systems they will have to comply with, and some of the many frictions that all of this inherently builds into the design and development of truly autonomous systems. This technical grounding makes this book an indispensable read for anyone looking to understand this area more fully.'*

Samuel Cranny-Evans, *Calibre Defence News*

'Examining the technological, legal, statutory, moral and ethical concerns that surround the deployment of autonomous weapons, War Without Oversight *offers a rigorous critique of current proposals to transfer the authority to kill people to automated systems in contemporary theatres of war. Given the undoubted relevance and urgency of these concerns, and the import of Walker's research into why we should question the apparent viability and efficacy of such systems, the volume is an indispensable reference point for discussing the impact of automated warfare in the 21st Century and beyond.'*

Professor Anthony Downey, Birmingham City University

'Paddy Walker explores the implications of humans being technically capable of handing over the decision to kill to a computer. Don't turn a blind eye, read this book; discover what you must do to make future generations thankful you lived.'

Air Vice-Marshal (Ret'd) Michael Harwood CB CBE, former Head of British Defence Staff United States

'Paddy Walker yet again uses his shrewd analytical skills to delve into one of the thorniest challenges posed by modern warfare – how to manage advances in technology so as to maintain control on the battlefield. Walker's approach is both practical and necessary so as to ensure that human decision-making stays central to current and future wars. A must-read for all students of military history and current conflict!'

Karin von Hippel, former Director General, RUSI

'An excellent book that emphasises the critical need for meaningful command and control in lethal engagements to address operational risks, accountability, and the evolving norms of warfare. It fully strengthens my view that interoperability is the cornerstone of modern warfare, enabling seamless collaboration, efficient resource utilisation and rapid decision-making among allied forces. It enhances operational effectiveness, strengthens coalition partnerships and ensures adaptability to evolving threats, making it essential for success in contemporary conflicts.'

Michael Holm, Founder, Owner and Chairman of Systematic

'An insightful and sharply focused critique of the development of autonomous weapons that is grounded in an in-depth contextual analysis of the complex technical, political, operational, economic, ethical and legal issues involved. This very timely and necessary primer develops compelling arguments for maintaining human oversight and accountability in contemporary and future warfare.'
Shona Illingworth, Professor in Art, Film and Media, University of Kent

'Some believe the future of war will look like SkyNet or the Matrix. But it probably won't. Paddy Walker offers a precise analysis of the feasibility and desirability of autonomous weapon systems waging wars for us. Walker's conclusions may surprise readers.'
Professor Sean McFate, Georgetown University, US National Defense University, and author of *The New Rules of War*

'Walker considers the drivers towards autonomy in weapons systems and the diverse technical, legal and moral challenges that different forms of autonomy might present. He sounds a stark warning – piece by piece, autonomy is already happening, with human agency being ceded incrementally. Whilst recognising the complexity of the challenge, this book makes a detailed case for rules to keep humans on the battlefield.'
Richard Moyes, Director, Article 36

'War Without Oversight is a clear-eyed view of the complexity which the emerging autonomous capabilities of weapons systems will bring to the battlefield (and beyond) in the coming decade. While dealing with anticipated technical advances in the coming decade, Walker manages to be grounded in the practical, drawing on and extrapolating from recent and ongoing conflicts. Even those readers who have been following the debates over autonomous weapons closely will find much to consider here.'
Laura Nolan, International Committee for Robot Arms Control

'The allied military establishment is giving too little constructive thought to artificial intelligence and autonomous weapon systems. This thoughtful and well-researched analysis of one of the greatest challenges now facing allied forces should be required reading for all diplomatic and military policy makers and their advisors.'
Major General Bill Robins CB OBE, former Director General, UK Defence Information and Communication Services, Senior Research Fellow, RUSI

'War has always been a human endeavour yet much of the clamour today predicts a wholesale change in that due to the arrival of autonomous systems on the battlefield. The reality, however, is that speakers and authors who parrot these claims have detached themselves from the detail of what needs to happen in order for their rhetoric to be made real. Paddy Walker provides us with a corrective to that: educating and informing us about the challenges and hurdles that must be met in order to make autonomous systems work in war. The requirement for supervision and oversight is clear: Humans may, just may, become less evident in combat over the coming decades but their actions and decisions will remain critical to employing systems effectively. Read it.'

Professor Peter Roberts, former Head of Military Science, RUSI, and Founder, Aurelius Labs

'War Without Oversight is a unique and much-needed contribution to the debate on autonomous weapon systems. I have not come across a text that manages so well to combine fine-grained technical detail with historical context and philosophical depth. Walker excels in spotlighting the challenging ways human-machine interactions are transforming war today. A must-read for scholars, activists and policymakers alike.'

Elke Schwarz, Professor of Political Theory at Queen Mary University of London

'This deeply researched analysis successfully balances both the technical and conceptual in outlining the applications of autonomous weapons on the battlefield. Painting a granular picture from modern day Ukraine to historic US naval actions, Walker captures a vast and interconnected technological landscape within a highly readable and compelling framework. This book raises essential questions for military practitioners, policymakers and researchers alike.'

Emily Tripp, Executive Director, Airwars

'This is a vital contribution to the debate on autonomous weapon systems. In painting the wider context within which these systems would sit, Paddy provides a well-informed argument on the need for human agency in warfare. There are those that would present the unsupervised use of autonomous technology in war as a utopian ideal. I challenge them to read this book.'

Nicholas Valentine, Lead Engineer – Human Machine Teaming, Defence Equipment and Support, UK MoD

War Without Oversight
Why We Need Humans on the Battlefield

PADDY WALKER

*A Primer on Autonomous
Weapons and Challenges to
Their Deployment*

Howgate Publishing Limited

Copyright © 2025 Paddy Walker

First published in 2025 by
Howgate Publishing Limited
Station House
50 North Street
Havant
Hampshire
PO9 1QU
Email: info@howgatepublishing.com
Web: www.howgatepublishing.com

Writer royalties go to the Imperial War Museums Institute

All rights reserved.

No part of this publication may be reproduced, stored in a retrieval system, or transmitted in any form or by any means including photocopying, electronic, mechanical, recording or otherwise, without the prior permission of the rights holders, application for which must be made to the publisher.

British Library Cataloguing-in-Publication Data
A catalogue record for this book is available from the British Library

ISBN 978-1-912440-58-0 (hardback)
ISBN 978-1-912440-59-7 (paperback)
ISBN 978-1-912440-63-4 (EPUB)

Paddy Walker has asserted his right under the Copyright, Designs and Patents Act, 1988, to be identified as the author of this work.

The views expressed in this book are those of the individual author and do not necessarily reflect official policy or position.

CONTENTS

Dedication		vi
Thanks		vi
Abstract		vii
Abbreviations		ix
1	Introduction	1
	1.1 The book's structure	12
	1.2 Introduction to key concepts	17
	1.3 Timelines around capabilities	20
	1.4 Contextual drivers	25
	1.5 Introduction to AWS feasibility	30
	Section One: An Analysis of Structural Challenges to AWS	34
2	Context: The role of context in the removal of weapon supervision	35
	2.1 Warfare's continuum of methods	38
	2.2 The role of context in the wider AWS debate	39
	2.3 Context's norms	41
	2.4 Defence planning	44
	2.5 Context's human angle	45
	2.6 The role of situational awareness and uncertainty	51
3	Drivers: Factors accelerating the removal of weapon supervision	54
	3.1 Current practice	57
	3.2 Technology creep and dual-use technology trends	65
	3.3 Structural and procurement drivers	69
	3.4 Ethical drivers	75
	3.5 Operational drivers	82

4	Deployment; Models for the removal of weapon supervision	95
	4.1 Capabilities, roles and use cases	101
	4.2 Planning tools	105
	4.3 Machine and human teaming models	110
	4.4 Developing models for autonomous weapons	116
	4.5 Flexible autonomy	118
	4.6 Swarming model for AWS deployment	123
	4.7 Operations and causes of failure in AWS models	130
5	Obstacles: General challenges to the removal of weapons supervision	136
	5.1 The Geneva Conventions and the Law of Armed Conflict	138
	5.2 Proportionality and distinction in AWS deployment	143
	5.3 Accountability in AWS deployment	150
	5.4 Article 36 and LOAC-compliant weaponry	152
	5.5 Behavioural constraints to AWS deployment	156
	5.6 Proliferation constraints and other lessons from Ukraine	166
	5.7 Ethical constraints to AWS deployment	172

Section Two: An Analysis of Practical Challenges to AWS Deployment ... 178

6	Wetware: Design challenges to AWS function	179
	6.1 Computational methods, software and intelligence	186
	6.2 Architectural approaches to AWS deployment	189
	6.3 The Delivery Cohort	193
	6.4 AWS learning architecture	197
	6.5 Missing pieces	207
	6.6 AWS control methodologies	211
7	Firmware: Embedded process challenges to AWS function	215
	7.1 Sources of technical debt	216
	7.2 Firmware ramifications of learning methodologies	223
	7.3 Reasoning and cognition methodologies	227
	7.4 Attention methodologies in AWS	229

8	Software: Coding challenges to AWS function	233
	8.1 Coding methodologies	236
	8.2 Coding errors	245
	8.3 Utility function	247
	8.4 Software processing functions	248
	8.5 Anchoring and goal setting issues	252
	8.6 Value setting issues	256
	8.7 Action selection issues	257
	8.8 Behaviour setting and coordination	259
9	Hardware: Build challenges to AWS function	262
	9.1 Hardware and sensor fusion issues for AWS	265
	9.2 Configuration and calibration issues	270
	9.3 Validation and testing	272
10	Oversight: Command and control constraints to AWS deployment	276
	10.1 The notion of meaningful human control	280
11	Conclusion	290

Selected Bibliography   309
Index   325
About the Author   337

DEDICATION

Knox Eustace Crawley
Born on 23rd May 2024
who will presumably have to sort this stuff out…

THANKS

Lloyd Clark
Sarah Gumb
Kirstin Howgate
Madeline Koch
Olive Reekie
Peter Roberts
Nicky Valentine
Elle Walker

The Humanities Research Institute, University of Buckingham
The Institute for the Public Understanding of War and Conflict, The Imperial War Museum, London
The Royal United Services Institute

ABSTRACT

An assault by Ukraine's 13th National Guard Brigade on Christmas Eve 2024 might be the first recorded example of a robot-only combined-arms manoeuvre against an adversary. This book considers the scenario where these robots are then unsupervised, programmed to make their own kill decisions. Autonomous weapon systems (AWS) are defined as robotic weapons that have the ability to sense and act unilaterally depending on how they are programmed. Such human-out-of-the-loop platforms will be capable of selecting targets and delivering lethality without any human interaction. This weapon technology may still be in its infancy, but both semi-autonomous and other precursor systems are already in service. There are several drivers to a move from merely automatic weapons to fully autonomous weapons which are able to engage a target based solely upon algorithm-based decision-making. This requires material step-change in both hardware and software and, once deployed, such weapons posit a significant change in how humans wage war. But complex technical difficulties must first be overcome if this new independent and, in time, self-learning weapon category can legally be deployed on the battlefield. AWS also pose basic statutory, moral and ethical challenges.

This book digs into the manifest complexity involved in fielding a weapon that can operate without human oversight while still retaining value as a battlefield asset. The book's aim is to shine a light on the practical and technical feasibility of removing supervision from lethal engagements. The subject's importance is that several well-tried concepts that have long comprised battlecraft may no longer be fit for purpose. In particular, legal and other obstacles challenge such weapons remaining compliant under the Laws of Armed Conflict. Technical challenges, moreover, include the setting of weapon values and goals, the anchoring of the weapon's internal representations as well as management of its utility functions, its learning functions and other key operational routines. While the recent development pace in these technologies may appear extraordinary, fundamental fault lines endure. The book also notes the interdependent and highly coupled nature of the routines that underlies these weapons' operation, in particular ramifications arising from its machine learning spine, and, in so doing, demonstrate how detrimental these compromises are across AWS deployment models. In highlighting AWS deployment challenges, the analysis draws on broad primary and secondary sources to conclude that meaningful human control should be a statutory requirement in all lethal engagements.

ABBREVIATIONS

AAR	After Action Review
AAV	autonomous aerial vehicle
ACL	autonomous control level
AI	artificial intelligence
AGI	artificial general intelligence
ANN	artificial neural network
APS	active protection systems
ATR	autonomous target recognition
AUVSI	Association for Unmanned Vehicle Systems International
AWS	autonomous weapon system
CACE	change anything changes everything
CCA	collaborative combat aircraft
CCW	Convention on Certain Conventional Weapons
CEP	central error probable
COA	course of action
COG	centre of gravity
COTS	commercial off the shelf
DARPA	Defense Advanced Research Projects Agency
DL	deep learning
DNN	deep neural networks
DoD	US Department of Defense
DOF	degrees of freedom
FPV	first-person view
GAO	US Government Accountability Office
GDP	gross domestic product
GGE	Group of Governmental Experts
GPS	global positioning system
GPU	graphics processor unit
HMC	human machine collaboration
HMT	human machine teaming

HLMI	high-level machine intelligence
HRW	Human Rights Watch
HVEC	high efficiency video coding
ICJ	International Court of Justice
ICRC	International Committee of the Red Cross
IDF	Israeli Defense Forces
IHL	international humanitarian law
IHRL	international human rights law
LAWS	lethal autonomous weapon systems
LLM	large language model
LOAC	Law of Armed Conflict
MHC	meaningful human control
ML	machine learning
MMC	machine-machine collaboration
MVP	minimum viable product
NATO	North Atlantic Treaty Organization
NGO	non-governmental organisation
OODA	observe, orient, decide, act
PED	processing, exploitation and dissemination
RMA	revolution in military affairs
RUSI	Royal United Services Institute
SIPRI	Stockholm International Peace Research Institute
THeMIS	Tracked Hybrid Modular Infantry System
UAS	uncrewed aircraft system
UAV	uncrewed aerial vehicle
V&V	validation and verification
WBE	whole brain emulation
WYSIATI	what you see is all there is

1
Introduction

Machines have long served as instruments of war, but it has traditionally been the relevant commander who has decided how such weapons are employed. Evolution of technology, however, has the potential to change that reality and the purpose of this book is to analyse the widespread implications of this development. Your author started his academic study into this area twelve or so years ago. At that time, it was easy to dismiss those technologists making wild claims about the reach of battlefield weapons a decade hence. Did they not understand that removing the human from these processes was simply impossible, technically infeasible and that uncrewed platforms were really some imagined things born out of science fiction? Precision-guided weapons may have first appeared above Vietnam more than half a century ago, but any wholesale jump to independent weaponry must, at its foundation, be little more than some annoying fancy being suggested by those who wished to grab headlines.

And yet today, as the world is well into its third year of Russia's invasion of its neighbour, the use of remote and increasingly unsupervised weapons is escalating throughout Ukraine's front lines. They are small, cheap and effective. Their airframe has been adapted from readily available consumer materials and, once fitted out with explosives to attack and electronic means to see, they are radically changing how war is undertaken. These drones slip into tank turrets or dugouts[1] and, notes the *Economist*, they can loiter and pursue targets before initiating lethality. The phenomenon is clearly an evolving norm of warfare with systemic implications for operations and logistics, extending

---

1 Economist, 'Killer Drones Pioneered in Ukraine are the Weapons of the Future', *Economist* (8 February 2024), https://www.economist.com/leaders/2024/02/08/killer-drones-pioneered-in-ukraine-are-the-weapons-of-the-future?

from the front lines to what used to be understood as safe rear areas. Although the drone does not yet constitute a revolution in how war is undertaken, its modus operandi matters because it certainly signposts a destabilising shift towards small, cheap and disposable weapons, a new harnessing of widely available and consumer technology, as well as a pathway from automatic to autonomous weaponry. This, then, is the context of this book.

But while this high-level narrative would have seemed far-fetched when your author first sharpened his pencil on the subject in 2012, it is also the surprising simplicity that underlies both the family and use-types of emerging drones which has rapidly catalysed their deployment. Developed from racing quadcopters, drone deployment has seen costs slide while new use cases arrive almost daily as combatants tailor newly ubiquitous technologies into novel means of waging war. Focusing on these platforms' current drawbacks is probably falling into the same trap as the author did years ago when playing Cassandra on any deployment of uncrewed assets on the battlefield.

> *An evolving norm of warfare with systemic implications upon operations and logistics, from the front lines to what used to be understood as safe rear areas*

Nor are the battlefields of Ukraine an isolated example. In Yemen, Houthi rebels have transformed a long-dated balance of power in the Red Sea by embedding cheap Iranian guidance kits into new generations of inexpensive and often home-grown anti-ship missile drones. As again noted by the *Economist*, Iran has similarly shown how an assortment of long-range strike drones can have a geopolitical effect that far outweighs their cost.[2] Indeed, the broad phenomenon is being accelerated by exactly the global nature of what is playing out on our television screens. Recently routed government forces in Myanmar have logged rebels' use of volunteers and 3D printers to assemble componentry for their drones in small and dispersed workshops. Criminal groups are reportedly close behind the militias. Even if these parties develop expertise to overcome drone countermeasures, uncrewed platforms have already proved to be transformationally cheap, accessible and impactful. These are very consequential developments in a field where previous progress has been incremental and piecemeal.

---

[2] Economist Editorial, 'Killer Drones Pioneered in the Ukraine Are the Weapons of the Future', *Economist*, 8 February 2024, https://www.economist.com/leaders/2024/02/08/killer-drones-pioneered-in-ukraine-are-the-weapons-of-the-future.

How, then, has the landscape been changed by this decade of democratisation in how precision effects are delivered? Is the march of technology unavoidable and are these AWS technically feasible? Should we extrapolate from the last five years of technical advances to conclude that it is already upon us? Here, it soon becomes clear that a considerable technical gap exists between weapons that are automatic and those that can operate with real autonomy, functioning without human supervision. The context, after all, is that a degree of autonomy has existed in high-end munitions for years and in cruise missiles for decades. But Ukraine also demonstrates that change is afoot. The novelty here is that cheap, reliable software sat within cheap, reliable microchips can enable machine intelligence to govern millions of previously low-end munitions. Capabilities also improve in every product cycle, posing new problems of regulation and ethics but also, of course, of obsolescence and the frictions of integrating new munitions into legacy force design. Your author is a director of NGO Article 36, a charity whose name derives from the obligation enshrined in the Geneva Convention that current and future arsenals should be compliant under the Law of Armed Conflict (LOAC); procurement's febrile timelines bring into question whether proper resources can really be allocated for the testing of next generation weapons. It is difficult to think that new means of violence will be verifiable to the extent required by current conventions and frameworks.

> *The novelty here is that cheap, reliable software sat within cheap, reliable microchips can enable machine intelligence to govern millions of previously low-end munitions*

An incontrovertible takeaway is that innovation is pushing all manner of parties towards autonomy. It is doing this so that parties can out-speed their enemy, outlast their enemy and surprise their enemy with new capabilities while mitigating pinch points such as people, training, jamming and other unwelcome interventions. It is also to out-budget the enemy by de-skilling and de-costing these systems. The aim is to incorporate all necessary intelligence (which need not be all that much, but which will include the weapon's tasking, its navigation, target identification and processing, enough to enable an engagement sequence) on board each individual weapon system, thereby divorcing it from human oversight. It is against this quick moving background that this book seeks to consider the challenges and consequences that thus arise.

The use of artificial intelligence (AI) in defence is a high-stakes undertaking. For the purposes of this introduction, an autonomous weapon is an armament in which the identification and selection of targets and the initiation of violent force are carried out under machine control. Lethal capacities are thus delegated by the weapon system to its subcomponents in ways that preclude deliberative and accountable human intervention. The subject's standing is illustrated by the House of Lords' 2023 Public Inquiry into weapon systems that are based on AI where an autonomous weapon includes (but is not be limited to) five fundamental characteristics. The first is lethality and having sufficient payload that the weapon can be deadly. The second is autonomous function. The third is a lack of veto whereby, once started, there is no way for a human to terminate the system's engagement sequence. The fourth characteristic concerns indiscriminate effect whereby the weapon platform executes upon its task regardless of condition, scenario or target. The final trait is more nuanced and covers the weapon's 'evolution', its ability to learn autonomously and expand its functions and capabilities in ways that exceed human expectations. Interestingly, much of this definition is taken directly from China's own designation and classification of these weapon systems.[3]

Some defining of terms is in order. As noted by Krepinevich in *The Origins of Victory*, AI really refers to developments across digital technology that create processes which are capable of performing tasks previously thought to require human intelligence.[4] ML is then a sub field of AI which in turn refers to specific digital systems that are

AI is not any instant 'out of the box' solution but instead a toolbox of capabilities that continues to evolve

able to *improve* their performance on a given task and do this over time and through experience. The lynchpin here is the new abundancy of data on which the US *National Institute of Standards and Technology* (NIST) provides useful context; from the beginning of time up 2003, some five exabytes of data (1,00,000,000,000,000,000 bytes), the equivalent of fifteen thousand times all of the content in the US Library of Congress, had been created. Seven years later, noted Eric Schmidt, that same amount of data was being created every two days.[5] This is the discontinuity that underpins ML and, without which, we would not even be considering the subject to hand.

---

3 House of Lords, 'Proceed with Caution: Artificial Intelligence in Weapon Systems', AI in Weapon Systems Committee, Report of Session 2023-24, HL Paper 16, p. 7 (Box 4).
4 Krepinevich, Andrew, Origins of Victory, *Yale*, 2023, p.88.
5 Schmidt, Eric, 'Every Two Days We Create as much Information as we did up to 2003', *Techcrunch*, https://techcrunch.com/2010/08/04/schmidt-data/.

This does not equate, however, to there being good clarity on just *how* AI will transform this landscape. It remains a general and poorly understood assumption that AI will enable autonomy which, in turn, will then see the delegating of decisions over life and death to some undefined authorised entity and within undefined boundaries.[6] Furthermore, any system that remains tethered by prescriptive rules (and routines that also allow few deviations) should instead be considered to be little more than automated. To be properly autonomous, the weapon must be able to identify and then choose among different courses of actions (and do this independently) in order to accomplish goals that are then based on its knowledge and understanding of the environment in which it is functioning.[7] This is by no means a proven set of circumstances on the current battlefield.

> [AI can] find patterns that no unaided human mind would identify [but] it can also make mistakes that no human brain would fall for, a phenomenon known as 'artificial stupidity'

Various observations arise. First, AI is not any instant 'out of the box' solution but instead a *toolbox* of capabilities that continues to evolve and, as such, would appear to be most valuable as a *complementary* skill set to the expertise and judgement of the Delivery Cohort. Second, the Cohort must continually challenge the argument that too much data is always a condition that is preferable to not enough data. Although AI might be able to sort through masses of data faster than armies of human analysts and, as noted by Krepinevich, 'find patterns that no unaided human mind would identify', it can also make mistakes that no human brain would fall for, a phenomenon known as 'artificial stupidity'.[8]

It should also be expected that a half-clever enemy will be able to corrupt data sets being used to train the AWS; given the specificity of data that is relevant to AWS' environments, such sets will likely be widely shared and even easier for adversaries to disrupt; a worrisome development here might be and adversary feeding a rival AI false data, polluting the Cohort's weapon system and leading, in the case of a large number of autonomous

---

6 Garfinkel, Ben and Allan Defoe, 'Artificial Intelligence, Foresight, and the Offense-Defense Balance', *War on the Rocks*, 19 December 2019, https://warontherocks.com/2019/12/artificial-intelligence-foresight-and-the-offense-defense-balance/.
7 Krepinevich, p.89.
8 Friedberg, Sydney, 'Big Bad Data: Achilles Heel of Artificial Intelligence', *Breaking News*, 13 November 2018, cited Krepinevich, p.104.

assets, either to simultaneous friendly force failure at scale or, notes the *Centre for AI and Digital Policy*, to AWS becoming weapons of mass destruction due to their scalable lethality.⁹ An adjunct risk would then be friendly autonomous assets launching unsupervised attacks that cross an enemy's 'red lines' and so triggering unintended escalation, instances of agent 'misbehavior' being difficult to identify and difficult to adjust.

Why then is it still so difficult to define the product set that occupies the coming chapters? What actually comprises the AWS? For the purposes of this book, such systems can be described schematically as a weapon with sensors, algorithms and effectors based upon robotic components that can be either stationery or mobile. Data collected by these sensors is processed computationally to enable independent detection, tracking and classification of objects. Target recognition and discrimination can then be achieved by comparing this sensed data remotely or, more likely in a communications-denied environment, with target types contained in that weapon's database or perception library. Finally, the system incorporates a weapon to engage selected targets.¹⁰ The processes and hardware that together constitute this broad platform are the subject of the book.

> *The subject is important precisely because autonomous weapons are the manifestation of handing over the decision to kill to a computer*

Why is this subject so important? First, autonomous weapons are the manifestation of handing over the decision to kill to a computer. Modern Western militaries reflect a fundamental principle of Western thought that

---

9 Centre for AI and Digital Policy, 'CAIDP Statement to the Meeting of High Contracting Parties to the Convention on Conventional Weapons', 15 November 2024, https://www.caidp.org/statements/. The research organisation's concerns centre around these weapons' unpredictability and lack of control, their exponential lethality and the inability of autonomous decision-making appropriately to factor for ethical and legal constraints. Its recommendations are for an immediate moratorium on AWS development until comprehensive regulations have been established, the classification of the weapon class as WMDs, the banning of non-compliant AI systems that cannot adhere to current laws, implementing a standardised form of reporting that permits independent oversight of AI in military operations and, finally, the appointment of a UN special rapporteur on AI and human rights.

10 This definition accords with that adopted by UN Special Rapporteur on Extrajudicial, Summary or Arbitrary Executions. See: Heyns, Christof, 'UN Document A/HRC/23/47, Report of the Special Rapporteur on Extrajudicial, Summary or Arbitrary Executions, United Nations', Human Rights Council, 23rd Session, Agenda item 3 (27 May 2013), https://www.ohchr.org/Documents/HRBodies/HRCouncil/RegularSession/Session23/A.HRC.23.47.Add.5_ENG.pdf. AWS is here defined as a 'robotic weapon system that once activated, can select and engage targets without further intervention by a human operator'.

every human being is valuable, and militaries will do everything possible to drive down risk to their individual soldiers. Second, AWS posit becoming the 'third revolution in warfare'. Once developed, the received wisdom is that autonomous weapons will permit armed conflict to be fought at distance and with precision, at disruptive scale and to a super-fast timetable. The past two decades, moreover, have seen a broad range of technological advances that have made this a practicable possibility and, in so doing, deployment of these systems will render obsolete several precepts of current battlecraft. And the removal of human oversight from targeting sequences creates a raft of consequences, both directly and as second order effects. It complicates battlefield command, battlefield control and current concepts of leadership, which may abruptly no longer be fit for purpose. It exacerbates the already stark asymmetries of so-called riskless warfare in which hazards are passed across to civilian populations in an opponent's territory.[11] Indeed, current debate on AWS is heavily influenced by an apparent certainty that unsupervised weapons are an inevitability and will work seamlessly. The matter's importance is also heightened by the absence of any international agency authorised to test such weapons, to regulate their use and to foster global protection of civilian interests.

There are, of course, several versos to this picture. Isolating AWS attributes is useful to demonstrate the difficulties involved in jumping from mere automaticity to autonomy given the significant barriers that remain before human supervision can be removed from weapon operation. These barriers are technical but also behavioural, legal, operational and ethical. Discrimination, proportionality and necessity, for instance, are all underlying legal requirements that will be reviewed later in the book but factor large when considering commanders' responsibilities in achieving goals. As an example, target discrimination in AWS concerns a system's ability to identify and differentiate between combatants who are legitimate military targets and those who are non-combatants, civilians or sheltered persons, all of whom are specifically protected under international humanitarian law (IHL). In this vein, one purpose of this book is also to review the degree of challenge in the nuanced, often subjective processes that rely upon complex mathematics

---

11 Kalmanovitz, Pablo, 'Judgement, Liability and the Risks of Riskless Warfare', cit. Nehal Bhuta and others, eds., *Autonomous Weapons Systems: Law, Ethics, Policy* (Cambridge University Press, 2016), p. 158. See also Anderson, Kenneth and others, 'Adapting the Law of Armed Conflict to Autonomous Weapon Systems', *International Law Studies*, 90, 386 (Stockton Center for the Study of International Law, 2014), pp. 389-390, http://www.dtic.mil/dtic/tr/fulltext/u2/a613290.pdf.

as the only available input to their complex decision-making. It is also to highlight sources of uncertainty that might affect this whole process and, later in the book, to suggest guardrails to the matter of supervision across lethal engagements.

As part this introduction, it is useful to provide the reader with some scene-setting. First, the list of required hardware (a combination of sensors, processors and actuators) is not particularly demanding. Ultrasonic sensors, odometry sensors and radar sensors are all well understood and readily obtainable. Motion cameras, the Global Positioning System (GPS), lidar and radar have long been available to facilitate capturing the data necessary for an autonomous car to understand its environment and, in the case of an AWS, that of its potential targets. What, however, involves hitherto unseen complexity is the intricate choreography of software tasks that must be undertaken sequentially and concurrently (and in wholly adversarial circumstances), and it is these that occupy much of this book's later technical sections.[12] Sensor fusion techniques, for instance, involve the processing of information from multiple sources to create comprehensive situational awareness for each weapon platform. Algorithms must recognise shapes and behaviours in ways that imitate the human brain in distinguishing the relevant from the irrelevant, the legal from the illegal. Wholly new combinations of capabilities (be they pattern recognition, inference and prediction, biometry or machine learning) must then be present if the platform is really to toggle between live feeds, archived intelligence, relevant rules of engagement and LOAC before deciding upon whether to engage a target with lethal force. All this should occur within an ecosystem based upon each weapon's ongoing improvement, with system 'learning' (in the widest sense of the word as it remains unclear such a capability will ever practically be possible) from prior operations to update algorithms to incorporate feedback and so enhance performance and minimise errors. This is the new world of unsupervised armaments.

Our scene setting does not finish here. As currently posited, the technical spine of the unsupervised weapon requires extensive (and precisely relevant) training data to get it ready to do its job effectively, predictably and safely. This entails collecting, labelling and maintaining very large data sets that

---

[12] The book is divided into two equal sections. The first (Chapters One to Five) consider the issue's 'softer' behavioural and legal challenges, while the second section (Chapters Six to Ten) look at the challenges for parties to deliver upon the matter's technical requirements. This second section, looking at the likely routines that will allow AWS operation, comprises the book's 'practical' analysis. Calling it a 'technical' analysis would be inaccurate as the technologies here are not yet set, are developing pell mell and anyway benefit from behavioural consideration in order to frame the challenges arising.

must cover all possible combat scenarios that the weapon may encounter. This must also be undertaken in a way that coordinates the weapon's training data with those likely circumstances. It is not by accident, therefore, that the fragility of this technical habitat is a recurring theme of this book. After all, exactly applicable data sets must by definition be limited; one target engagement is very different from another, the arising data is messy and very rarely recorded in any suitable sense. Furthermore, current machine learning models likely as not perpetuate all the biases that present themselves in both this training data and the execution steps that must all be coded by human individuals and which alone comprise these weapons' decision processes.

Other challenges are similarly systemic. Abstraction (here, the ability to isolate individual parameters, different values and ideations) is properly difficult. Without being able to ascribe reliable boundaries to each 'happening' on the battlefield, the notion that predictable interpretation can be done without human involvement quickly falls apart. Humans innately undertake evaluation.

> *The intention is that the weapon system learns from prior operations and incorporates feedback in order to enhance performance and minimise errors. This is the new world of the unsupervised armament*

They readily appreciate delineation and learn from these outputs. For this reason, the book's later chapters focus upon coding challenges in moments of ambiguity where those very happenings are vague, inconclusive and unrecognisable (likely given the chaos of the battlefield) and can each be framed in multiple ways depending upon the context used. That the local commander then struggles to understand why an artificial agent has chosen a particular course of action also scotches both the precept of accountability and the compliance of that engagement.[13] The nub here is that that commander should no more expect an unsupervised machine to transfer knowledge from one context to another than from one independent machine to a colleague machine. These are key weaknesses if weapons are to move from being human operated to conditionally autonomous.

Two sets of tools are useful for readers to navigate this space. First is a framework of questions to test the uncertainties that are thrown up along the

---

13 The book's illustrations demonstrate the point. With the exception of the hand-drawn frontispiece, all the sketches have been produced by ChatGPT's Dall-E programme. The point here is that their generation is inexplicable and cannot be repeated. Using the exact same language to request an image returns unaccountably different sketches for each iteration. The more detailed the user's request, the more different the program's output.

book's path. After all, it has long been the human combatant making the split determinations that will now be fulfilled by algorithms should the human no longer be in that loop. In this vein, therefore, consider how a machine program might break down the significant happenings that would trigger those human's processes. Remember too the consequences of a bad decision, either a tragic mistake or a tragic omission. Can the often random-walk of a battlefield really be processed digitally into relevant sequences for onward decision-making, and what guardrails are required to ensure that all such happenings have been considered? The reader must be satisfied that these outputs are tested and mediated to safeguard accuracy and relevance. A corollary is to reckon upon how this new means might change the way that humans wage war.

Assuming for a moment that procurement executives have squared away these technical challenges, the reader's second tool kit should then examine AWS' likely operational challenges. Can parties really shoehorn these technological developments into existing engagement rules in ways that still safeguard compliance with legal frameworks? International criminal law holds that commanders who fail to avoid non-negligible risks to civilians and other protected persons can be liable for negligence or recklessness. Similarly, how can these machines factor for all the contextual components that underpin battlefield processes such as command, politics and social vectors that make up an engagement? Finally to this point, defence planning (and, indeed, this book's analysis) involves more than usual guesswork. After all, it is absolutely not clear how such weaponry will be deployed on the battlefield. Readers should remember that ambiguity is everywhere, whether across technical advances, priorities or resources, whether across the deployment of autonomous decision aids as seen in Israel's Gaza 2024 campaign or, in time, whole weapon systems where multiple components have independent agency.[14] Remember also that

> *Can the often random-walk of a battlefield really be processed by a machine into manageable sequences for onward decision-making?*

---

14 Groom, Geoffrey, 'Artificial Intelligence Applications for Automated Battle Management Aids in Future Miliary Endeavours', Naval Post Graduate School USA, thesis, June 2019, https://apps.dtic.mil/sti/pdfs/AD1080249.pdf. A decision aid here relates to a 'colleague' weapon that provides varying levels of digital assistance. This book primarily considers the deployment of weapons autonomy from the perspective of a sophisticated, resourced state and, generally, one of the one hundred and ninety-six signatories to the Geneva Conventions. This assumption is

defence planning will become orders of magnitude more challenging once humans are not around.

A further consideration is whether weapon independence constitutes a Revolution in Military Affairs or, failing that, some slightly more nuanced discontinuity in how war will be waged. AWS, after all, would seem to fit the bill: foundational, irreversible changes in battlefield processes brought about by novel, unsupervised processes. But this relationship is complicated by both the incremental and cumulative nature of these challenges and, often, by these challenges' unanticipated secondary effects. Agency, after all, is being ceded incrementally. Although humans currently retain primary control across the whole engagement sequence (the notion of humans-in-the-loop with the human operator at least approving or vetoing recommendations from the weapon system), the likely pathway is that human involvement will be diluted in stages as technology substitutes processes hitherto overseen by humans (the notion, then, of humans-on-the-loop and, eventually, humans-out-of-the-loop). While it may not seem beyond human capacity to retain final sanction over these systems, it is the degree and workings of this sanction that concerns this book.

This 'Delivery Cohort' is a recurring device for this book, responsible for deciding processes, delivering practices, tuning new skills and refining command in order to remove human supervision from lethal engagement

At its highest level of autonomy, the weapon system is operating independently with humans playing a minimal or no role in decision-making as the mission begins. And while readers juggle with the degree to which AWS deployment upends current practices, they must also factor for the role of these weapons' 'Delivery Cohort' who, together, comprise all the many constituencies and parties deploying unsupervised weapons and are detailed in the accompanying footnote. This Cohort is therefore a recurring device throughout this book as the collection of humans responsible for deciding processes, delivering practices, tuning new skills and refining command in order to remove supervision from lethal engagement.[15]

---

valid as the Conventions confer an obligation to comply with, inter alia, the Laws of Armed Combat and other responsibilities and commitments set out in Chapter Five (Obstacles).

15 For the purposes of this book, the term 'Delivery Cohort' is used as a device to convey the parties involved in delivering the deployment of AWS and will include, inter alia, the following taskings: neurophysiologists to coordinate AWS networks; psychologists to coordinate learning and cognition; biologists for adaption strategies; engineers for control routines; logisticians; roboticists; electrical specialists; behaviorists; politicians, civil servants and diplomats; people

1.1 The book's structure

In December 2016, the Fifth Review Conference of the Convention on Certain Conventional Weapons (CCW)[16] agreed to formalise discussions that had first begun in 2013. Its terms were to 'explore and agree on possible recommendations on options related to emerging technologies in the area of lethal autonomous weapons systems, in the context of the objectives and purposes of the Convention, taking into account all proposals – past, present and future'.[17] Eight years later, the CCW has now hosted its twelfth such Review Conference. The structure of this book is therefore driven in part by what has already been written on the subject but more by the pace of change across the subject's environment and relevance. In this vein, the premise of removing supervision is examined through the prism of recently published research, first considering behavioural constraints and then, in the book's second section, on the nuts-and-bolts architecture of AI. An important assumption here remains that the core methodology of AI has not changed materially over the past quarter century.[18] The backbone for AI agents is similar today to years past. This may

---

working in non-governmental organisations; sociologists; lawyers; company directors; weaponists; military tacticians; manufacturers; professionals involved in miniaturisation, simulation, configuration, coding, power supply and modularity; specialists in sensors, in distributed and decentralised routines; ethicists; data scientists and data engineers; and specialists in tooling and calibration.

16 The CCW is the United Nations body that is tasked to 'prohibit or restrict further the use of certain conventional weapons in order to promote disarmament' and the 'codification and progressive development of the rules of international law applicable in armed conflict'. See: UN Libraries, 'Preamble, 1980 Convention on Prohibitions on the Use of Certain Conventional Weapons Which May Be Deemed Excessively Injurious or to Have Indiscriminate Effects', United Nations Treaty Collections, 22495 (2 December 1983), https://treaties.un.org/doc/Treaties/1983/12/19831202%2001-19%20AM/XXVI-2-revised.pdf.

17 The wording is taken from the 'Final Report of the 2016 Informal Meeting of Experts on Lethal Autonomous Weapons Systems, Geneva, UN Documents Publishing (10 June 2016), https://www.unog.ch/80256EDD006B8954/(httpAssets)/DDC13B243BA863E6C1257FDB00380A88/$file/ReportLAWS_2016_AdvancedVersion.pdf. Convention here relates to the Geneva Convention. A detailed analysis on international humanitarian law, international human rights law, the tipping point between these two frameworks and their role in the LOAC can be found in Chapter Five (Obstacles), specifically 5.1 (Geneva Conventions and Laws of Armed Combat).

18 Hammond, K, 'Why Artificial Intelligence Is Succeeding: Then and Now', *Computerworld, Artificial intelligence Today and Tomorrow* (2015), para. 4, http://www.computerworld.com/article/2982482/emerging-technology/why-artificial-intelligence-is-succeeding-then-and-now.html. See also: Guszcza J and N Maddirala, 'Minds and Machines: The Art of Forecasting in the

appear counterintuitive but recent developments in AI, albeit narrow AI[19], have occurred not because any methodological discontinuity has surfaced to disrupt practices but because of new computational capacity, a disruptive expansion in the available volume of applicable data upon which these agents can be trained, as well as the availability of wholly novel processing speed.

Although the central premise behind 'network training' has existed for more than thirty years, early AI efforts were based on a relatively tiny universe of just a few thousand examples being applied to poorly differentiated problem sets. Today this same technique remains the default but now involves billions of examples and runs on machines with bespoke chip sets that allow them to learn much faster from these iterations. The same dynamic holds true for how an autonomous weapon might operate, although considered impossible as recently as a decade ago. Commentators now regularly debate their feasibility, fuelling a general 'revolution of expectation' based upon this new marriage of massive datasets and processing speeds that enables extraction of meaningful signal.

A goal for this book is then to provide equilibrium to the AWS debate, by explaining the subject's several versos, literally the left-hand side of a manuscript leaf that is meant to be read second but often articulates the subject's obverse or counter factual. Progress in the development of AI has been anything but linear, its recent past being 'full of hype and disappointment'.[20] Bubbles in this space have involved hype about expert systems, neural networks, and hard and fuzzy logic models as well as a dependence upon (and then subsequent relegation of) complex statistics that enable machine reasoning. This book's technical analysis therefore relies on a broad historiography of research to identify both best practice but also chokepoints in AI as they might relate to unsupervised

> *An important assumption here remains that the core methodology of AI has not changed materially over the past quarter century*

Age of Artificial Intelligence', Deloitte University Press, *Deloitte Review*, 19 (2016), paras. 3-4 of 27, https://dupress.deloitte.com/dup-us-en/deloitte-review/issue-19/art-of-forecasting-human-in-the-loop-machine-learning.html.
19 AI relates here to non-sentient computer sequences that are focused on one narrow task (or a combination of several such narrow tasks) enhanced by access to massive training data sets. Discussion on the interface between AI and artificial general intelligence is set out in Chapter Six (Wetware), specifically: 6.1 (Software versus intelligence).
20 Brooks, Rodney, 'The Seven Deadly Sins of AI Prediction', *MIT Technology Review* (6 October 2017), paras. 3-4 and generally, https://www.technologyreview.com/s/609048/the-seven-deadly-sins-of-ai-predictions/.

violence. These sources are global. They come from diverse academic institutions, defence establishments, practitioners and third sector parties. This second historiography is, furthermore, typically less than five years old and reflects current empirical work rather than what is theoretically possible in the science.

Examination of AI capabilities as they relate to this battlefield context unsurprisingly lags, as negligible historiography really exists for the exercise. Research may be available on autonomy and its accompanying technologies, but there has not been much effort to knit them together with issues relating to human supervision in the uncertain, changeable world of the conflict realm. For the purposes of this introduction, identifying high-level use cases provides useful context. First, the independent weapon will require intricate routines to administer onboard goals and values. Second, it will require dynamically managed utility functions, namely the set of mathematical routines that rank alternative courses of action according to their worth to the weapon system (and determined solely by its programming).[21] Third, all of this requires an anchoring mechanism to ensure the platform learns correctly and that its subsequent actions do not stray inappropriately from its intended purpose. All of these capabilities may have theoretical foundations when studied on a university workbench but involve magnitudes of complexity if they are to be deployed together (usefully and legally) in a real battlefield setting within a weapon that is making decisions without human intervention.

> *These capabilities may have theoretical foundations when studied on a university workbench but involve magnitudes of complexity if they are to be deployed together (usefully and legally) in a real battlefield setting*

The issue then becomes whether the removal of that supervision is *undoubtedly* technically feasible. This is not an idle question given the

---

21 For a discussion on these topics, see Chapters Six, Eight and Nine. Chapter Six (Wetware) isolates complexities arising from likely fundamental architectures in autonomous weapons. Chapter Eight (Software) identifies specific fault lines that might result from particular operational routines likely to underpin weapons-directing AI. Chapter Nine (Hardware) then highlights difficulties stemming from the physical properties of such systems. The point of these three chapters is to demonstrate the cumulative technical complexity that underlies AWS deployment. Taken together, the book's review of coding and architecture pinpoints discrepancies that exist between capabilities that are feasible and tasks that are expected within these platforms' function.

backdrop here of breakneck development across the technical competencies required by the Delivery Cohort to realise this aim. An obvious danger for a book on this subject is being blindsided by the unforeseen, allowing practitioners in this field to posit very credible scenarios based on mashups of the latest technologies.[22] Technical progress, however, tends to be chaotic with technologies evolving incrementally in stops and starts rather than arriving fully formed. After all, a 'certain' dead-end today can be tomorrow's ubiquitous breakthrough solution. Nor is it necessarily clear which grouping of technologies might overturn how battlecraft is undertaken. This is an important observation, the more so given that an autonomous weapon need not to be very intelligent and, notes Laura Nolan, is empirically unlikely to be capable of real learning over the medium term. It is simply a weapon that selects and engages targets on its own. A third block of required analysis comprises arguments against the introduction of autonomy in weapon systems. This concerns existing supranational LOAC and, on a more granular level, corresponding Rules of Engagement. It includes consideration of moral, ethical, and economic arguments against the withdrawal of meaningful human control (MHC) across engagements including the key role of context and situational awareness in the mix.

Of course, most of these challenges themselves have versos, and the book must then consider arguments that together advocate *for* weapon autonomy, issues such as political dividends, procurement advantages, operational benefits and rewards related to dual-use technology and basic commercial interest that act as an accelerant to fielding these new means. For example, AWS likely provide commanders with force multiplication. They may remove soldiers from harm's way. In this vein, the book's fourth historiographical section reviews where machine control might lead to better ethical performance in combat situations, where machines might make battlefields safer for humans, especially civilians. To conclude, the book's final segment reviews possible solutions in this AWS debate and the matter of MHC.

All of this must be judged against the book's central question on feasibility. How intractable are AWS' technical, operational and contextual

---

22 Oberhaus, Daniel, 'Watch "Slaughterbots": A Warning about the Future of Killer Robots', *Motherboard* (2017), https://motherboard.vice.com/en_us/article/9kqmy5/slaughterbots-autonomous-weapons-future-of-life. Professor Stuart Russell, director of computer science at University of California, Berkeley, concludes that 'this is not speculation. It is the result of integrating and miniaturizing technologies that we already have'. Russell's *Slaughterbots* provided a useful baseline at the time of this book's original conception and informs much of its contextual analysis.

impediments? What might interim and transitional autonomous systems look like? How will states adopt such weapons and how will they be fitted into battlecraft? A final line of inquiry then asks whether a role exists (and the shape of that role) for MHC to become an overarching control mechanism and, conceivably, a statutory umbrella for the deployment of unsupervised weapons. This is doubly relevant in the case of what is a new means of warfare that has yet to be deployed. This may appear uncomfortable given that computing was little more than a fringe activity less than two generations ago, prompting, therefore, a predominantly humanities-based analysis of the emerging weapon class. The analysis deliberately focuses upon concepts rather than, say, specifics of AWS coding. It is a conceptual review of these determinants and their consequences.

Clearly, there are degrees of weapon autonomy, a continuum that underpins the discussion in later chapters. The notion, however, has already driven several of the UK Ministry of Defence's Joint Concept notes on Human-Machine Teaming where 'future force design must find the optimal mix of manned and unmanned platforms, and balance employment of human and machine cognition for various tasks'.[23] The challenge, however, is that the idealised 'centaur model' of seamless human-machine teaming breaks down the moment that its processes require inputs faster than the human operator can provide. These half-way models also rely on sufficient

> *The AWS must understand its status relative to its mission, grasp what is important and then decide upon an action sequence (including, presumably, on when and how to be lethal), and to do this independent of external guidance*

(and empirically unlikely) means of communication being in place between machine and the human; the machine, after all, is operating at some part of this continuum on its own. At the very least, it must understand its status relative to its mission, grasp what is important and then decide upon an action sequence (including, presumably, on when and how to be lethal), and to do this independent of external guidance.

Two final elements of context are relevant to this preamble. First, the book's analysis is limited to the current era of narrow AI. Should artificial

---

23 UK Ministry of Defence, 'Human-Machine Teaming', Joint Concept Note 1/18 (2018), p. 44, https://assets.publishing.service.gov.uk/media/5b02f398e5274a0d7fa9a7c0/20180517-concepts_uk_human_machine_teaming_jcn_1_18.pdf.

general intelligence (AGI) and machine sentience come to pass, then many of the deductions of this book will likely no longer be valid. Second (and in recognition of the pace of innovation in the space), the author needs to set an end-date for the book's study: the middle of 2024. Developments and advances after this date must wait for a second edition. To isolate key precepts, the book's focus is upon whole-weapon independence. Unless stated, the analysis usually assumes deployment of sophisticated and wide-task AWS that are intended to operate in isolation.

One further piece of context is useful to frame the book. It builds upon and updates the author's earlier PhD dissertation, *Challenges to the Deployment of Autonomous Weapon Systems*.[24] That work included some two thousand footnotes which have generally been removed here in order not to overwhelm the narrative. Readers looking for more detail on the book's original source material (in particular for the technical analysis of the book's later chapters) should access the 2019 dissertation as set out in the footnote below. This book, moreover, stays close to that paper's original structure. It also sets out some of these more relevant sources in its bibliography that follows Chapter Eleven.

1.2 Introduction to key concepts

This introduction now undertakes two tasks. It provides an overview to the argument's central concepts in order to ground the reader. It also provides an appraisal of important themes that comprise the work's individual chapter headings to signpost the book's overall structure. A key conjecture of the book is that the right of combatants to choose their means and methods of warfare[25] is not unlimited.[26] This highlights a key principle but also exposes a potential

---

[24] Walker, Paddy, 'Challenges to the Deployment of Autonomous Weapons Systems', University of Buckingham (2019). The dissertation is available for download on various platforms including Academia.edu (see https://www.academia.edu/74687808/Challenges_to_the_deployment_of_autonomous_weapons). An abbreviated version of the same appeared in *Royal United Services Institute Journal* (July 2020), see https://www.tandfonline.com/doi/epdf/10.1080/03071847.2021.1915702.

[25] 'The Protocol Additional to the Geneva Conventions' (August 1947) and relating to 'The Protection of Victims of International Armed Conflicts (Protocol I)' (8 June 1977), hereinafter referred to as 'Additional Protocol I'; the document refers alternately to 'methods and/or means of warfare', 'means and methods of attack' and 'weapon, means or method of warfare', https://ihl-databases.icrc.org/ihl/INTRO/470.

[26] This principle is variously stipulated in, for instance, 'Article 22' of the Hague Regulations (1907); see: 'Respecting the Laws and Customs of War on Land' and 'Article 35(1)' of Additional Protocol I, https://ihl-databases.icrc.org/ihl/INTRO/470.

fault line. An overarching assumption is that humans should exercise control over combat operations and also, crucially, over individual attacks.[27] And, although this is of course the basic tenet of IHL, a central pillar of LOAC, together the laws of war that underpins much of this book's analysis[28], it does not reflect the matter's loud verso about the degree to which parties actually adhere to these conventions. Readers must all the time question how much these 'rules' really factor in the decision-making and actions of, for instance, autocracies, non-state parties and non-peer adversaries.[29]

The issue demonstrates the topic's maze of contradictions. Replacing humans with machines might reduce casualties and costs, but conceivably lowers the threshold for going to battle in the first place and reduces subsequent risks of an arms race in new technologies. It is not an accident that the Future of Life Institute conflates how humans resolve conflict with the society in which those humans live, the consequence being that

> *All of a sudden, the decision of only a very few people (those in a position to authorise AWS deployment) may be the tipping point for a party to drift into conflict*

if machines are making decisions about who and when to kill it clearly changes this equation.[30] It is also the availability of independent weaponry to any political constituency that has broad constitutional ramifications. All of a sudden, the decision of only a very few people (those in a position to authorise AWS deployment) may be the tipping point for a party to drift into conflict.[31] Unlike nuclear devices, after all, AWS technology requires no specialised or hard-to-obtain materials.

---

27 As defined in US Department of Defense, 'DoD Dictionary of Military and Associated Terms' (2017).
28 International Committee of the Red Cross (ICRC), 'A Guide to the Legal Review of New Weapons, Means and Methods of Warfare: Measures to Implement Article 36 of Additional Protocol I of 1977', *International Review of the Red Cross* , 88, 864 (2006), p. 931, https://international-review.icrc.org/articles/guide-legal-review-new-weapons-means-and-methods-warfare-measures-implement-article-36.
29 Walker, Paddy and Peter Roberts, *War's Changed Landscape: A Primer on the Forms and Norms of Conflict* (Howgate Publishing, 2023), pp. 110 and 225.
30 Future of Life Institute, 'Autonomous Weapons: An Open Letter from AI and Robotics Researchers', Future of Life (2015), https://futureoflife.org/open-letter-autonomous-weapons/. The difficulty here, after all, is that AWS deployment becomes normalised for certain grey zones and deniable tasks such assassinations, destabilising neighbouring states and other polities, subduing particular populations or even selectively killing a particular ethnic group.
31 Suarez, Daniel, 'The Kill Decision Shouldn't Belong to a Robot', Ted.com (2013) https://www.ted.com/talks/daniel_suarez_the_kill_decision_shouldn_t_belong_to_a_robot.

It is also useful then to review certain terms that will appear throughout this book. Autonomous weapons have long been divided by the US Department of Defense (DoD) into three subtypes, as noted by NGO Human Rights Watch (HRW) in its first November 2012 report, *Losing Humanity*, according to the amount of human involvement.[32] Human-in-the-loop weapons include robots that can select targets but only deliver force with a human command.[33] Human-on-the-loop weapons can select targets and deliver force under the oversight of a human operator who can override the robot's actions. HRW's third definition covers fully autonomous robots that can select targets and deliver force without any human input or interaction. This book concerns this third category of autonomous, human-out-of-the-loop weapon systems, although, as evidenced below, the original heuristic is no longer wholly helpful as it does little to identify elements of the argument about what is generally risky, dangerous or prone to unintended consequences.

It is important to understand the difference between machines that are autonomous (machines that, to varying degrees, are self-learning and therefore 'evolving', the focus of this book), automated (machines that are nevertheless complex, deterministic and rules based but are not able to learn through assimilated feedback) and merely automatic machines that are simply based on programmed thresholds. It is the level of human control and intervention (and, conversely, the degree of machine freedom) that is the

---

32 Docherty, Bonnie, 'Losing Humanity – The Case Against Killer Robots', Human Rights Watch (2012) http://www.hrw.org/reports/2012/11/19/losing-humanity-0, p. 2. Both the CCW and UN states-parties were slow to voice concerns over possible use of fully autonomous weapons outside of armed conflict (for example, policing and law enforcement and border control). HRW subsequently investigated concerns in this area in its May 2014 'Shaking the Foundations' report.
33 These definitions have been developed by HRW. Its paper, *Killer Robots* published in 2012, proved to be an early alert on the dangers posed by AWS and is often cited in this book. For a discussion and timeline on reports on AWS, see: HRW and Harvard Law School International Human Rights Law Clinic, 'Reviewing the Record: Reports on Killer Robots from Human Rights Watch and Harvard Law School International Human Rights Law Clinic (2018), generally http://hrp.law.harvard.edu/wp-content/uploads/2018/08/Killer_Robots_Handout.pdf. The extent of NGO advocacy against AWS deployment is best evidenced by the broad composition of civil society organisations including coalitions such as the Campaign to Stop Killer Robots (incorporating, inter alia, HRW, Amnesty International, the Association for Aid and Relief Japan, the International Committee for Robot Arms Control, Mines Action Canada, Pugwash Conferences on Science and World Affairs, PAX Netherlands, the Nobel Women's Initiative, Article 36, the Latin America Human Security Network (SEHLAC), and the Women's League for Peace and Freedom). An analysis of civil society's early AWS debate is provided by University of Sheffield's Research Excellent Framework: Impact Case Study (Ref 3B), 'Shaping International Policy and Stimulating International Public Debate of Autonomous Weapon Systems', Sheffield University (2014).

key distinction for this book. It is also the whereabouts in a weapon system where autonomous function and agency is to be found. Simply referring to autonomy as a general weapon attribute is far too imprecise. Instead, it is the *task* nature that is being undertaken autonomously (at subsystem or function level) that matters to this analysis. Several autonomous capabilities, moreover, such as navigation, visual cueing and system-level reporting may of course be quite uncontentious, while others, such as targeting, engagement and recurrent action, may present enduring difficulties.

Three other classifications require discussion. While semi-autonomy involves human oversight in target selection, supervised-autonomy involves machine-selected targets with humans subsequently confirming any lethal engagement. Here, it is human and communication shortcomings that might lead to full weapon autonomy whereby the machine is selecting and engaging targets without human supervision.

Readers should also understand the difference

> *It is this unavoidable indivisibility of the human (as programmer, enabler, coordinator, user and victim) and lethal machine that is so often lost when considering these systems' deployment*

between 'capability-independence', the weapon's ability to accomplish a task, and 'organisational independence', the ability of the weapon to achieve that task within a wider battleplan within the components that make up that battleplan's sociotechnical infrastructure (mission priorities, other assets, constraints and context).[34] In this sense, the categorisation of weapons into lethal, non-lethal or less lethal is unhelpful as it masks the inconvenient fact that weapon effects are never solely a function of that weapon's design but depend, of course, upon its use and the vulnerabilities of those affected by it. Indeed, it is this unavoidable indivisibility between the human (as programmer, enabler, coordinator, user and victim) and lethal machine that is so often lost in discussions on these systems' deployment.

1.3 Timelines around capabilities

Weapons, of course, can already identify, track and engage targets and a purpose of this book is therefore to look even further ahead and focus on the

---

[34] Sartor, Giovanni and Andrea Omicini, 'The Autonomy of Technological Systems and Responsibilities for Their Use', in Nehal Bhuta and others (eds.), *Autonomous Weapons Systems: Law, Ethics, Policy* (Cambridge University Press, 2016), p. 44.

future battlefield. In this vein, recent battlecraft in Ukraine has clearly hastened the deployment of uncrewed assets, already prevalent in Nagorno-Karabakh, in Yemen, Libya and Syria.[35] Indeed, the Ukraine conflict has widely provided an incubator to uncrewed technologies in space and in the air, on the ground and under the sea. No longer is the uncrewed weapon system merely stationary or simply designed to repeat pre-programmed routines within set parameters and time frames. Although still constrained, the use of uncrewed assets in Ukraine has demonstrated beyond doubt ever greater platform independence and ever greater breadth of tasking.

The Scottish poet Andrew Lang quipped that statistics are often used in the same way that a drunkard uses a lamppost ('for support rather than illumination') but, even in an arena as fast moving as drone warfare, examples are useful to frame these developments and to highlight emerging capabilities. First, the Ukrainian government expected to field more than one million indigenous first-person-view drones during 2024. That is a staggering number, both in its absolute value but also relative to months (yet alone years) earlier. By way of context (and thinking of Mr Lang's statistics), this is also double the number of artillery shells supplied to Ukraine by the entire European Union in 2023.[36] The phenomenon has driven both warring parties. Russia's KUB-BLA unit, for instance, is a loitering drone developed jointly by Kalashnikov and Zala Aero Group.[37] Its manufacturers claim it can precisely hit ground targets and deliver specific payloads to target coordinates that are specified manually or acquired from payload targeting imagery. The system, they assert, incorporates AI and uses visual identification in order to facilitate real time recognition and classification of targets. Ukraine has similarly fielded new drones (variants on the Bayraktar TB2 and Phoenix Ghost) and loitering missiles (Switchblade), both purporting to use remote, independent navigation and object recognition capabilities.

Second, although much of this book's later analysis considers Ukrainian advances in uncrewed platforms, Russia has successfully played

---

35 Meaker, Morgan, 'Ukraine's War Brings Autonomous Weapons to the Front Line', *Wired Magazine* (24 February 2023), https://www.wired.co.uk/article/ukraine-war-autonomous-weapons-frontlines.
36 Zafra, Mariano and others, 'How Drone Combat in Ukraine is Changing Warfare' (26 March 2024), Reuters, https://www.army-technology.com/projects/zala-kyb-strike-drone-russia/.
37 See, for instance, *Army Technology* (12 January 2023), https://www.army-technology.com/projects/zala-kyb-strike-drone-russia/.

catch-up. Its Lancet drone is 'a smart, multi-purpose weapon capable of autonomously finding and hitting a target without requiring ground or sea-based infrastructure'.[38] This 'us versus them' narrative highlights the febrile pace of developments. In doing so, it also signposts a clear trend that weapons are becoming ever smarter. The context here is twofold. On the one hand, weapon development is playing out that 'revolution in expectations' discussed above. Observers' predictions ('I've seen it and therefore it exists') increasingly reflect soldiers' own experiences in the field, stoking ever more outlandish expectation.[39] On the other, emergent means are, as a rule, tending to accelerate parties' embrace of autonomy, a further step to normalising unsupervised weapons in future arsenals.[40]

Versos, however, exist to much of this narrative. First, the deployment of broadly capable AWS still basically remains a hypothetical construct[41] and states remain wary of defining too clearly their positions on the matter (in particular relating to statutory instruments that might fetter their options).[42] Battlefield integration, moreover, is rarely seamless: analysis in February 2024 suggested that the Russians, at least, had turned off their Lancet's touted AI-guidance, its automatic target recognition having proven

> *Observers' predictions ('I've seen it and therefore it exists') increasingly match solders' experiences on the ground, stoking yet more outlandish expectation*

---

38 Editorial, *Automated Decision Research*, Weapon Systems with Autonomous Function Used in Ukraine (28 June 2022), https://automatedresearch.org/news/weapons-systems-with-autonomous-functions-used-in-ukraine/.
39 For a useful (notwithstanding the piece is now a decade and a half old) discussion on 'the battlefield of the future' see, generally: Wood, R, 'The Technical Revolution in Military Affairs' (2010). As Wood comments, 'innovation in weapons design is driven by the absolute need for survival and, thus, will always advance aggressively… The study of innovation is crucial since the technology of war interacts with the actual practice of fighting'. See also: Rubenstein, R and others, *Practicing Military Anthropology: Beyond Expectations and Traditional Boundaries* (Kumarian Press, 2013), generally.
40 Russell, Stuart, 'AI Weapons: Russia's War in Ukraine Shows why the World Must Enact a Ban', *Springer Nature*, Vol. 614 (23 February 2023).
41 In particular, the task type and capability set that should characterise AWS. See, generally: Center for a New American Security, 'Autonomous Weapons and Human Control' (2016), pp. 4-5, https://www.files.ethz.ch/isn/196780/CNAS_Autonomous_Weapons_poster_FINAL%20(1).pdf.
42 The position of the UK Foreign Office is an example. See: John Templeton Stroud, UK UN delegation, Fifth Review Conference, Convention for Conventional Weapons, HRW/Article 36 side event (December 2016). UK's negotiating position regarding the CCW has been to point to both the current absence and long-term unfeasibility of AWS to define its hands-off stance on AWS deployment policy.

to be unacceptably unpredictable.⁴³ Humans, after all, struggle to understand what is happening when they are under stress, in danger and facing deliberate deception and, in a similar vein, Ukrainian decoys (be they fake HIMARS rocket launchers or replica anti-aircraft radar) routinely trick Russian drone operators and their artillery teams into wasting ordnance while leaving well-camouflaged assets untouched.

In assessing autonomy, the challenge is neatly framed by Arthur C. Clarke's aphorism from 1962: 'It is impossible to predict the future and all attempts to do so in any detail appear ludicrous within a few years.'⁴⁴ For the purposes of this book, 'near term' is reckoned very generally to be within the half-decade or so to 2030. 'Medium term'

> *'It is impossible to predict the future and all attempts to do so in any detail appear ludicrous within a few years'*

then relates to developments that may occur in the fifteen years thereafter to 2045. But any framework around timing comes full of caveats, the more so given that opinions on when readers might expect weapons-directing AI are as confident as they are diverse.⁴⁵ By way of context, Nick Bostrom was estimating in 2015 that high-level machine intelligence (HLMI) had a fifty per cent chance of arriving by 2040 and a ninety per cent probability by 2075. Public understanding of AI and AGI is generally muddied by what is often called the 'temporal framing mechanism', that common misunderstanding on timescales for promised technological developments and the blurring of distinctions between visionary research goals and what are their more realistic short-term outcomes.

The implausibility of this is exactly evidenced by the wide range of expectations of AGI's arrival; Asian experts forecast HLMI a telling forty-four years earlier than their US counterparts. Indeed, a paper published In 2024 surveying almost three thousand published AI researchers now estimate the date for when AI could beat humans at every possible task is 2047, seven years after Bostrom's earlier half: half call. This is also a dramatic move from just twelve months earlier when the anticipated date was thirteen years later,

---

43 Freedberg, Sydney, 'The Revolution that Wasn't: How AI Drones Have Fizzled in Ukraine (So Far)' (20 February 2024), *Breaking Defense*, https://breakingdefense.com/2024/02/the-revolution-that-wasnt-how-ai-drones-have-fizzled-in-ukraine-so-far/.
44 Bawden, David, 'The Nature of Prediction and the Information Future: Arthur C. Clarke's Odyssey Vision', *Aslib Proceedings*, 49, 3 (1997), pp. 57-60.
45 Bostrom, Nick, *Superintelligence; Paths, Dangers, Strategies* (Oxford University Press, 2014), p. 19.

in 2060.[46] We are reminded of Amara's law whereby we tend to overestimate the effect of a technology in the short run and underestimate its effect in the long run. As an aside, social change is always even slower than technological change.

What becomes apparent is that predictions in this space are fraught with very large uncertainty but also with methodological error as well. We have no idea if we are five per cent or twenty per cent or 0.05 per cent along the way to developing machine sentience. Experts' simulations often assume exponential growth, a similar mistake to that made by economists and investors much of the time.[47] Instead, the likelihood adopted by this book is that advances will, at some point, slow as progress becomes ever harder to achieve, and the goals of AGI move yet further away. In this, it is useful to remind ourselves that we are assuming three quite separate and individually tricky characteristics for independent weapon platforms: a capacity to deduce and decide courses of action; an awareness of surroundings; and, based on coordinated analysis of those surroundings and tasking, action sequences that contribute to mission performance rather than simpler mission execution.

> *The uncertainty is considerable because we have no idea if we are five per cent or twenty per cent or 0.5 per cent along the way to developing AGI*

Another characteristic might appear obvious but also requires restatement. An AWS must obey the same *physical* laws that govern all the physical entities. It cannot change shape and size arbitrarily; it must use effectors to move itself around and do this remotely based upon recognised rules; it requires an energy source to think, sense and move and when moving (regardless of dimension); it will take time to speed up and slow down. Physical laws must dictate its internal model of the world, its locus and place within its passing environment. And with this in place, each machine must then be able to search through broad possible solutions that are based solely upon this statistical framework as it relates to those physical laws. Performance, moreover, will be based upon its inherent ability to solve problems and to do this by toggling between its then-current internal model (the 'representation' of its passing place in its milieu) and its dynamic updating of these models.

---

46 Mollick, Ethan, 'Signs and Portents' (6 January 2024), One Useful Thing, https://www.oneusefulthing.org/p/signs-and-portents/comments.
47 Joachim Klement, 'When Will Skynet Arrive?' (14 November 2023), Klement on Investing, https://klementoninvesting.substack.com/p/when-will-skynet-arrive.

Predictions here have long been prone to revision. Procurement, after all, depends upon the scope of capabilities being envisaged; a portfolio of incremental changes is easier than all at once removing human involvement from these loops. In this vein, the Chief Robot Scientist at the US DoD confidently suggested in 2015, nearly a full decade ago, that autonomous machines would be deployable by militaries within ten years, with unsupervised convoy applications coming available 'sometime after'. Other parties have been more conservative, each one reinforcing what is clearly a lack of consensus on timetables for deploying AI and AWS. In an open letter (since reinforced and published in 2023[48]), leading AI practitioners warned that the lethal autonomous systems would be feasible 'in years, not decades', the context being that the letter's signatories reckoned that AI had already reached a point where the deployment of these systems was practically (if not legally) feasible. This prompts a further requirement to define the capabilities that will constitute AWS. Here, it is useful to consider the weapon class within the context of a continuum which, at one end, sit narrow-task (but still independent within the parameters of that tasking) weapon platforms to, further along, much broader commission systems with mission as well as task competences.

> We are reminded of Amara's law whereby we tend to overestimate the effect of a technology in the short run and underestimate its effect in the long run

1.4 Contextual drivers

Individual AWS components must work independently[49] but act together. The tipping point between weapon automaticity and autonomy thus depends upon the combination of platform and weapon which together create a system that is lethal and that on its own can select and engage targets

---

48 Edwards, Benj, 'OpenAI Executives Warn of Risk of Extinction from Artificial Intelligence in New Open Letter', *Ars Technica* (30 May 2023), https://arstechnica.com/information-technology/2023/05/openai-execs-warn-of-risk-of-extinction-from-artificial-intelligence-in-new-open-letter/.

49 See: 'Unmanned Systems Integrated Roadmap, FY2013-2038', Department of Defense: 'Unmanned systems that have the option to operate autonomously are typically fully pre-programmed to perform defined actions repeatedly and independent of external influence or control', p. 66.

 without further intervention.⁵⁰ But this is also to mask the complexity of what is required. The inconvenient truth is that a system which relies simply for its operation upon the prior knowledge of its designer (rather than on its own precepts) is not capable of independent operation. This important distinction really captures the whole debate and the tipping point at which the weapon has autonomous choice regarding target selection in the use of lethal force.⁵¹

Given this pace of innovation, what then is the best way to consider the broad issue of battlefield autonomy?⁵² A useful starting point should be to map the field into which the Delivery Cohort is deploying these systems. In

---

50 Heyns, Christof, 'Report of the Special Rapporteur on Extrajudicial, Summary or Arbitrary Executions', p. 8. This distinction informs the position of NGO Article 36. See: Article 36, 'Structuring Debate on Autonomous Weapon Systems', Memorandum for delegates to the Convention on Certain Conventional Weapons (CCW), Geneva (14 November 2013). Similarly, HRW's Mary Wareham points to the definition of weapons 'able to select and attack targets without any human intervention' being the broadly adopted working definition for AWS (in conversation with the author, June 2014); see also: Campaign to Stop Killer Robots, 'Urgent Action Needed to Ban Full Autonomous Weapons', CSKR London (23 April 2013), generally, http://stopkillerrobots.org/wp-content/uploads/2013/04/KRC_LaunchStatement_23Apr2013.pdf.
51 Analysis on the subject was first provided by Hanon, Leighton, 'Robots on the Battlefield – Are We Ready for Them?', *American Institute of Aeronautics and Astronautics* (2004), p. 7, http://arc.aiaa.org/doi/abs/10.2514/6.2004-6409. AWS technology is already subject to a defined set of recognised institutional measurements. These greyscales of autonomy range from level one (simple remote guidance) to level ten (full autonomy). Autonomous control level (ACL) six will, for instance, allow multiple uncrewed weapon systems to recognise multiple targets and allocate those targets between systems. A further example is useful. ACL 9 is intended to enable groups of automated systems to assess the battlefield, the number and the location of targets. ACL 9 incorporates AWS' analysis of targets' threat potential in order to allocate overall mission priorities. It even envisages the skipping between low- and high-value targets. For a popular analysis of the subject, see: *Drone360 Magazine*, http://www.drone360mag.com. It is noteworthy for this analysis that the US Military's Global Hawk AUV has autonomous take-off and landing, and can self-determine destinations, adjust and set speeds, altitude, roll, pitch and yaw, but only ranks at 2.5 on the ACL scale of one to ten.
52 An overview is as follows: first, it looks to establish a contextual framework to frame the breadth of factors that influence the broader question of weapon control (Chapter Two, Context). It then identifies drivers influencing the removal of human oversight in lethal engagements (Chapter Three, Drivers). Material obstacles exist, however, for any material shift towards independent weapons. Chapters Five through Nine (Obstacles, Wetware, Software, Firmware and Hardware) seek first to identify and then to evaluate the significance of technical and other related fault lines that exist regarding AWS deployment. These chapters together consider AWS' technical feasibility. Chapter Ten (Oversight) then analyses the concept of MHC as a key pivot to the AWS debate.

one corner are politics, culture and society.⁵³ But it is also necessary to factor for the military perspective, and how generals reconcile the promise of technological advances with generations of under-delivery to arrive at a position based squarely upon what is practically achievable. There are a few unhelpful contextual drivers here. First, it is not unusual for military organisations to succumb to 'presentism', that universal human condition whereby commentators consider their time as a period of unprecedented turbulence while simultaneously attributing an exaggerated tranquillity to the past. Similarly, 'neophilia' is the belief that what is being experienced is entirely novel.

Other issues require thought. Setting aside the notion of improbable and never-before-seen military discontinuities, three fifths of this book is taken up with considering technical fault lines that constrain this march towards autonomy. Second, the very complexity of AWS deployment deserves consideration, as it too undercuts any notion of 'planning certainty'. While generals and politicians both like assurance, readers should recall Bismarck's point that the statesman is 'like a wayfarer in the forest who knows which direction he is walking but not at what point he will emerge from the trees'. For this reason,

> *The inconvenient truth is that a system which relies simply for its operation upon the prior knowledge of its designer (rather than on its own precepts) is not capable of independent operation*

the book must consider the why and to whom deployment of AWS might appeal and the several drivers that press for autonomy in states' (and others') arsenals. This is again an involved relationship. A broad move towards robotics is indisputable. The consensus view today is that the global market for robotics will increase each year by fifteen per cent from thirteen billion dollars in 2021 to twenty-five billion dollars in 2026.⁵⁴ These statistics, moreover, cover seventeen robot types and nineteen robot-performed tasks, and the dual-use nature of robots is clearly an important catalyst to their being harnessed for battlefield tasks⁵⁵, a phenomenon that is reviewed in

---

53 See: Gray, Colin, *Strategy and Defence Planning: Meeting the Challenge of Uncertainty* (Oxford University Press, 2014). Also: Gray, Colin, *The Future of Strategy* (Polity Books, 2015) and *Another Bloody Century* (Phoenix, 2005). In particular, Chapter Two (Context) in this book seeks to repurpose several of Gray's arguments within the AWS debate.
54 See, generally, 'Robotics: Technologies in Global Markets Report, 2021–2026', Research in Markets (January 2022), https://researchandmarkets.com.
55 The term 'battlefield commander' does not relate in this work to any particular rank but is used throughout to convey a nomenclature of the superior who is controlling the in-theatre

Chapter Three (Drivers) in its analysis of technology creep and investment across machine autonomy by the commercial sectors.[56]

What, then, is this 'portfolio' of advantages that autonomy promises? Removing human supervision suggests quicker reaction times to adversarial threats, another accelerant towards quicker decision cycles and ever faster data processing.[57] It posits better persistence and better endurance, and doing all of this while reducing humans' exposure to enemy fire. It is therefore little wonder that these enablers coalesce into a general 'revolution in expectation' whereby, all of a sudden, weapons autonomy appears to be an inescapable development.

> *'Presentism' occurs when commentators consider their time as a period of unprecedented turbulence while attributing an exaggerated tranquillity to the past*

In this vein, the US DoD's *Unmanned Systems Integrated Roadmap FY 2013–2038* plots an unambiguous future where 'the prevalence and uses of unmanned systems [will] continue to grow at a dramatic pace',[58] requiring that the department update again its guidance on AWS in order to reflect the pell-mell progress of the past half-decade.[59] The broad procurement assumption is now that robotics are ideal for 'dull' missions[60] (long-duration undertakings with mundane tasks that are ill-suited for crewed systems), 'dirty' missions (exposure to hazardous conditions), and deep (behind enemy lines) and 'dangerous' missions.

---

deployment of AWS. This may or may not be separate from other political authorities in the decision process. It also relates to a level of accountability and legal responsibility.

56 Throughout this work, the term 'driver' is used to refer to accelerators, prompts and catalysts for further development and eventual deployment. Source: Dictionary.com, http://www.dictionary.com/browse/drive: i) definition 25, 'vigorous onset or onward course towards a goal or objective' and ii) coincidentally, definition 26 'a strong military offensive'. For a pictorial review of recent combat robotics, still relevant in 2024, see: Melendez, S, 'The Rise of the Robots: What the Future Holds for the World's Armies (2017), Fast Company blog, https://www.fastcompany.com/3069048/where-are-military-robots-headed.

57 The human fighter pilot needs at least 0.3 seconds to respond to a simple stimulus and more than twice as long to make a choice between several possible responses.

58 US Department of Defense, 'Unmanned Systems Integrated Roadmap FY 2013-2038', p. 20.

59 US DoD, Directive 3000.09, Autonomy in Weapon Systems (25 January 2023), https://media.defense.gov/2023/Jan/25/2003149928/-1/-1/0/DOD-DIRECTIVE-3000.09-AUTONOMY-IN-WEAPON-SYSTEMS.PDF, and announcement, DoD announces update to DoD Directive 3000.09, 'Autonomy in Weapon Systems', https://www.defense.gov/News/Releases/Release/Article/3278076/dod-announces-update-to-dod-directive-300009-autonomy-in-weapon-systems/.

60 The relevance of autonomous uncrewed machines for dull, dirty, and deep and dangerous missions has long been acknowledged and is well covered by Neal, PJ, 'From Unique Needs to Modular Platforms: The Future of Military Robotics', US Naval Institute (2010), pp. 1-7.

It is also necessary to consider how AWS deployment might raise *ethical* standards on the battlefield. The notion is that machines will all but 'remove' humans from the combat frontline. The subject of ethics is always fraught but doubly so when in the context of machine operation, machine learning and a machine's unsupervised behaviour. The verso, of course, is that it is unrealistic to assume humans will always adhere to the legal essentials of armed combat, especially in moments of stress. Soldiers, according to fellow professionals at arms, 'are, at best, a variable tool in waging war'.[61] Emerging practice across drone operations does little to help this, prompting West Point's Lieber Institute to issue a paper entitled 'The Legal and Practical Challenges of Surrendering to Drones'.[62] An upshot might be that AWS deployment is somehow justified on a practical level by recurring human lapses in battlefield *jus in bello*. Instead, a code-

> *Good soldiers with lesser equipment always trump poor soldiers with the latest technology*

based 'ethical governor' might edit its host machine's actions in advance of lethal engagement. The thinking might also be that these unsupervised platforms could one day operate more appropriately than human soldiers if built without need for self-protection.

For the purposes of this introduction, several continua would appear to exist between, at one end, the notion of independent 'killer robots' (presumably roaming the battlefield with a degree of sentience) to, at the other, the more likely advent of task-specific, human-machine teaming based upon hardware with specific autonomous capabilities to be used in specific applications.[63] Chapter Four therefore considers autonomy through the lens of it being an enabler. Wholly new mission types are possible without humans in the loop, particularly in areas such as cyber and electronic warfare where decision speed is critical to success, the more so given that removing human supervision is likely to be both incremental and from a wide variety of

---

61 Maloney, Col S, 'Ethics Theory for the Military Professional', *Air University Review*, 32, 3 (1981), p. 55.
62 Biggerstaff, William Casey and others, 'The Ukraine Symposium: The Legal and Practical Challenges of Surrendering to Drones', Lieber Institute and West Point, 8 February 2023.
63 See, generally: HRW, Arms Division, *Losing Humanity – the Case Against Killer Robots* (Washington, 2012), http://www.hrw.org/reports/2012/11/19/losing-humanity-0. The organisation is widely recognised to have been the first NGO to highlight the issues posed by the deployment of autonomous weapons.

currently supervised battlefield tasks[64]; from covertly deployed networks of smart mines to stand-alone systems controlling the rapid-fire exchange of cyber weapons; from swarming autonomous machines intended to disrupt enemy operations to unmanned aircraft targeting an enemy's positional, navigational and timing capabilities. Another continuum then becomes whether the platform is 'command executing' (whereby the weapon receives its order, carries it out and then awaits the next command) or 'sovereign executing' (the weapon is deployed without task or time constraints and is guided instead by less defined objectives).

1.5 Introduction to AWS feasibility

The purpose of Chapters Five to Ten is then to unpick systemic challenges to these weapons' operational deployment. First, Chapter Five (Obstacles) identifies non-technical constraints, reviewing the legal framework into which such weapons must fit. This book adopts the broad assumption set out by the International Committee of the Red Cross that procuring parties (here, states rather than non-state players) place weight on their weapons being LOAC compliant.[65] How might AWS fit into existing frameworks that require finely nuanced calculations regarding existing law? Here, there are four legal hurdles that comprise LOAC. Each requires compliance. Is each specific attack proportional? Is it militarily necessary? Has due process been demonstrably undertaken to ensure that the selected target is a *bona fide* combatant or conforming asset? LOAC also requires that appropriate action first be undertaken to prevent unnecessary human suffering arising from that (and every) lethal engagement. This is a complex set of prerequisites that confound the human, let alone a machine.

Other 'soft' complexities exist. Given the laws' indistinct nature, a clear challenge is that weapon compliance depends entirely upon the lines of code driving its sequences notwithstanding the imprecision that exists

---

64 US Defense Study Board, 'Summer Study on Autonomy' (June 2016), https://dsb.cto.mil/reports/2010s/DSBSS15.pdf.
65 ICRC, 'The Use of Armed Drones Must Comply with the Laws of Armed Combat' (2013), https://www.icrc.org/eng/resources/documents/interview/2013/05-10-drone-weapons-ihl.htm.

between international humanitarian law and international human rights law and between regional and national rules on warfighting. Compliance is then further hamstrung by obstacles presented by the interpretation of even the most basic components of engagement rules.[66] This complexity[67] means it is rarely straightforward to determine which legal framework applies in each engagement.

A further immutable principle is that weapons and their consequences must be controllable, a fundamental requirement for the moral acceptability, political legitimacy and general legality of 'organised violence'.[68] Expecting a weapon to select its own targets is problematic on several levels, not least for the responsible human commander to predict and understand every specific target, every precise engagement moment, the location where violence is administered and the environment within which violent effects are to be undertaken. Without this direct link between commander and outcome, deployment of AWS is *prima facie* unlawful, the more so as irregular warfare clearly dilutes conventional designations across battle zones and competencies. Nor is this balance between supervision and autonomy a zero sum whereby increasing the one results in a corresponding decrease of the other.

Wholly new mission types are possible without humans on the loop, particularly in areas such as cyber and electronic warfare where decision speed is so critical to success

The relationship is also muddied by having to engineer predictability into processes, the subject of this book's practical analysis and considered in Chapters Six to Nine. A requirement to predict is an enduring challenge to commanders and the Delivery Cohort alike. As the behaviour of automated systems becomes more complex, the less predictable these systems must become. This contradiction is a key driver to the book's findings, from AWS' 'temporal' dislocation (given the dynamic mix of a battlefield's time

---

66 Throughout this book, the UK rules of engagement are taken from JSP 383, 'The Joint Service Manual of the Law of Armed Conflict', Ministry of Defence (28 August 2013), https://www.gov.uk/government/collections/jsp-383.
67 ICRC, 'Handbook on International Rules Governing Military Operations' (2013). The publication runs for 1,464 pages and covers general obligations during combat including targeting and command responsibilities.
68 The harmful effects of weapons must be foreseeable and must at all times be under the control of those who employ them. See: ICRC, Draft Rules for the Limitation of the Dangers Incurred by the Civilian Population in Time of War, Article 14, ICRC (1956). See also: International Law Commission, Articles on the Responsibility of States for Internationally Wrongful Acts, Article 8 and 23 (1), UNGA Res 56/83 (2001).

and space and domain) and the brutal narrowing of timeframes to the erosion of slack and situational awareness. Nor is abstraction helped by the blurring of boundaries between weapon types, effects and outcomes as subcomponents become less distinct, thoroughly interconnected and even geographically distributed.

Finally to this point is the widespread creation of 'technical debt', the subject of Chapter Seven (Firmware) and an unavoidable byproduct of this layering. After all, these weapon processes must rebase moment-by-moment including the platforms' anchoring, their goal and value setting and the updating of the machines' utility functions. This may look like a set of regular verification routines but it actually equates to dynamci tuning, all of which must now be undertaken without third-party direction. This, moreover, is not just about maximising performance but is the only means available to the Delivery Cohort to ensure even basic compliance with passing norms, technical conformance and adherence to LOAC.

Foundational changes in practices will be needed if AWS are deployed. A key proposition of this book is that several well-tried concepts that have long been used in battlecraft may no longer be fit for purpose. These concern capabilities, processes, the frictions of integration as well as wholesale developments in doctrine and training that must first be in place. In this vein, Chapter Ten (Oversight) considers the enduring role of humans across this piece and whether MHC might provide an ongoing control mechanism for AWS deployment. This remains a key agenda item in the CCW and, although progress in that body continues to be patchy, the historiography for this book's final section continues to develop quickly, in particular on the overarching significance of target selection as the critical control function of a weapon. Here, therefore, the book generally uses the playbook of NGO Article 36 and that organisation's definition of MHC encapsulating the 'when, where and how weapons are used; what or whom they are used against; and the effects of their use'.[69]

MHC must operate at three distinct layers: *ante bellum*, *in bello* and *post bellum*. The distinction is important because it points to MHC imbuing all phases of battlecraft where each of these conditions must inform and then shape each of the constituents in the control debate. It also needs to do this as deployment relates to the design, acquisition and use of such tools.

---

69 Article 36, 'Killing by Machine; Key Issues Understanding for Meaningful Human Control' (2015), generally.

MHC is therefore not a single test in an engagement sequence. Instead, it is an overarching framework to govern the use of violence and one which must be evident right down to the level of individual direct attacks. Here, its obligations require the operator and relevant command chain to evaluate, understand and then factor for the expected outcome of that specific weapon in that specific context and with the obligation being in place for each and every lethal engagement. Within this framework, the AWS must then be deployed for a certain purpose and for a certain time. Without the efforts of the Delivery Cohort, these systems are devoid of agency and intentionality and, on this basis, it is for good reason that this book next considers the role of context as a key challenge to replacing human oversight.

SECTION ONE

An Analysis of Structural Challenges to AWS Deployment

2
Context

The role of context in the removal of weapon supervision

Context is the frame of reference within which this book's study (both behavioural and technical) should be read. It is an analysis of the circumstances that form the setting for weapon independence. A useful starting point is provided by Stuart Russell, professor of computer science at UC Berkeley, whose video *Slaughterbot* portrays a 'fictional near-future in which autonomous explosive-carrying microdrones are killing thousands of people around the world'.[1] Russell's imaginary narrator quips that autonomy allows the user 'to separate the bad guys from the good... watch the weapons make the decisions... take out your enemy virtually risk-free'.[2]

Russell's portrayal is just one of several scenarios that contemplates the world of combat transformed by the adoption of lethal autonomous weapons. Together, these pictures weave a quite plausible narrative on

---

1 See: Ackerman, Evan, 'Lethal Microdrones, Dystopian Futures, and the Autonomous Weapon Debate', *IEEE Spectrum*, 15 November 2017, https://spectrum.ieee.org/automaton/robotics/military-robots/lethal-microdrones-dystopian-futures-and-the-autonomous-weapons-debate. The eight-minute video by Russell is available at https://www.youtube.com/watch?v=9CO6M2HsoIA and https://www.youtube.com/watch?v=ecClODh4zYk. Russell's portrayal suggests AWS with multiple roles. It notes that AWS deployment might come with a high risk of failure but with little consequence being attached to that failure. It also posits real expansion to the scope and daring of AWS, the likelihood of an AWS arms race, a lower bar for future engagements and, finally to this point, fast cross-proliferation of lethal autonomy between military services.
2 Russell, Stuart, *Slaughterbot*, YouTube, 0.15 minutes/1.50 minutes/2.47 minutes, https://www.youtube.com/watch?v=ecClODh4zYk. Russell concludes the video in person by stating that 'this short film is more than speculation. It shows the results of miniturising technologies that we already have [and that] allowing machines to choose to kill humans will be devastating to our security and freedom' (7.15 and 7.34 minutes).

how autonomous weapon systems (AWS) might be integrated into parties' battlecraft.[3] Indeed, the utility of these pieces is to demonstrate a portfolio of use-sets for these platforms, both tactical (area denial, defence in depth, flank security, deep operations, kill-boxes and asymmetric operations) and strategic (high-profile impact and other 'morale sapping actions'). Russell's *Slaughterbot* echoes century-old anxieties that deployment of a new weapon system is a harbinger of war on a vast scale. The video quite deliberately passes over the fact that war is cruelty and cannot be refined. Instead, he

> *Autonomy allows the user 'to separate the bad guys from the good... watch the weapons make the decisions... take out your enemy virtually risk-free'*

paints a picture of remote, surgical warfare, and it is the gulf here between these two positions (on the one hand, AWS' promise of targeted, refined violence; on the other, the gritty and fundamental viciousness of war) that captures what is an enduring norm.

It also starts to explain the widely varying conclusions reached by those trying to reconcile specific combat technologies with specific battlefield outcomes. As I argued in *War's Changed Landscape*, the pace of war is clearly quickening. Similarly, the means for combatants to undertake mischief (be that conventional or liminal, above or below the threshold of outright war) are expanding as analysts use already enlarged data sources to base their decision-making. The corollary here is that it is only through understanding context that the reader can properly calibrate all the developments discussed in this book's following chapters.

Russell's scenario-setting also underlines the degree of transformation

> *Context here is an analysis of the circumstances that form the setting for weapon independence*

that AWS deployment posits in how fighting will be undertaken. As parties embrace autonomy to gain advantage on the battlefield, this realisation is likely to be complicated by parties' smokescreening – by the strategic benefits of surprise and the time lag between commercial development of a technology

---

3 See, generally, Russell, Stuart, 'Take a Stand on AI Weapons', *Nature*, 521, 7553, 27 May 2015, https://www.nature.com/articles/521415a. Russell opines that 'the capabilities of autonomous weapons will be limited more by the laws of physics (constraints on range, speed and payload) than any deficiencies in the AI systems that control them... One can expect platforms deployed in the millions, the agility and lethality of which will leave humans utterly defenceless'. An overview is also usefully provided by Sufge, Erik, 'What Might a Killerbot Arms Race Look Like?', *Popular Science*, 28 May 2015, https://www.popsci.com/what-would-killerbot-arms-race-look/.

and its implementation as a battlefield capability. The difficulty, of course, is that AWS have yet properly to be deployed, and contextual analysis must therefore remain a matter of conjecture. Ukraine and other current conflict zones also suggest that it is recent battle experience

> *Given that there can be no objectively correct answer, it is context that provides a consistent baseline to the exercise*

which really provides hard evidence of a weapon's immediate utility. Not only are the consequences uncertain, but they will also vary significantly for each affected party. In the case of AWS, uncertainty also arises from the number of deployment models, weapon specifications, platform configurations and other exogenous deployment factors that combine to complicate the analysis. Here context becomes the common prism through which to weight these components.

This chapter is divided into three sections. An introduction to context's significance is followed by consideration of the defence planner, the contextual ramifications of ambiguity and how situational awareness is critical to unsupervised engagements. All of these factors are complicated by the challenge of establishing cause and effect: drivers are rarely correlated between each promising improvement, whether that be from newly autonomous processes, from operational performance, cost reduction, expanded combat options and force multiplication to the advantageous expansion of the battlefield into new spaces.[4] A danger then becomes circularity between all of those consequences and what then becomes the 'new' context (those new norms and forms whereby the effects of AWS deployment alter what was that battlefield's

> *Commentary is curiously monotonal, skipping questions about technical uncertainties but also avoiding discussion on how autonomous weapons will actually help commanders achieve goals*

'preceding' context). After all, there is almost no public discourse on the impact that AWS will have on command and how loosening supervision affects general security. Also lost in the debate are those human factors which

---

4 Francis, David, 'How a New Army of Robots Can Cut the Defense Budget', *The Fiscal Times*, 2 April 2013, http://www.thefiscaltimes.com/Articles/2013/04/02/How-a-New-Army-of-Robots-Can-Cut-the-Defense-Budget. The article stated that, at that time, the Pentagon was already spending eight hundred and fifty thousand dollars per annum 'just to keep each soldier on the battlefield'. This compared to the cost of AWS such as TALON (two hundred and thirty thousand dollars), the SGR-Ai (two hundred thousand dollars) and the 710 Warrior (three hundred and fifty thousand dollars).

have previously defined success in war (will, fear, human genius and front-line élan).

2.1 Warfare's continuum of methods

If human supervision is the ballast in lethal engagement, Martin van Creveld's *Technology and War* provides a useful stepping stone for readers to navigate the issue. Matters of oversight need to be viewed with a long contextual lens. Battlecraft's continuum is complicated, morphing from practices characterised by workforce mass to ones of firepower mass (van Creveld's age of machines) and then to practices instead defined by the adoption of technology (analogous to AWS, his age of systems).[5] AWS deployment seems to sit neatly within this framework, but, if AWS deployment is really to upend battlefield practices, their technical means must be at least as seamless as the practices that they replace.

Revolutions in military affairs (RMAs) are rare.[6] This is not to suggest that an RMA can only be driven by the advent of a new technology. Instead, disruptions today should be judged in light of the battlefield's broad context, what Antoine Bousquet terms the 'wider socio-technical milieu'. We need to think about technologies and the battlefield changes that they empower through this wide aperture.[7] In the case of AWS, how does the cradle-to-grave effect of an 'independent' weapon play in the media, across social media, in politics and then defence planning? Diluting the human's role prompts a long list of consequences, not least because few AWS characteristics are either intrinsic or exclusive to the autonomous weapon. Instead, they depend

---

5 van Creveld, Martin, *Technology and War: From 2000 BC to the Present Day* (Simon & Schuster, 2010), pp. 217-219, 311. See also: Worcester, Maxim, 'Autonomous Warfare: A Revolution in Military Affairs', *ISPSW Strategy Series: Focus on Defence and International Security*, Issue 340 (April 2015), pp. 2-3, https://www.files.ethz.ch/isn/190160/340_Worcester.pdf.
6 Mansoor, Peter, 'The Next Revolution in Military Affairs', *Strategika*, Hoover Institute, Issue 39 (15 March 2017), para. 1 of 4, https://www.hoover.org/research/next-revolution-military-affairs. In considering the ramifications of AWS deployment, Mansoor discusses here the relative development of the first firearms, the socket bayonet, the dreadnought battleship, carrier aviation and blitzkrieg.
7 Bousquet, Antoine, 'A Revolution in Military Affairs? Changing Technologies and Changing Practices of Warfare', *Technology and World Politics*, McCarthy, D (ed.) (Routledge, 2017), pp. 2-3, https://www.academia.edu/34469743/A_Revolution_in_Military_Affairs_Changing_Technologies_and_Changing_Practices_of_Warfare.

upon how each weapon is fielded and, counter-intuitively, upon other quite human interventions.

The paradox is clear. Although the Delivery Cohort may deploy its AWS with quite particular purpose, the platform's purportedly self-learning basis means that the more independence is given to these systems, the less their capabilities will correlate to the intentionality of their creators.[8] Analysis here is often distorted by technological determinism whereby perceived changes in war's conduct are retroactively (and carelessly) attributed to one emerging technology or another.

Adequate contextual weight is rarely given to traditionally soft factors such as logistics, passing tactics and day-to-day

The platform's purportedly self-learning basis means that the more independence is given to these systems, the less their uses will correlate to the intentionality of their creators

operational changes (vicissitudes, for instance, in the tooth-to-nail ratio of combat troops to support staff) and how these affect adoption, integration and deployment of new means.[9]

2.2 The role of context in the wider AWS debate

Context also fosters confidence by dialling down the risk of being spectacularly incorrect. The role of the defence planner, much covered below, is exactly not to be too wrong. Context provides the tools for common analytical foundation. It underwrites continuity across the debate in what, after all, is still largely a future-orientated phenomenon. But that is to ignore that same 'milieu' and its many threads

that must feed into how we understand the place and direction of these developments. Another pointer comes from the pages of the British Army's Action Centred Leadership and the explicit priority that it gives to the

---

8 For a discussion on changing capabilities posited by AWS see also: *Economist*, 'Autonomous Weapons Are a Game-changer' (25 January 2018), https://www.economist.com/special-report/2018/01/25/autonomous-weapons-are-a-game-changer.

9 Tamburrini, Guglielmo, 'On Banning Autonomous Weapon Systems: From Deontological to Wide Consequentialist Reasons', cit. Nehal Bhuta and others, *Autonomous Weapons Systems: Law, Ethics, Policy* (Cambridge University Press, 2016), p. 126. Any human detected in the prohibited area is classified as a legitimate target.

role of context in military operations.[10] 'Understanding Context' is given unambiguous precedence in the Army's three leadership silos, 'Achieve the Task', 'Build Teams' and 'Develop Individuals'. Context, states Army doctrine, is the key factor in how militaries should conduct their mission, 'the collection of circumstances that form the setting for an event in terms of which it can be fully understood'. All of this is then complicated by the inescapable temporal component of context.[11] At every point of the AWS deployment equation, timescales are certain to become ever shorter[12], the context being that technologies rarely come fully formed. They arrive piecemeal, upending their procurement and integration, the more so as the competencies for autonomous function must be available synchronously, working first time and every time.

> *Historical perspective is the only protection against undue capture by the concerns and fashionable ideas of today*

A further purpose of this chapter is to help the reader judge how setting and context practically affects AWS deployment. After all, frictions abound, whether from popular pressure and special interest groups, from institutional inertia, from behavioural considerations or from the discomfiture of earlier sunk costs. Friction also arises from political expediency, tight budgets and other commercial pressures. Context should remind us of a long tradition of human ingenuity trumping newly minted technology as well as economic considerations when scarce resources have been allocated elsewhere. It also warns us of the stymicing effects of organisational resistance, of insufficient skills or available means for those tasked with these practices' integration. The list may be long but it reflects the sinuous, empirical role of context and, in the case of AWS, how the public eye deals everyday with the matter of

---

10 Royal Military Academy Sandhurst, Centre for Army Leadership, 'Army Leadership Doctrine', Edition 1 (2016), p. 17. See also: https://www.army.mod.uk/media/24335/20210923_army-leadership-doctrine-web_final.pdf).
11 Marshall, Michael, 'Timeline: Weapons Technology', *New Scientist* (7 July 2009), https://www.newscientist.com/article/dn17423-timeline-weapons-technology/. We should remember that the first decade alone of the twenty-first century saw the advent of national defence shields, active denial systems, high energy lasers, pulsed energy projectiles, neuroscience-based human enhancement and, in Metal Storm, a gun capable of firing several million rounds per minute. As defined in this book's introduction, near term is reckoned generally to be within the period to 2030. Medium term then relates to developments that may be expected to occur in the fifteen years thereafter to 2045.
12 Scharre, Paul, 'Making Sense of Rapid Technical Change', Center for a New American Security (17 July 2017), generally https://www.cnas.org/publications/commentary/making-sense-of-rapid-technological-change. See also: Latiff, Robert, 'How Technological Advancements Will Shape the Future of the Battlefield', *Signature* (13 October 2017).

artificial intelligence (AI). In August 2017, for instance, the press carried a coordinated statement from the world's top academics and industry leaders warning the United Nations about the dangers of autonomous weapons. Alerting the UN to 'a third revolution in warfare', the letter warned that 'once this Pandora's box is opened it will be hard to close'[13], a theme that has since been echoed and amplified by others, including the Future of Life Institute.[14]

As usual, however, there is a verso. At the other end of this contextual lens sits General McMaster, erstwhile National Security Adviser to the US White House, who prefers instead to scold those divorcing war from its political nature, in particular where developments 'promise fast change and efficient victories through the application of advanced military technologies'.[15] In this vein, this book's analysis stays quite close to Michael Howard's aphorism that 'the roots of victory and defeat often have to be sought far from the battlefield',[16] as it is just as likely that social, political and cultural factors will surpass technical considerations in the argument. It is entirely intentional that the British Army's Leadership Doctrine separates the *nature* of conflict (an immutable, unchanging contest of will, characterised by uncertainty, chaos and friction) from the *character* of conflict (dynamically changing and shaped by means, technology and chance). Any substitution of human agency by an algorithm concerns conflict's character. It is fundamentally just another way of killing one's adversaries.

> *Once this Pandora's box is opened, it will be hard to close*

2.3 Context's norms

Use of force by one individual against another has long been an intensely personal affair. While human beings may be ever more remote from the point of violence, a human somewhere in the kill chain has up to now unequivocally taken the decision to take life. Indeed, the ethical and legal norms for this

---

13 For the text of the open letter: Future of Life Institute, 'An Open Letter to the United Nations Convention on Certain Conventional Weapons' (20 August 2017), https://futureoflife.org/open-letter/autonomous-weapons-open-letter-2017/.
14 Gronland, Kirsten, 'AI: Artificial Intelligence, the Military and Increasingly Autonomous Weapons', Future of Life Institute (9 May 2019), https://futureoflife.org/resource/state-of-ai/.
15 McMaster, Lt Gen HR, 'On the Study of War and Warfare', Modern War Institute (24 February 2017), https://mwi.usma.edu/study-war-warfare/, para. 6 of 10.
16 Clark, Lloyd, 'Blitzkrieg: Myth, Reality and Hitler's lightning war – France 1940' (UK: Atlantic Books, 2016), p. 2.

 behaviour have developed over millennia to determine when one human may use force against another, both in peace and in war. Norm violation is usually quite recognisable and methods to rectify most norm breaches are universally understood even if they are not universally agreed.[17] The conditions that come together to define these norms, some defined but most customary, also provide us with useful context to understand the decisions in front of the Delivery Cohort and it is important, therefore, to consider this framework.

Norms are either enduring (existing, long-dated and persistent behaviours that underpin the conduct of warfare), evolving (behaviours in flux, usually because of recent advances in a technology or practice) or new (a behaviour or rule set that has recently undergone significant change but is now cemented and understood as having captured a permanent alteration in war's conduct).[18] The issue, of course, is that AWS deployment is likely to grossly destabilise existing norms. Importantly, all of this is a matter of degree: advanced militaries (as evidenced, for instance, by Russia's tactics in Ukraine) already undertake violence without necessarily using their full range of available technological capabilities.[19]

For the Delivery Cohort, culture behaves much like norms in framing its decision, also splitting between enduring, evolving and new. Enduring culture is immutable, set in stone and very slow to change. It is a bedrock and reflected in a state's strong traditions and customs. However, whereas norms are generally persistent, the happenings that make up new elements of culture occasion rapid, inexplicable change, often accelerated by catalysts that are exogenous and unexpected. Furthermore, any one cultural influence will likely not correlate with another. Contextual components here include demographic and social considerations, gender and religious considerations and, across geographies, the presence of quite individualistic customs and

---

17 Heyns, Christof, 'Autonomous Weapon Systems: Living in a Dignified Life and Dying a Dignified Death', cit. Bhuta and others, '*Autonomous Weapons Systems: Law, Ethics, Policy*, p. 2.
18 Walker, Paddy and Peter Roberts, *War's Changed Landscape? A Primer on the Forms and Norms of Conflict* (Howgate Publishing, 2023), pp. 2-3.
19 Franke, Ulrike and Jenny Soderstrom, 'Star Tech Technologies: Emerging Technologies in Russia's War on Ukraine', European Council on Foreign Relations (5 September 2023), https://ecfr.eu/publication/star-tech-enterprise-emerging-technologies-in-russias-war-on-ukraine/. For overview, see also: Strachan, Hew, *The Direction of War – Contemporary Strategy in Historical Perspective* (Cambridge University Press, 2013), p. 167.

frictions that only complicate easy pigeonholing. Context's cultural influences are just as hard to abstract as our traditional norms. Together, however, they comprise the drivers that inform political decisions around using new means and, for our purposes here, whether AWS find their way into battlecraft. They also make for a volatile mix and one which is liable to change at any point in time. These trends are invariably difficult to predict, arrive in bunches and interact erratically, conforming to Colin Gray's useful conclusion that 'the course and outcome of war is shaped by many factors, not least of the human, the cultural, and the political, in addition to the possibilities opened by machines'.[20]

> *Substitution of human agency by an algorithm concerns conflict's character. It is fundamentally just another way of killing one's adversaries*

In considering AWS, it is also the pace of any change that is important. But pace creates contextual difficulty for determining which and when each new norm might next prevail. Reaction to novel means has often been inconsistent; public disquiet against Spitfire aircraft flying over Southern England in 1937 turned very quickly to public adulation in advance of the Battle of Britain; the context here for AWS is that all regimes are eventually attentive to public opinion. Whilst certain constituents might momentarily abhor autonomous weapons on ethical grounds, others might support them as a way to influence neighbours and avoid casualties. Here it should be reiterated that this book's commentary is largely directed to deployment of state-sponsored AWS. The emergence of these weapons under entirely new fighting paradigms, perhaps by an as-yet undefined state system or, more obviously, by a non-state party, will likely have contextual ramifications outside the scope of this book.[21]

> *Context fosters confidence by dialling down the risk of being spectacularly incorrect. The role of the defence planner is exactly not to be too wrong*

---

20 Gray, Colin, *Another Bloody Century* (Phoenix, 2005), p. 22.
21 Geneva Centre for Security Policy, 'Perils of Lethal Autonomous Weapon Proliferation: Preventing Non-State Acquisition' (2018), generally. See also: Wallach, W, 'Towards a Ban on Lethal Autonomous Weapons: Surmounting the Obstacles', *Communications of the ACM*, 60, 5 (2017), p. 28.

2.4 Defence planning

Context both anchors and complicates the processes of the defence planner[22] whose role is always be informed by strategic, political and environmental analysis but also by more granular factors that relate to the mission in hand. These are all then focused in a unified effort to develop options given available capabilities and resources. Context is thus key in providing planners with a framework for their own allocation decisions as well as a baseline to shape defence in volatile times. It is the first responsibility for every state to undertake appropriate planning, the lowest hurdle being the pursuit of a course of minimum regret. This highlights an interesting conundrum. In considering AWS, it does not follow that proving a technology's feasibility automatically leads to that technology's adoption given the meld of factors (political, economic, operational, ethical and cultural) that really make up that adoption's context. With AWS, furthermore, its disruptive capabilities create unique challenges to that planning process, calling into question states' first principles on identity and red lines, on laws and ethics, and on the future of military command and organisation, as well as national assumptions on how defence is to be carried out.

> *It is the first responsibility for every state to undertake appropriate planning, the lowest hurdle being the pursuit of a course of minimum regret*

Moreover, the defence planner must only be broadly correct to mitigate (as best as possible) outright error. Long-term defence planning does not need to be right in some absolute sense; instead, it needs only to be correct enough regarding its principal purpose to deter aggression and to ensure homeland integrity. Planners, like economists, remain in 'the dismal position' of having certainty neither about the effects of their output nor the worth of their choices. But the further into the future that context is stretched, the less confident must also be its influence. As Gray further concludes, reputations are dented when vanity seduces its owner to venture a guess too far.[23]

---

22 Walker and Roberts, *War's Changed Landscape*.
23 Gray, Another Bloody Century, p. 41.

Sidestepping miscalculation also requires that defence planners factor for economic considerations. Resources are finite and 'treasure' must be wisely allocated. Regardless of exactly correct figures, the costs of warfare between the US and China have variously been calculated to be some thirty per cent of China's gross domestic product (GDP) and seven per cent of US GDP.[24] Studies note the truism that full state-upon-state war

> 'The course and outcome of war is shaped by many factors, not least of the human, the cultural, and the political, in addition to the possibilities opened by machines'

invariably costs the protagonists 'substantial military losses to bases, [military assets]; significant political upheaval at home and abroad; and huge numbers of civilian deaths'.[25] So what, then, is the context here? It should be that disproportionate costs encourage instead a role for alternative means of warfare including, of course, non-lethal methods; intelligence, diplomacy, financial practices, public relations, covert and other means that fall short of conventional warfare.

2.5 Context's human angle

A key driver in AWS deployment remains the ever more layered role of the human in current battlefield processes. The context, after all, is that all manner of industrial and scientific advances have failed thus far to sideline the sailor, soldier or pilot and, in considering whether the decision to kill can ever be delegated away from the human agent, it is context that returns the reader to the central role of people in this equation. This observation has several constituents. Humans lead. They create and experiment and judge. They rouse and inspire. Motivation covers a portfolio of traits from incentive and impetus to catalyst and drive. While it is these characteristics that have traditionally won the fight, the Cohort's decision set is now about fighting remotely and over greater distances and likely with assets that are increasingly outside human oversight.

---

24 Foot, Rosemary, 'Constraints on Conflict in the Asia-Pacific: Balancing the 'War Ledger'', *Political Science*, 66, 2 (2014), p. 119.
25 Jones, Seth, 'Much "Political Warfare" in Our Future', *Breaking Defense* (2 February 2018), paras. 6-13 of 19.

Much of public debate on AWS is framed by cultural forces, none more powerful that the Terminator franchise. Created by James Cameron and Gale Anne Hurd, a Terminator (or T-unit) is the fictional weapon class of hunter-killer robots built by an artificially intelligent computer called Skynet. Their franchise focuses on a post-apocalyptic war between this synthetic intelligence and a resistance effort carried out by surviving humans. Appearing in 1984, the storyline starts at 'Judgement Day', the date on which Skynet becomes self-aware and its creators attempt to deactivate the network. Skynet immediately judges humanity to be its threat, building what becomes a series of autonomous machines for surveillance and assassination missions. The franchise continues to be an unlikely but disproportionately strong catalyst to shaping public perception around robotic autonomous weapons.

Kyle Reese, the fictional soldier sent back in time to stop these sentient robots, describes his unsupervised antagonists as 'an infiltration unit, part man-part machine,

Combat distils down to competencies, whether of the fighting soldier and the skills needed for killing people and damaging things or, perhaps, a machine undertaking these things

underneath a hyper alloy combat chassis, microprocessor controlled, fully armoured [and] very tough'. Cameron's units are self-healing, their central processing unit being an artificial neural network able to learn and adapt; 'the more contact I have with humans, the more I learn'. Terminators can withstand firearms and explosions, albeit with consequences to their organic disguise layers. 'It can't be bargained with. It can't be reasoned with. It doesn't feel pity or remorse or fear'.

The relevance of the franchise today is that it continues to fire expectation around AWS, reinforcing the mythology that surrounds these platforms in the public eye. First, Terminators adapt and advance; prior to the development of the T-600, Skynet uses non-humanoid assets which proved poor infiltrators, later models therefore introducing shifts in tactics where more nuanced behaviours are more important than widespread destruction. Second, technical advances directly equate to enhanced capabilities and impact; 'nanotechnological transjectors' allow one machine to hack another machine while a robot's endoskeleton can split to allow its facsimile to act independently, so making Terminators twice as difficult to take down. It is this same combination of adaption and advance that frames perceptions around machines' deployment across the battlefield; a low-cost drone jet engine already costs less than twenty percent of its predecessor technology, new construction methods requiring seventy percent fewer mechanical components.[26] In line with the Terminator franchise, drones are already capable of multiplying mid-air; a multi rotor UAV can split into two, three or even six smaller drones depending on operational requirements; LLM-based protocols mean that a swarm can independently divide tasks and do this in real time; layered resilience strategies mean that uncrewed assets are ever less susceptible to jamming and countermeasures. Just as Terminators operated continuously without break, AWS designers are similarly focused upon creating interconnected lethal machines that work in concert and which are networked for coordinated operations. This then is the new context for the *near-term* introduction of AWS into modern battlecraft.

Narratives tell us that it will be robots and other remote means which will increasingly undertake that fight. But the task of context is to remind us that it is humans that must still seal the battle. As long noted in US Army Handbook, 'however much the tools of war may improve, only soldiers willing and able to endure war's hardships can exploit them'.[27] Indeed, those viewing autonomy through the lens of technical determinism should remember that it is usually much easier to predict technological change, even to understand how it should work, than it is to comprehend what it will mean. Combat here still distils down to competencies, whether of the fighting soldier and the skills needed for killing people and damaging things or, perhaps, a machine undertaking these things. After all, defence planners are plying their trade in an age when the human has not been totally in charge of that machine's effects for some time. We long ago reached a point where we no longer really know what a whole set of machines are actually doing for us. And this raises an additional (even unexpected) conundrum: while humans remain responsible for AWS design, its testing and validation, its coding, calibration and maintenance, humans are abrogating themselves from the moment when that weapon kills.

How might AWS deployment skew the long-dated relationship between human soldier and the battlefield? Changes on a human level are rarely obvious. It will always involve risk for the Cohort to identify and exploit such adjustments. Moreover, the only really relevant human context is that autonomous, automatic and manual means of violence each remain means for dispensing lethality. It is of secondary importance to those on

---

26 AlpineMacro, 'The Drone Arms Race', *Innovation Themes and Strategy*, 25 September 2024, p.3.
27 US Army, *Serving a Nation at War* (Army Strategic Communications U.S. Army Public Affairs, 2004), p. 8.

the receiving end how they are being attacked. Indeed, in state systems of warfare, it is only the state's political (and therefore human) authorities that can authorise use of lethal force[28], buttressed by social and cultural forces that reflect each state's communities (which are similarly derived from entirely human interactions).

The contextual corollary is that conflict will persist as a universally human activity long after AWS deployment and, in pretty much every sense, fielding machines that kill without human intervention can unfortunately be considered as just another component to humans' initiating violence for political purposes. And so, although we may tinker with equivalences regarding personal danger and effect, weapons that are independent do not in themselves affect the unchanging nature of war. AWS deployment fundamentally represents

Humans are then abrogating themselves from the moment when that weapon kills

just another way for humans to kill each other. But similarly, not all wars are settled by violence and so it is important that the reader does not undervalue other non-lethal types of warfare such as non-state, economic, electronic and other means of coercion, all of which again are intrinsically human activities.

Other elements of human context have an impact on deployment. An example is in the intensity of AWS adoption where humans either embrace the speed of change brought about by a disruptive new practice or resist such transformation. An early survey, dating back to 2007 (but all the more important as it dates from the early adoption phase of drone technology), mapped attitudes towards uncrewed weapons within the US Air Force. It found that more than one third of pilots had 'wrapped up their professional identity so tightly around the act of flying' that they would rather leave the service than fly a remotely piloted aircraft.[29] Whether or not this is unsurprising, it nevertheless illustrates the behavioural challenges confronting seamless integration of novel practices. There are several

---

[28] See Harelson, Lonnie, 'The Principles of War: Valid Yesterday, Today and Tomorrow', Joint Forces Staff College, Norfolk US (2005), pp. 24-26 ('The Future Has a Need for the Principles of War'). Fighting wars is not a science and context must therefore factor in the unpredictable nature of human involvement in its processes. For discussion on the use of force and relevant case studies, see: International Committee of the Red Cross, *The Use of Force in Armed Conflict: Interplay between the Conduct of Hostilities and Law Enforcement Paradigms*, pp. 13-43 (November 2013).

[29] Cantwell, Houston, *Beyond Butterflies: Predator and the Evolution of Unmanned Aerial Vehicles in Air Force Culture* (School of Advanced Air and Space Studies: Maxwell Air Force Base, AL, 2007), pp. 81-85.

elements to this. Suspicion about new means comes from the long-held precept that every change in a military context is a destabilising source of risk. It lessens the planner's ability to predict. Caution, moreover, is a deeply embedded planning trait which, in military circles, has long been reinforced by soldiers' lack of trust in unproven technology and by unease at how new means might affect their areas of responsibility. It is fed both by organisational stasis and by the Cohort's general 'busyness' with participants' overarching focus on current operational priorities. It is always easier to undertake a well-understood task that relies on muscle memory than to be innovative and pioneering, the more so when personal safety and reputation might be at stake.

In this vein, the broad traits of AWS are unlikely to inspire rank-and-file confidence. After all, what is being suggested here are composite and inherently complex systems, driven by impenetrable software that the local operator cannot readily fix and which, as later detailed, are likely to deliver fundamentally unpredictable outcomes. This scepticism has long been reinforced by day-to-day technical lapses, small but also big, and, as noted in the US Army's Army Equipping Strategy, the requirement that these complex systems undergo a long and necessarily unstable period of adoption. A further point of friction arises. Empirically, technical suspicions are difficult to refute once lodged in the soldier's mind, the more so as the weapon becomes ever more isolated from that soldier (the case with AWS and its basis of operational logic upon which nearby humans have no influence). Nor will the rank and file understand weapon autonomy as a concrete, visible object but instead as a diffuse, remote set of capabilities corralled together into an independent ecosystem. In a further contextual twist, it is exactly this imprecision that has also complicated the role of civil society in its efforts to ban such technologies; the same features that make autonomous weapons enticing to military procurement also make placing restrictions on them difficult.

> Not all wars are settled by violence and so it is important that the reader does not undervalue other non-lethal types of warfare, all of which are intrinsically human activities

Context is also informed by the ambiguities and biases that accompany the introduction of new measures. It is shaped by the operational realities faced every day by each new practice's users and influencers. The extent and success of a deployment are largely determined by the priority given by each service to embedding whatever methods that it sees being most useful to its future practices. How much departure is required by the incoming technology? What degree of disruption is likely for each party's role, responsibility and prospect? Does deployment require a rewrite of doctrine and new training methods, and who in the organisation is responsible for delivering these changes? What oversight is allocated to the process? Prejudices further harden if budgets are tight and when acquisition of untried hardware is deemed less attractive than buying readily available and proven alternatives. If a long period of experimentation is then required, it may be that the new means does not deliver capabilities for several years. In this vein, context prompts parties towards more cautious and incremental paths.

Suspicion about novel means comes from the long-held precept that every change in military context is a destabilising source of risk

This conforms to the enduring norm that it is often not until the occurrence of crisis or hot conflict that adoption of new technologies really accelerates and can properly be embedded into that party's battlecraft.

A further contextual drag arises from the degree of transformation required in these new practices as parties familiarise themselves with incoming methods. Institutional muscle memory is always challenging to shift. The argument is generally that emerging practice fosters initial paralysis but, once overcome, can tip over into enthusiasm, experimentation and acceptance. But the trajectory of this path is unknowable, and three additional factors are worth consideration. First, planners' stasis from 'knowledge explosion' is a characteristic of there being actually *less* appetite for parties either to make sense of the present or to forecast the future. The Cohort is either overwhelmed or, just as likely, shuts itself down waiting for the next new development to supersede the one being implemented. A second generality is that planners reckon that embrace of new means always requires that they run to catch up. Third, an institution that already reckons that it has learned lessons (perhaps from an earlier campaign or experiments with a prior technology) often enters a period of reduced creativity and inadequate internal questioning. This certainly feeds into that same 'sociotechnical milieu' but, in this case, one in which context actually slows the pace of battlefield change.

Still other nuances influence planners' action. The *physical* components of a new technology (here, perhaps its effects, its range, accuracy and cost) often appear easier to identify and embrace than less definable traits such as training, morale, organisation, doctrine and the quality of leadership that each new method will entail. It is always easier for the planner to focus on tangibles, giving a technology's physical footprint outsized heft in decision-making, often at odds with the age-old aphorism that 'good men with poor ships are better than poor men with good ships'.[30]

The point is usefully laboured in the *US Army Handbook*, written back in 2004, by its exaggerated statement that 'however much the tools of war may improve, only soldiers willing and able to endure war's hardships can exploit them'.[31] Although weapon innovation is clearly a constant, its impact may yet be blunted by emulation, by parallel discovery, by adaption and evasion. Indeed, the inescapable fact is that newer and better devices are always just around the procurement-corner and there is never a point of rest for the defence planner.

2.6 The role of situational awareness and uncertainty

To complete this chapter's analysis, its final section considers context from the perspective of the weapon system. What evaluation tools should reasonably comprise AWS targeting and decision processes? Key to this is the notion of 'situational awareness', the understanding of the platform's operational environment in all of its dimensions. This is a considerable ask given context's largely subjective makeup.[32] Situational awareness may be a broad and dynamic concept, but it is also an essential element for commanders to adhere to the legally accountable rules of conduct in armed conflict. The double complication here is that hard-to-abstract context must reconcile with hard-to-abstract legal constructs, all exacerbated by the uncertainty of what is taking place on the battlefield.

Uncertainty certainly impacts the confidence with which commanders can task algorithm-driven assets. Although it is unrealistic for an algorithm to assign perfect certainty to an outcome (in this case, 'I am one hundred per cent sure that this is a tank'), uncertainty estimation is nevertheless the key

---

30 Mahan, Admiral Alfred Thayer, The Influence of Sea Power upon the French Revolution and Empire 1793-1812, Volume one (1898), p. 102.
31 US Army, Serving a Nation at War, p. 8.
32 Such awareness incorporates judgements that have been based upon, *inter alia*, classification of persons, escalation/de-escalation of force, definition and ramifications of protected places, battlefield responsibilities and permissions, appropriate selection of armaments as well as the three principles of distinction, military necessity and proportion.

precept for neural networks.³³ Absent human oversight, compliant operation relies upon the weapon amending its passing understanding of its wider condition (its situational awareness) and, in the case of deployed AWS, to do this based only upon data that has been polled by that device's sensors. This masks an intricate process. First, those inputs must be mediated. Each stimulus is effectively competing with an almost limitless number of adjacent signals to provide its host with its version of a moment-to-moment picture on the machine's relevant environment. Human perception is an effortless process (human eyes, after all, do not tell the brain what objects they see any more than a camera informs what objects it is capturing), but the challenge here is the adjudication of those sensory signals as well as the management of their intensity. Second, inputs must be filtered according to threshold values (the pivot point that triggers a weapon response) and also by a test that determines 'match', 'mismatch' and 'novelty' relationships that may or may not exist between these sensed signals and the machine's initial representations. This must also be 'right first time, every time'.³⁴

Weapon autonomy cannot be understood as a concrete, visible object but instead as a diffuse, remote set of capabilities corralled together into an independent ecosystem

Even at a theoretical level, achieving situational awareness in machines is surprisingly difficult. First, there is a subtle difference between incoming data being categorised a mismatch (no corresponding fit with the machine's training) and being understood as a 'novelty' (the same test outcome but also without a corresponding feedback signal). Routines, moreover, where discrepancies are so nuanced are unusually prone to spoofing, error and paralysis. A further challenge is that these process types must be based squarely upon feedback models, each subsequent (and yet still intermediate) output requiring its own filter before being reintroduced, weighted and, as necessary, acted upon (and all

However much the tools of war may improve, only soldiers willing and able to endure war's hardships can exploit them

---

33 iMerit Post, 'A Comprehensive Introduction to Uncertainty in Machine Learning' (1 September 2022), https://imerit.net/blog/a-comprehensive-introduction-to-uncertainty-in-machine-learning-all-una/.
34 Han, Meghan, 'Lethal Autonomous Weapons and Info-Wars: A Scientist's Warning', *Medium* (6 July 2017), paras. 7, 12 and 21 of 27, https://medium.com/@Synced/lethal-autonomous-weapons-info-wars-a-scientists-warning-cc95798bc302.

in keeping with each machine's goals). For compliant function, the weapon must also predict (and search) from the widest universe of possible contextual representations before it can accept alteration of its immediately prior state.

This opens a Pandora's box of challenges that provides the basis of this book's later technical analysis. Having only code to separate 'wanted' from 'unwanted' associations (in what, after all, is a chaotic battlefield environment) creates intractable complexity given the adjacent tasks that must now be undertaken if the process is to alter or suppress a behaviour.[35] First, the weapon must identify and factor for associative meanings.

> *Routines where discrepancies are so nuanced are unusually prone to spoofing, error and paralysis*

Are these contextual associations to be pre-set and immutable, or rules based and varying? Within what framework might they then be improved through subsequent learning steps and processes? Second, it is these driving fundamentals (data's 'which', 'when', 'how often' and 'degree') that have the uncomfortable feature of both arising from the weapon's system architecture but simultaneously dictating the effectiveness of the weapon that can then be deployed.

The aim of this chapter has therefore been to provide an introductory prism through which self-directing weapons can be judged, with context providing the framework with which to weight these challenges. It is because of the very wide art of the possible (at one end, the dystopian scenario put forward by Professor Russell) that context acts as ballast to what is otherwise a quickly evolving sets of circumstances.

> *Context acts as ballast to what is otherwise a very quickly evolving set of circumstances*

It becomes a fundamental tool in this book's overall investigation, driving the book's subsequent chapter headings in order for us to understand why (together, 'drivers'), how (together, 'deployment') and why not (together, 'obstacles') AWS deployment may or may not be a viable construct over the short fifteen years or so until 2040.

---

35 'Its grammar, indeed, may be its own, but not its logic' (Clausewitz, cit. Robert Cassity and Jacqueline Tame, 'The Wages of War Without Strategy: Beyond the Present – A Call to Clausewitz and to Conscience', *Strategy Bridge*, Abstract and generally (23 August 2017), https://thestrategybridge.org/the-bridge/2017/8/23/the-wages-of-war-without-strategy.

3
Drivers
Factors accelerating the removal of weapon supervision

This book now turns to consider the many drivers that contribute, either singularly or in concert, to deploying automated weapon systems (AWS). These catalysts are rarely isolated and tend instead to overlap. They have the broadest genesis, different intensities and, as detailed below, are usually behavioural (from the political, social, ethical and moral to economic, operational and doctrinal) in how they influence our landscape. An important caveat is that the world still remains in the early stages of developing robotic battlefield technologies and that fully autonomous weapons represent the far end of this continuum.[1] Two observations arise. First, states' arsenals are in considerable flux as parties grapple with the appeal of integrating robotics into their plans and, second, the gap between what is currently deployed and what might be envisaged has never been wider or more volatile.

Push and pull factors compete to influence state adoption of new means. These include militaries' goals to improve processes and stay ahead of likely adversaries. Here, robotics and the removal of human oversight promise a portfolio of benefits, and our purpose is therefore to review and critique this potential. At its most basic, ramping up uncrewed battlefield practices can sidestep circadian shortcomings to human performance such as fatigue,

---

1 Several continua (relevant to AWS deployment) are reviewed in this book including those related around weapon tasking and assignment, procurement practices, and defence planning. Continua also exist in battlefield practices (firepower mass to manpower mass to technology mass), in weapon capability and the move from automation to autonomy as well as the sequences within target acquisition, selection and dispatch. For a discussion on this relationship, see: Sharkey, Noel, 'The 'Evitability' of Autonomous Robotic Warfare', *International Review of the Red Cross*, 94, 886 (Summer 2012), generally https://www.icrc.org/eng/assets/files/review/2012/irrc-886-sharkey.pdf.

the often-inconvenient human requirements for training, upkeep and other support. There is no debate, for instance, that robots should have a main role in army logistics, the payload of Boston Dynamics' three-hundred-and-forty-pound robotic BigDog already being three times greater than that which can be hauled by the regular infantry soldier and all without the tiresome requirement of weekend rotas, pensions and medivac.

Equally important, however, are push factors, that long list of contextual drivers to AWS deployment that includes economic, social, environmental and legal factors. It is these more sinuous components that provide the common thread to this chapter's analysis, the migration from robotics to machine independence being influenced by a meld of factors: the political landscape; the ramifications of an ageing population (with, perhaps, a decreasing tolerance for casualties); the quickly evolving character of warfare evidenced in Ukraine as well, of course, as the pervasive use of robotics across the commercial and domestic domains.

> *A period of humdrum development is all at once inflected by a tipping point in the product's adoption after which the system's implementation switches from linear to exponential*

Some formality and structure are therefore useful if these catalysts are to be understood. Stanford futurologist Tony Seba provides such a framework which, once repurposed, may be useful to understanding these phenomena.[2] In this vein, a disruptive weapon system should exhibit four characteristics. First are permanent changes in the cost curves of underlying technologies. Second, there should be dramatic increase in the weapon's capabilities brought about the convergence of these technologies. Third, these changes should prompt innovation in how these technologies can be fielded as well as their integration into current practices. Finally, the disruption process should be subject to what Seba identifies as an S-curve adoption whereby a period of humdrum development is all at once inflected by a tipping point in the product's adoption after which implementation switches from linear to exponential. As evidence for his model, Seba uses the cost curve disruption in lithium batteries, in computing power, in lidar as well as the acceleration that arises from more systemic implementation models such as open-sourcecomputing.[3]

---

2 Seba, Tony, 'Clean Disruption: Why Conventional Energy and Transportation Will Be Obsolete by 2030', Presentation to Swedbank (17 March 2016).
3 Ibid. Seba identifies a fourteen per cent improvement in lithium battery performance per one dollar from 1995 to 2010 and a twenty per cent improvement thereafter. In 2000, one teraflop

Context, of course, is again at hand as the necessary foundations for building autonomous weapons have been in place since the late 1990s. More than two decades earlier, researchers at University of California, Los Angeles and Hewlett-Packard had succeeded in building microscopic integrated circuits using single molecules as building blocks. James R Heath, the professor leading that project, suggested at the time that a molecular computer with the processing power of one hundred conventional personal computers would be about the size of a grain of salt. The procurement implication of this has only accelerated given developments in smart manufacture, more widespread supercomputing and almost unlimited memory capacity in devices so small that they are on the scale of insects.

These developments then coincided with emerging notions of an 'automated battlefield'[4], best captured in US defense secretary Chuck Hugel's 2016 Defense Innovation Initiative (otherwise referred to as Third Offset Strategy) and its redoubled focus on innovation to overcome operational challenges.[5] That programme had already been preceded by a number of quadrennial defence reviews with similarly heroic labels such as 'Reconnaissance-Strike Complex', 'Transformation', 'Air and Sea Battle' and 'Anti-Access/Area Denial'.[6] Together, these reviews shape how this chapter is organised. It first considers factors encouraging the embrace of autonomy in how we fight before looking more broadly at technology creep

---

of processing power cost forty-six million dollars and required housing in one hundred and fifty square metres' space. In 2016, 2.3 teraflops cost just fifty-nine dollars and were available as a personal hard drive. By 2019, twenty teraflops, the performance criterion for autonomous vehicles, was being deployed. The lidar unit that cost one hundred and fifty thousand dollars in 2012 costs less than one hundred dollars in 2024.

4 Long, Jeffrey, *The Evolution of US Army Doctrine: From Active Defense to the Airland Battle and Beyond* (US Army Command and General Staff College, Fort Leavenworth, Kansas, 1991), p. 122 and generally. Also: Romjue, John and others, *Prepare the Army for War: A Historical Overview of the Army Training and Doctrine Command, 1973-1993* (TRADOC Historical Series, Office of the Command Historian, Virginia, 1993), pp. 44-48.

5 Louth, John and Christian Moeling, 'Technological Innovation: The US Third Offset Strategy and the Future of Transatlantic Defense', Armaments Industry European Research Group, Policy Paper (December 2016), pp. 3-6, http://www.iris-france.org/wp-content/uploads/2016/12/ARES-Group-Policy-Paper-US-Third-Offset-Strategy-December2016.pdf. See also: Pellerin, Cheryl, 'Third Offset Bolsters America's Military Deterrence', US Department of Defense (31 October 2016), https://dod.defense.gov/News/Article/Article/991434/deputysecretary-third-offset-strategy-bolsters-americas-military-deterrence/. Pellerin's analysis of the US 'Third Offset Strategy' highlights the role of artificial intelligence (particularly machine learning) as a key component in improving the strengths and cost-effectiveness of its forces.

6 US Department of Defense (DoD), 'Unmanned Systems Integrated Roadmap FY2013-2038'. Here, DoD states that autonomy in uncrewed systems will be critical to future conflicts 'that will be fought and won with technology'. The document highlights that 'the special feature of an autonomous system is its ability to be goal-directed in unpredictable situations'. It also sets out a 25-year vision for the development, production, test, training, operation and sustainment of uncrewed systems technology throughout the DoD.

and the phenomenon of dual-use in many of the technologies underpinning unsupervised practices.

3.1 Current practice

A useful starting point to understand these drivers comes from the Stockholm International Peace Research Institute (SIPRI). By 2020, its dataset on weapon systems exhibiting degrees of autonomy already comprised three hundred and eighty-one different examples. That universe, however, does not of course equate to three hundred and eighty-one wholly autonomous systems.[7] Two points arise. The battlefield has long been experiencing broad automation across its processes, from crewed and hybrid and all the way to fully autonomous systems. The phenomenon, moreover, is evident almost regardless of platform, from transition examples in decision-aids and air defence, across drone systems, even in main battle tanks.

What then are the drivers accelerating innovation, a new breadth of tasking and an emerging democratisation in how parties procure these assets (including quicker accreditation of suppliers and more agile commercial collaborations)?[8] Here, states' adoption of uncrewed aircraft systems (UAS) provides a useful case study. A decade ago, by July 2013, the number of UAS procured by the US Department of Defense (DoD) already exceeded ten thousand units, mirroring the inaugural requirement announced in DARPA's Robotics Challenge that year for machines that must undertake complex automated tasking (here, the negotiation of rough terrain, removal of debris and dealing with multipart maintenance tasks). A decade forward and the ecosystem is unrecognisable; from Ukrainian software engineers shuttering their offices to deploy their laptops adjacent to field artillery units, data scientists can now retrain target recognition models just minutes after a drone strike in order to improve accuracy. Software applications (such as Ukraine's Uber for Artillery) have reduced the time between target identification and engagement from twenty minutes to just thirty seconds. Those same developers have written classifiers in order to detect Russian tanks hidden behind natural and artificial camouflage and to do this better than the human

---

7 The point here is that such systems include an element of autonomous function in their operation. SIPRI is therefore identifying unsupervised components in the weapon rather than that whole weapon being capable of autonomous function.
8 Richardson, William and others, 'Towards Agile Procurement for National Defence: Matching the Pace of Technological Change', Canadian Global Affairs Institute (June 2020), https://www.cgai.ca/toward_agile_procurement_for_national_defence_matching_the_pace_of_technological_change.

eye simply using video feeds.⁹ Innovation is clearly upending battlefield practices.

Key to this analysis is therefore to understand the pace and depth of transformation. By 1995, University of Texas students had demonstrated an UAS that was able to take off autonomously, locate and identify biohazardous material, map the location of each such barrel and return to its start point.¹⁰ Three years later and still some quarter of a century ago, a largely autonomous UAS no bigger than a model airplane successfully crossed the Atlantic to land at a predetermined landing point in mainland Europe. The timeline throughout has been accelerated by initiatives within the DoD. By 2001, the US Senate Armed Services Committee had hardwired an aggressive schedule to develop autonomous and uncrewed means.

> *'Abandoning the use of... [artificial intelligence] in weapon systems is akin to abandoning electricity and internal combustion engines'*

This was soon backed by the funding of two central goals: within ten years, one third of all deep-strike aircraft should be uncrewed and, within fifteen years of that date, one third of all ground combat vehicles should operate without human beings on board. Such 'institutional purpose' was critical then and remains critical today, the introduction of autonomy also being pursued across services. A main tenet of its oft-repeated Task Force Report, The Role of Autonomy in DoD Systems, has been to promote autonomy's operational benefits (segmented by domain, scenario and environment) and the 'capability surprise' that these innovations offer.

A question for this chapter is also to evaluate the phenomenon's persistence and the degree to which weapon autonomy is here to stay. This was certainly the case when your author started his studies a decade past but, as evidenced by developments right across uncrewed assets, this is no longer really a relevant line of inquiry. To quote Paul Scharre, 'abandoning the use of ... [artificial intelligence] in weapon systems is akin to abandoning

---

9 Richardson, Michael and others, 'Battlefield Autonomy: A Proven Way Forward, Defense Acquisition University (February 2024), https://www.dau.edu/library/defense-atl/BattlefieldAutonomy.
10 Reyes, Arthur and others, 'Overview of the University of Texas and Arlington's Autonomous Vehicles Laboratory', Department of Computer Science and Engineering, Technical Report CSE-2003-13 (2013), generally. By way of subsequent narrative, in the four years to fiscal year 2010, flight hours for UAS had increased from 165,000 hours to more than 550,000 hours and the inventory from less than 3,000 to 6,500.

electricity and internal combustion engines'.[11] This book's strapline may be that we need to retain humans on the battlefield, but the narrative here is an almost singular assumption that weapon development is unstoppable. This is the new context, the new norm and one that is unlikely to change, commentators pointing to the same processes that allow self-driving cars to avoid pedestrians instead enabling weapons to hunt and attack targets on their own.

The industry's global trade association reckons that some three thousand companies have been involved for more than a decade across the UAS procurement chain. Today, the global drone market alone is valued at more than twenty-five billion dollars with, interestingly, several of the larger participants (3D Robotics, AeroVironment, BirdsEyeView, Delair, DroneDeploy and PrecisionHawk) founded as long ago as the distant 2010s.[12] The 2023 Las Vegas Drone Show had more than two hundred trade stands with representatives from seventy-five countries. An adjacent driver, of course, is that all the building blocks upon which autonomy depends can be procured from well-diversified sources, chiming with Seba's disruption model whereby rapid scaling up usually precedes a tipping point in a technology's general adoption. It is then only a short hop before morphing into a fully fledged driver to that technology regardless of underlying industry.

> *A finding from Seba's disruption model is that rapid scaling up usually precedes a tipping point in a technology's general adoption*

But hidden versos exist, both technical and behavioural. An uncomfortable truth is the counterintuitive ramping-up in personnel that is required to support those very technologies that are intended to free weapons from this human supervision. An inescapable nub is that these technologies still depend upon human-centric endeavour. US Air Force estimates from 2016 suggest that it was already taking seventeen people to fly each uncrewed aircraft.[13] Surprisingly, this ratio has worsened in the intervening half decade. Although a thirty-strong company supports an Apache attack helicopter, the

---

11 Schreiner, Max, 'AI in War: How Artificial Intelligence Is Changing the Battlefield', The Decoder (21 January 2023), https://the-decoder.com/ai-in-war-how-artificial-intelligence-is-changing-the-battlefield/.
12 IMARC Analysis, 'Top 11 Drone Manufacturers in the World', https://www.imarcgroup.com/top-drone-manufacturing-companies, accessed February 2024.
13 Norton, Travis, 'Staffing for Unmanned Aerial Systems (UAS)', pp. 43, 89 and 91. See also: Hambling, David, *Swarm Troopers: How Small Drones Will Conquer the World* (Archangel Ink, 2015), p. 34.

suggested crew for the US Grey Eagle drone is currently a company of one hundred and thirty-five individuals.[14] And all of this at a time when the US military fell one whole quarter short of its 2022 recruiting target. While these numbers will change as the underlying technologies evolve, the human 'cost' of tasking for uncrewed aerial vehicles (UAVs) must still include those in the Delivery Cohort responsible for logistics and preparation, weapon control and calibration as well as all of those tasks related to piloting the platform. Given the coordination required of planning and maintenance, launch and recovery, surveillance

An uncomfortable truth is the counterintuitive ramping-up in personnel required to support those very technologies that are intended to free weapons from human supervision

and PED (processing, exploitation and dissemination), it should be no surprise that the number of people supporting each more sophisticated UAV amounts to more than two hundred professionals.

It is also the portfolio of capabilities that catalyses deployment of these assets. Matching specific battlefield tasks with specific enabling technologies promises useful (and often novel) means for the commander.[15] It is also in the interests of commercial parties to broaden AWS use cases, usually by repurposing existing means in these assets' procurement. BAE Systems' press release for its Taranis UAV was early to tout its weapon system's 'autonomous brains'[16], the pitch for their platform suggesting military advantage of surgical engagement but also a cost benefit relative to its crewed equivalents. The trend has accelerated over the past half decade with parties developing teaming and wingman initiatives. The Australian Air Force, for example, is building upon its Boeing's MQ-28 Ghost Bat platform to include unsupervised striking of enemy targets, remote surveillance, the jamming of enemy signals and the ability to act as decoys.[17] A similar initiative is the

---

14 Freedberg, Sydney, "'Unmanned' Drones Take Too Many Humans to Operate, Says Top Army Aviator', *Breaking Defense* (27 February 2023), https://breakingdefense.com/2023/02/unmanned-drones-take-too-many-humans-to-operate-says-top-army-aviator/.

15 In order to define 'battlefield activities', a useful starting point herewith is still provided by US Army Field Manual, *Intelligence Preparation of the Battlefield*, Section 1 (FM 34-130, July 1994), https://fas.org/irp/doddir/army/fm34-130.pdf. The document covers battlefield environment, effects, threats, management as well as battle execution, space and time as it relates to AWS deployment.

16 Smith, Chris, 'What Is Taranis? Everything You Need to Know About Britain's Undetectable Drone', BT website (21 November 2017).

17 Losey, Stephen, 'New in 2024: Air Force Plans Autonomous Flight Tests for Drone Wingmen', *Defense News* (30 December 2023), https://www.defensenews.com/air/2023/12/30/new-in-2024-air-force-plans-autonomous-flight-tests-for-drone-wingmen/.

drive towards collaborative combat aircraft, a low-observable platform that can escort and coordinate with crewed aircraft.

When it comes to defence, the number of buyers is small. For large and complex systems, governments' defence departments are really the single customer, defence markets generally being a closed monopsony.[18] The type of defence manufacturer, however, is rapidly evolving as digital horizons expand; in January 2024, for instance, even OpenAI quietly removed language from its usage policy forbidding the use of its products for 'military and warfare'.[19] At the time of writing, the US Department of Defense had eighty three active contracts for generative AI work, each contract value ranging between $4 million and $60 million. Defence startup Anduril raised $1.5 billion in late 2024 to develop its 'Arsenal' manufacturing platform that will purportedly be capable of producing "tens of thousands of autonomous weapons" each year.[20] Anduril's move is in line with the US Pentagon's Replicator initiative intended to resource companies capable of producing thousands of attritable autonomous systems, an early recipient being AeroVironment that makes Switchblade drones.

Two deployment challenges arise from these developments. First, buying decisions occur infrequently and still have long purchase cycles. Second, buyers rightly fear that manufacturers lack appropriate incentive to keep improving and iterating their products once that contract has been awarded. This is important because of the fundamental difference between AI and software; 'AI is monolithic', notes Rebesco and Manganiello, 'Unlike software, where deterministic outputs to the same input are a core principle, the performance of AI models is highly contextual and time-dependent. Models drift, data changes and even so-called foundational models [must be] fine-tuned, retrained and repurposed to new users and new use cases'.[21] There is only the best model for a particular data point at a particular time. In autonomous systems, the 'fit' between a data point and the model being procured is ephemeral and unique and requires comprehensive management from the moment it is deployed.

---

18 Rebesco, Jim and Anthony Manganiello, 'Rething the Role of a Systems Integrator for artificial Intelligence', *War on the Rocks*, 20 August 2024, https://warontherocks.com/2024/08/rethinking-the-role-of-a-systems-integrator-for-artificial-intelligence/, p.3.
19 Lazzaro, Sage, 'OpenAI is Quietly Pitching its Products to the US Military and National Security Establishment', *Fortune*, 17 October 2024, https://fortune.com/2024/10/17/openai-is-quietly-pitching-its-products-to-the-u-s-military-and-national-security-establishment/.
20 Knight, Will, 'Palmer Luckey's Defense Startup, Anduril, Raises $1.5 Billion to Produce AI-Powered Weapons', *Wired*, 8 August 2024, https://www.wired.com/story/anduril-palmer-luckey-funding-ai-drones-arsenal-factory/.
21 Rebesco, Jim and Anthony Manganiello, p.4.

Just as the arrival of novel capability creates that tipping point in adoption identified by Seba, it also fosters an institutional 'fear-of-missing-out'.[22] Examples are useful to illustrate these new possibilities. From fifteen miles' distance, the US Predator UAV can already work with image features that measure just a handful of inches across. Coherent change detection then allows that machine to note differences between the scene under observation and one that had been recorded earlier, allowing those on the ground to identify, for instance, disturbances left by an improvised explosive device on the side of the road. The point here is to highlight the cumulative effect of individual technical advances whereby quite specialised developments in componentry can quickly catalyse wholly new means of battlecraft that parties are compelled to adopt.[23]

> *There is only the best model for a particular data point at a particular time. The 'fit' between a data point and the model is ephemeral and unique*

Similarly, innovation can also accelerate refinement to these models. Ukraine has been an active testing ground for commercially originated technology, from satellite systems to the broadest variety of drones. Starlink's deployment to Ukraine was the first time a commercial company provided the backbone for a 'country's military capability during wartime. Not only does the system illustrate interesting technical innovation, but it also signposts institutional innovation on the battlefield, evidencing disruption in how states are sourcing technology outside traditional procurement methods. Procurement, so often a question of incremental upgrading and amalgamation, has quickly become a matter of substitution and the integration of novel systems into legacy force design.[24]

The adoption of first-person-view (FPV) drones provides a further case in point. Offering cheap, accurate firepower[25], the platform innovation matters precisely because the weapon class embodies important new trends in how

---

22 Ling, Justin, 'To Beat Russia, Ukraine Needs a Major Tech Breakthrough', *Wired* (4 January 2024), https://www.wired.com/story/ukraine-russia-future-war-tech/.
23 In order to deliver its service, the Predator must then deploy multiple cameras featuring various levels of zoom from a 45° wide-angle view down to an ultra-narrow 0.2° tunnel view. On a standard 35mm camera, the equivalent lenses at these extreme ends would be a 55mm wide-angle lens and a 12,000mm telephoto. See: Hambling, *Swarm Troopers*, p. 45.
24 Walker, Paddy, 'Agile Procurement? Norms and Challenges to the Integration of Novel Systems into Legacy Force Design', Ares & Athena, Centre for Historical Analysis and Conflict Research (Summer 2022).
25 Detsch, Jack, 'Ukraine's Cheap Drones Are Decimating Russia's Tanks', *Foreign Policy* (9 April 2024), https://foreignpolicy.com/2024/04/09/drones-russia-tanks-ukraine-war-fpv-artillery/.

war is being fought: the shift towards small, cheap and disposable weapons; the increasing use of consumer technology; and a drift towards autonomy in battle. These incremental changes can soon add up to a quite material shift in war's conduct, even if they do not constitute any individual revolution in themselves. In this case, innovation is also important to states' treasuries. An entry-level FPV drone costs between four hundred and fifty dollars and six hundred and thirty dollars.[26] As noted by the *Economist*, unguided artillery shells cost between eight hundred dollars and nine thousand dollars,[27] and the budget for a GPS-guided munition is some one hundred thousand dollars. For the purposes of comparison, a planner can purchase more than five hundred of those same FPV drones for what is costs to buy a two-hundred-and-twenty-five-thousand–dollar Javelin anti-tank missile. These are extraordinary economics.

> *Procurement, so often a question of incremental upgrading and amalgamation, has quickly become a matter of substitution and the integration of novel systems into legacy force design*

Innovation is also transforming parties' operational domain whereby a typical Ukrainian assault group of a dozen soldiers is now accompanied by almost the same number of drone operators, half of whom are FPV drone pilots with the remainder undertaking reconnaissance tasks using other drone variants.[28] As a pre-cursor to less supervised munitions, FPV drones are also attractive for closing the firepower gap. In early 2024, Ukraine faced a critical shortage of conventional artillery, firing some three thousand shells each day, just one quarter the figure that Russia was able to field. Drones have then provided an agile addition to Ukrainian artillery efforts. Because they can relay pictures to their operators in real time, these technologies also provide detailed context on the developing battlefield. As evidenced by internet footage of drones chasing vehicles or being flown into buildings or trenches, the platforms offer novel capability to artillery assets. The weapon class also offers psychological advantage, whether loitering after an artillery

---

26 Cook, Ellie, 'Ukraine's Cheap FPV Drones More Efficient than Prized Artillery', *Newsweek*, 20 December 2023, https://www.newsweek.com/ukraine-fpv-drones-mykhailo-fedorov-russia-avdiivka-1853646.
27 Economist, 'How Cheap Drones Are Transforming Warfare in Ukraine', *Economist* (5 February 2024), p. 2.
28 Hambling, David, 'What Does Ukraine's Million-Drone Army Mean for the Future of War?', *New Scientist* (19 January 2024), https://www.newscientist.com/article/2413260-what-does-ukraines-million-drone-army-mean-for-the-future-of-war/.

barrage or forcing adversaries to disperse into small groups that are only able to conduct operations under the cover of darkness.

Many of these enabling capabilities are already diluting the human element required to dispense lethality. In particular, autonomous object recognition should in time enable AWS to complete tasks whether or not the weapons are subject to adversarial jamming. Recognition tools, moreover, are already a widely available and inexpensive commercial add-on for uncrewed assets. America's Switchblade 300 already fields a hardened version of the capability at a reported cost of fifty-three thousand dollars while Russia's Ovod (Gadfly) FPV drone has been trialing a similar 'terminal guidance' system 'rooted in AI' since 2022. The double discontinuity here is the disruption to the technology's cost structures and also, at the same moment, in what these platforms can do. As a result, the battlefield is suddenly awash with variant systems, each new one being a step up from its immediate forerunner. The *Economist* reports, for instance, that the Ukrainian Scalpel drone costs as little as one thousand dollars and can lock onto a target designated by its pilot. Its AirUnit drone, still in prototype at the time of writing, promises an autonomous version that is likely to be even less expensive.

This snowball effect is itself a driver, both within weapon types and also in the user acceptance of these assets. Again, several small adaptions have coalesced into what is now a key piece in how fighters fight. Replacing previously bespoke componentry with programmable chips has, for instance, made frequency hopping a much easier task, reducing the efficacy of enemy jamming and the compromising of communications. Adding electro-optical means to GPS navigation has then enabled remote terrain-tracking to be undertaken

> *The weapon class embodies new trends in how war is being fought: the shift towards small, cheap and disposable weapons, the increasing use of consumer technology and a drift towards autonomy in battle*

from on board the platform without having to rely upon remote guidance; already in commercial production, they are proving cheap to introduce, easy to engineer and compliant under the US International Traffic in Arms Regulations. War is getting sneakier and more customised. From the Houthis' use of guidance kits in Yemen driving anti-ship missiles in the Red Sea to Iranian development of long-range strike drones, parties are realising that they can generate geopolitical effects that far outweigh the cost of those initiatives.

These remote platforms have long been extraordinarily lethal. The US and UK's large variant UAV currently carry up to fourteen Hellfire missiles.

They are also versatile. The Reaper can be dispatched carrying just four such missiles but with a pair of laser-guided five hundred-pound bombs. These drivers are also agnostic to tasking, whether primarily as a defence system (responding either automatically or autonomously to incoming munitions, or both), as an anti-personnel system tackling incursions into a defined area or as a weapon tasked with offensive seek-and-destroy capabilities. Indeed, the difference between a machine that can do these things and one that can make its own attack decisions is increasingly only a matter of programming, with improvements in one such capability prompting expectations across that platform's greater capabilities. This is particularly the case in uncrewed assets. Aerial drones have played an outsized role in the war in Ukraine, but uncrewed ground vehicles are similarly becoming a battlefield staple. Typically, four or six wheeled, uncrewed machines are already kitted out for multiple purposes, from supplying isolated front-line positions, evacuating injured combatants, placing or destroying landmines, and otherwise delivering explosive ordnance.[29] At the time of writing, unit range remains a limiting factor to their deployment, still between two and three kilometres and still requiring human supervision from targeteers and others providing support. But their place on the battlefield is certain, a general quip being that 'no one really cares if the ground vehicle gets destroyed, except the accountants'.

3.2 Technology creep and dual-use technology trends

Machine independence is not, of course, the preserve of military applications and significant blurring exists between commercial and military use of autonomous technologies. This creates a dichotomy. On the one hand, incoherence complicates arguments around banning self-directed weaponry. On the other, we have seen that commercial interests are pivotal in creating technical solutions for military use; those same autonomous capabilities found in a remotely operating search-and-rescue vehicle now underpin functions across a lethal autonomous weapon. The technology pathways of civilian aviation provide clear precedent. The Airbus

---

29 Burgess, Matt, 'Robots and Fighting Robots in Russia's War in Ukraine', *Wired* (30 January 2024), https://www.wired.com/story/robots-are-fighting-robots-in-russias-war-in-ukraine/.

A320 can take off and land itself and a study by the Humans and Autonomy Laboratory reported that pilots already spend less than three minutes per flight with their hands on cockpit instruments.[30] Boeing 777 pilots similarly report that they are in control for less than ten minutes each flight.

System autonomy has long been pioneered by industry. The US credit, debit and prepaid card industry is a case in point. It monitors billions of transactions per second with autonomous tools that identify fraudulent transactions within milliseconds, all undertaken on a customer base of several billion cards. The norm here is that autonomous processes have been managing data velocity and mass without human supervision for decades. For decades, IBM's Watson for Oncology has provided cancer doctors with recommended courses of treatment within seconds from its unstructured universe of medical documents. Similarly, Google's machine-learning password classifiers authenticate users, reducing the chance of being scammed by ninety-nine per cent.

Autonomous agents have been similarly disruptive across other trades, and it is therefore to be expected that a similar shift will now take place in battlefield practices. Autonomy, moreover, particularly features across civilian uses of drone technology.[31] Non-military autonomous practices already decide actions in uncrewed assets used across US police and border security as well as by parties involved in agriculture, maritime and forestry. Civilian markets investing in autonomous aerial means include the coastguard, parties involved in oil and gas, in electricity grids, in climate modelling and across the communications industries. Commercial drone applications that are regularly advertised by the likes of Amazon reinforce the narrative of what appears to be a single technology that is already ubiquitous and inevitable. UAV and autonomous means are becoming ubiquitous, cites the 2024 website for the Association for Uncrewed Vehicle Systems International

---

30 See also: Bray, Hiawatha, 'You've Heard about Self-Driving Cars. What about Self-Flying Planes?', *Boston Globe* (30 May 2024), https://www.bostonglobe.com/2024/05/30/business/self-flying-planes.

31 Examples abound. Drone technology is used to provide promotional video for real estate agencies and field inspection for farmers. As above, budgets are the principal driver given such UAV services can be delivered at a fraction of the cost of a helicopter. For sources, see: Lantigua, Joshua, 'Commercial Use of Drones Has Already Taken Flight', phys.org (11 December 2013), http://phys.org/news/2013-12-commercial-drones-flight.html and Global Research, 'Drones: From "Military Use" to "Civilian Use". Towards the Remote UAV Policing of Civil Society?', Centre for Research on Globalization (16 May 2012), http://www.globalresearch.ca/drones-from-military-use-to-civilian-use-towards-the-remote-uav-policing-of-civil-society/30876. See also: Walker, 'Agile Procurement', pp. 65-68. Also: Simonite, Tom, 'Sorry, Banning 'Killer Robots' Just Isn't Practical', *Wired* (22 August 2018), https://www.wired.com/story/sorry-banning-killer-robots-just-isnt-practical/.

(AUVSI), reckoning that some two thousand different drone types are in production at the time of writing.³²

Unsurprisingly, all this quickly feeds into military procurement. Adoption of both uncrewed and generally unsupervised platforms has long been considered a priority by the US House of Representatives' Committee on Oversight and Government Reform.³³ The pace and sophistication of implementation have increased exponentially, leading Christof Heyns, 'United Nations Special Rapporteur, to conclude in 2013 that while 'bright lines are difficult to find',

> *The refitting of equipment often signals the embrace of new capabilities, further diluting human supervision in these combat assets, whether through updates to those programmes or through changes in use-case*

lethal automated weapons are 'combinations of underlying technologies with multiple purposes'.³⁴ The landscape is being transformed by incremental 'technology creep' evidenced by rolling equipment refits. Creep is both the rapid iteration of individual technologies and also the combination of developing technologies in a manner that moves them away from their first intended specification. The refitting of existing equipment often signals the embrace of new capabilities, further diluting human supervision in these combat assets, whether through updates to those programmes or through changes in use-case. The challenge for civil society is that this same incrementalism can camouflage quite quick movement along a weapon's pathway to autonomy; a seemingly innocuous innovation (perhaps a small mobile robot newly programmed to respond autonomously to protect peacekeepers from local ensnarement) might actually mark a material discontinuity in how war is waged.

Statistics reinforce the issue. Some eighty per cent of Ukrainian-generated targets in 2022 were likely derived from drone-produced intelligence. But more than ninety per cent of those Ukrainian drones were destroyed in just

---

32 Source: Auvsi, http://www.auvsi.org/home.
33 Anderson, Kenneth, 'Rise of the Drones: Unmanned Systems and the Future of War', testimony submitted to the Subcommittee on National Security and Foreign Affairs, Committee on Oversight and Government Reform, US House of Representatives (23 March 2010), http://digitalcommons.wcl.american.edu/cgi/viewcontent.cgi?article=1002&context=pub_disc_cong, generally.
34 Human Rights Council, 'Report of the Special Rapporteur on extrajudicial, summary or arbitrary executions, Christof Heyns' (9 April 2013), https://www.ohchr.org/sites/default/files/Documents/HRBodies/HRCouncil/RegularSession/Session23/A-HRC-23-47_en.pdf.

the first quarter of that year.[35] During that period, the average life expectancy of a fixed-wing drone was thought to be six flights. A less sophisticated quadcopter was expected to last just three flights. These assets' survivability will doubtless improve as further innovative features come on stream, which will also accelerate adversaries' efforts to counter the weapon class. Indeed, effective measures to defeat drone attacks have seen the main battle tank to become a frequently requested combat asset by brigade commanders in Ukraine in 2024. At the same time, however, these have been matched by technical advances that both morph and increase UAV tasking, including new iterations of video cameras, thermal imagers and radio antennae in order to process targets, manage and then distribute data to the best placed 'shooter'. It is these advances that continue to push planners to embed autonomy across their battle systems.

A seemingly innocuous innovation in technology might actually mark material discontinuity in how war is waged

Each side of the Ukrainian conflict has employed somewhat different approaches to embedding artificial intelligence (AI) into its weapon systems. Ukraine and its Western allies have tended to focus upon fast identification, tracking and targeting while their adversary has depended upon loitering munitions and improving intelligence, surveillance and reconnaissance capabilities to undertake precision targeting. Put simply, Western AI-enabled systems appear to focus 'on the left side of the observe, orient, decide and act loop',[36] prioritising faster targeting and generally better war fighting capabilities, while Russia has adopted programmes to automate the entire kill chain. An adjunct challenge is that communication has become increasingly fraught given advances in jamming, spoofing and closing down pockets of transmission.

This has complicated drone operations, the more so given the new collection of vast amounts of video footage, far too much either to be sent (given insufficient bandwidth and adversarial meddling) or processed (given insufficient intelligence capacity). The release valve here is that such data management moves instead on board the uncrewed platform, a further driver towards autonomous operation. This also accords with the use of small,

---

35 Economist, 'The War in Ukraine Shows How Technology Is Changing the Battlefield', *Economist Special Report, Lessons from Ukraine* (3 July 2023), p. 4.
36 Bendett, Sam and Jane Pinelis, 'How the West Can Match Russia in Drone Innovation', 25 January 2024, War on the Rocks, https://warontherocks.com/2024/01/how-the-west-can-match-russia-in-drone-innovation/.

low-powered chips which can already figure out rudimentary information about potential targets from the drone's sensed data, a job that would once have been done only on a distant cloud server.

3.3 Structural and procurement drivers

In considering structural drivers, it is useful to understand the high-level conditions that must be in place in advance of AWS adoption. First, there needs to be broad trust in these weapons' capabilities. There also needs to be general *cultural* acceptance of those technologies and for this to be evident across all organisational, institutional and societal parties. There then needs to be in place reasonable availability of (and hence familiarity with) these new capabilities. Trust here is clearly driven by the weapon's fitness for purpose and the repeatability of machine outcomes, with confidence also becoming a catalyst if users are really to delegate decisions to an algorithm. Finally to this point is a clear process of assessing benefit: to be adopted, autonomy must reduce the burden for both soldier and commander, with identifiable knowledge transfer between machines following repeated testing (and this taking place in both collaborative and non-collaborative environments).

Institutional drivers come in different guises. In 1958 US president Dwight Eisenhower created the Advanced Research Projects Agency with its purpose of undertaking research and development projects to expand the frontiers of technology and science. That role had been created in response to the Soviet launch of Sputnik two years before. Today, that mission has been taken on by the Defense Advanced Research Projects Agency (DARPA) to ensure that US military technology retains an edge over its potential enemies and so preserve decisive military advantage for the US Armed Forces. As an accelerator, therefore, DARPA's programmes provide useful insight into America's passing military priorities (as well, of course, as its potential weaknesses) and, in identifying drivers to AWS, give context to how the US intends autonomous platforms to develop. Indeed, a recent theme for DARPA has been adaptive systems, autonomous decision aids and intelligent battlefield processes that are increasingly independent of human oversight.

This is also reflected in US weapon procurement. By 2016, the DoD was already spending three billion dollars on uncrewed aircraft comprising, by number, forty per cent of all US aircraft.[37] By 2023, twenty-nine programmes were fully dedicated to the procurement of UAS, from uncrewed carrier mission vehicles, control assets, counter-drone products to, interestingly, uncrewed aerial targets.[38] The scope of deployment models, each with different degrees of supervision and human oversight, has similarly expanded. When the American Federal Aviation Administration introduced its national framework for registering uncrewed aircraft in late 2015, more than one hundred and eighty thousand drones where were registered in the first two weeks. As evidenced in Chapter Five's discussion on proliferation, the US is not alone in this phenomenon; by 2018, more than seventy countries were operating uncrewed aircraft with thirty armed UAS programmes established or in development.

Trust here is clearly driven by the weapon's fitness for purpose and the repeatability of machine outcomes

Other deep-seated trends act as institutional drivers. The first of these is run-away budgets and the escalating costs of having 'boots on the ground' (or, more importantly, in the air). Between 2001 and 2012, US compensation costs per active-duty service member grew nearly sixty per cent.[39] Today, it costs the DoD more than one hundred thousand dollars each year across its services to fund each combatant's health, accommodation and well-being costs.[40] Pensions, pay, training and equipment are additional. Adjusted for inflation, this equates to an annual (and unsustainable) growth in the high single digits over the past decade.[41] Crewed systems, moreover, are costly in any military environment: the human fighter must be able to breathe, eat and take care of bodily functions. His and her safety must be addressed

---

37 Gettinger, Dan, 'Drones in the Defense Budget', Center for the Study of the Drone, Bard College, October 2017, generally. Also: Gertler, J, 'US Unmanned Aerial Systems', Congressional Research Service, CRS R42136 (3 January 2012) and J Gertler, 'How Many UAVs for DoD?', Congressional Research Service, CRS IN10317 (2015), generally.
38 McNabb, Miriam, 'How Much Will the US Department of Defense Spend on Drones in 2023? AUVSI's Report', *Drone Life* (17 January 2023), https://dronelife.com/2023/01/17/how-much-will-the-u-s-department-of-defense-spend-on-drones-in-2023-auvsis-report/.
39 Work, Robert and Shawn Brimley, '20YY; Preparing for War in the Robotic Age', Centre for a New American Security (January 2014), p. 20, https://www.cnas.org/publications/reports/20yy-preparing-for-war-in-the-robotic-age.
40 Shaughnessy, Ian, 'The Ethics of Robots in War', *NCO Journal* (21 February 2024), https://www.armyupress.army.mil/Journals/NCO-Journal/Archives/2024/February/The-Ethics-of-Robots-in-War.
41 Over the same period, the share of the base DoD budget for military personnel-related costs rose from thirty per cent to thirty-four per cent and was expected to consume forty-six per cent of the budget by 2021 even with a 2.6 per cent historically normal real annual rate of growth.

with armour, redundant control apparatus and escape systems. In the case of ground troops, transportation must be available to and from a contact zone as well as being tailored to support each combatant while in theatre. Each subordinate system must then be underpinned by its own complex logistics chain. In addition, medical supplies, facilities and staff must be immediately available to evacuate and treat the injured. The human component of these chains requires extensive management, training and generous benefits programmes for life.[42] Simply to recruit a soldier costs fifteen thousand dollars in the US while the cost of treating an injured US soldier is reckoned to average two million dollars. It is against this familiar context that calls intensify to remove humans from the battlefield.

By comparison, the cost of robotic systems continues to reduce given increased component commonality and innovative means of manufacture. Considered below, swarm robotics alone are expected to grow by a compound annual rate of more than thirty per cent to 2028.[43] Even by 2015, studies undertaken by Duke University were concluding that the fully loaded crew cost per hour of a UAV was ninety per cent less (one hundred and fifty dollars compared to two thousand dollars) than a crewed aircraft.[44] This cost differential may be subject to the vagaries of accounting but the clear trajectory is towards the embrace of robotics and uncrewed battlefield assets.[45] Fully costed, parties

The human fighter must be able to breathe, eat and take care of bodily functions. His and her safety must be addressed with armour, redundant control apparatus and escape systems

---

42 The National Audit Office's analysis provides useful context here in demonstrating the current costs of manpower assets. twenty-nine per cent of the UK's 2017 budget of 35.3 billion GBP was accounted for by service personnel, four per cent by civilian contractors and four per cent by administration. Pensions to service personnel comprised two per cent of the UK's defence budget in 2017. See: UK National Audit Office, 'A Short Guide to the Ministry of Defence' (2017), p. 9, https://www.nao.org.uk/wp-content/uploads/2017/09/A-short-guide-to-the-Ministry-of-Defence.pdf.
43 Werner, Pieter, 'Swarm Robotics Market Set for Explosive Growth by 2028', *RockingRobots* (9 October 2023), https://www.rockingrobots.com/swarm-robotics-market-set-for-explosive-growth-by-2028/.
44 Jaffe, Ian, 'Former Fighter Pilot, Duke Prof Missy Cummings Talks Drones', *Duke Chronicle* (15 September 2015), http://www.dukechronicle.com/article/2015/09/former-fighter-pilot-duke-prof-missy-cummings-talks-drones. Note, however, that sensor multiplicity on UAV is likely to dull this advantage if further analysts are then required to process additional data.
45 For Russian military cost inflation, see: Global Security, 'Russian Military Budget' (last modified 4 September 2023), http://www.globalsecurity.org/military/world/russia/mo-budget.htm. For Chinese military cost inflation, see: Perrett, B, 'China's Inflation-Adjusted

appear to spend more than thirty thousand dollars per hour to fly their F-35 while a Reaper costs less than four thousand dollars to operate for each hour of use.[46]

The differential is also mirrored in the relative *capital* costs of uncrewed versus crewed assets. It is expensive to have humans in and around these platforms and spiraling hardware costs create their own institutional driver in the face of squeezed procurement budgets. It is with reason that defence inflation and planners' attachment to very expensive platforms has prompted the tongue-in-cheek aphorism that, fifty years hence, 'the entire US defence budget will be able to purchase just one aircraft'.[47] Advocates unsurprisingly suggest that weapon autonomy provides an answer to exactly these escalating costs, in particular the procurement of increasingly 'exquisite' and multi-role platforms, prompting the Centre for a New American Security to suggest that escalating hardware costs mean that US armed forces are no longer able to replace front-line combat systems on a one-for-one basis.[48]

The example of the F-35 provides an appropriate case study on procurement drivers. 'When an aircraft has a pilot on board', notes Christian Enemark, 'there is a need to accommodate and protect frail human flesh in the engineering, construction and use of that aircraft'.[49] By 2014, eight years into production, the F-35 cost some one hundred and ninety million dollars'[50], more than validating the calculation in 2012 by Congress' watchdog agency that the average price for an F-35 aircraft had already doubled since the programme's inception.[51] The total cost of ownership, including maintaining

---

Defense Budget Up 7.5%,' *Aviation Week* (18 March 2013), http://aviationweek.com/awin/china-s-inflation-adjusted-defense-budget-75.
46 Hambling, *Swarm Troopers*, p. 66.
47 Wolf, Katharina, 'Putting Number on Capabilities: Defence Inflation versus Cost Escalation', European Institute for Security Studies, Brief Issue, 27 (July 2015), p. 1 and generally https://www.iss.europa.eu/sites/default/files/EUISSFiles/Brief_27_Defence_inflation.pdf.
48 Centre for a New American Security, cit. Work and Brimley, p. 22. Surviving against steadily improving guided munitions already requires that all such crewed platforms incorporate costly stealth technology, stand-off ability and highly capable active and passive defences. In the case of the F-35 aircraft, see: https://www.washingtonpost.com/sf/brand-connect/the-f-35-how-it-works/.
49 Enemark, Christian, 'Armed Drones and the Ethics of War: Military Virtue in a Post-Heroic Age' (Routledge, 2014), p. 98.
50 A useful summary is provided by: Macias, Amanda, 'The Pentagon Is Trying to Figure out the True Cost of Its Costliest Weapon System, the F-35', CNBC (28 February 2018), https://www.cnbc.com/2018/02/28/pentagon-wants-to-know-true-cost-of-f-35-system.html. An uncrewed UAV, for instance, requires no cockpit pressurisation or temperature control and may have more space and payload capacity for fuel allowing it to stay in the air for longer (long-dwell, high altitude capabilities).
51 Shalal-Esa, Andrea, 'Insight: Expensive F-35 Fighter at Risk of Budget 'Death Spiral'', Reuters (15 March 2013).

and supporting the F-35 over its lifetime, is forecast to be more than three hundred million dollars per unit.[52] Longer production runs mean that this up-front figure per unit (depending, of course, upon variant) moves but, by 2023, it was averaging one hundred and nine million dollars per airframe while the lifetime cost of the whole programme still hovered around 1.7 trillion dollars.[53]

Nevertheless, the F-35 remains the critical platform for high-intensity conventional warfare with the US Air Force at various times planning to buy more than seventeen hundred of these machines. The extraordinary expectation is also that these aircraft will remain in service through to 2070.[54] But these jets are still quite specific in their use case. And they are certainly too expensive to lose in great numbers, because of their cost and also because of their very long replacement cycle and the potential to lose pilots and consequent public relations ramifications. Developments in Ukraine, furthermore, would suggest a marked *narrowing* in possible tasks as well as a lengthening list of assignments to which F-35s are wholly unsuited, from countering low-cost enemy drones in the air littoral to engaging low-value, dispersed targets. This stark economic mismatch is not going away, the cost of a single Chinese Sunflower suicide drone being less than thirty thousand dollars, just the number to cover sixteen thousand of these assets for just one F-35. This realisation is already affecting procurement, as the US Army cancelled its future reconnaissance helicopter in 2024; fielding a costly crewed scouting platform makes little sense given the advent of inexpensive drones that do not require trained, expensive pilots at the controls.

Fifty years hence, 'the entire US defence budget will be able to purchase just one aircraft'

Reducing complexity across the purchasing process is also a driver for AWS. The F-35's first development contract was signed in 1996 and, while the first units actually flew in 2006, the platform did not finally arrive into service until 2015, nearly twenty years after initial consent.[55] The complexity

---

52 Hambling, *Swarm Troopers*, p. 94.
53 Haywood, Justion, 'How Much Does An F-35 Cost?', *Simple Flying* (20 December 2023), https://simpleflying.com/how-much-does-an-f-35-cost/.
54 Barno, David and Nora Bensahel, 'Drones, the Air Littoral, and the Looming Irrelevance of the US Air Force' (7 March 2024), *War on the Rocks*, https://warontherocks.com/2024/03/drones-the-air-littoral-and-the-looming-irrelevance-of-the-u-s-air-force/.
55 This has certain procurement ramifications. Only two consortia were of sufficient size to compete for the contract to construct the high value aircraft. Cost and commercial risks

of crewed platforms also reduces the physical number of units that a state can afford. As platforms become fewer and dearer, a simple numbers game of how many aircraft you can get into the sky itself becomes a driver to the deployment. An emerging norm is that AWS promise cheaper

> *Escalating hardware costs mean that US armed forces are no longer able to replace front-line combat systems on a one-for-one basis*

and more numerous systems that can still deliver appropriate lethality. While these drivers may not be new (there were similar discussions taking place around mechanisation and aerial warfare in the 1930s), there is little argument made against weapon inflation.

Operational restrictions around current crewed weaponry also act as a catalyst. First, protecting the high value F-35 pilot through system duplication and physical shielding may account for up to sixty per cent of that platform's capital outlay. Second, this level of duplication and work-around may also degrade [operational] performance by up to eighty percent.[56] Unlike in its uncrewed, autonomous alternative, the combat aircraft's pilot remains the most critical (and most vulnerable) component of the F-35 ecosystem.[57] In addition to creating procurement and design constraints, the pilot is also a priority for adversarial targeting, becoming an increasing burden on that platform's performance in terms of additional armour, life support and operational limitations.

> *The stark economic mismatch is not going away, the cost of a single Chinese Sunflower suicide drone being less than thirty thousand dollars, just the number to cover sixteen thousand of these assets for just one F-35*

Augustine's Law XV holds that building in additional layers of redundancy compromising reliability as well as increasing costs exponentially: 'the last 10% of performance generates one-third of the cost and two-thirds of the

---

empirically limit business competition (and any subsequent chance for market forces to reduce a programme's build cost) as well restricting the benefits of possible collaboration. See: Hambling, *Swarm Troopers*, p. 94.

56 Zenko, Micha, 'The Coming Future of Autonomous Drones', Council on Foreign Relations (4 September 2012), https://www.cfr.org/blog/coming-future-autonomous-drones. Zenko was early to discuss the use of UAVs tasked with actions that remain short of warfare but which are nevertheless designed to secure military advantage.

57 As much as forty-five per cent of the costs of the A-4 Skyhawk is accounted by the crew protection, system redundancy and other outlay on shielding which would be avoided if the platform was uncrewed.

problems'.[58] Another risk to these expensive platforms is that the next technical shift may in short order erode their relevance. The Ukraine war has proved that the adoption of low-cost, agile UAV makes the presumption (let alone the guarantee) of airspace superiority ever more doubtful.[59] This is only likely to be upended further by the arrival of multiple, low-cost swarming systems.

3.4 Ethical drivers

It is also true, however, that several ethical considerations endorse diluting the human's role in lethal engagements, notwithstanding that their often intangible nature can make evaluation here a frustrating exercise. It may also seem a conclusion that is more based on heuristics than upon worked-through evidence. A strong theoretical notion is that machine-based safeguards can be built into autonomous weapon systems to ensure compliance with international humanitarian law (IHL). Supporters of roboticist and roboethicist Ronald Arkin, erstwhile professor of interactive computing at the Georgia Institute of Technology, have long promoted the general viability of what he terms an 'artificial ethical override'.[60] A driver for this argument comes from a 2006 Mental Health Advisory Team report out of the US Surgeon General's office in 2006 asserting that uncrewed combat systems could obviate certain ethical challenges in combat conditions.[61] According to that study, soldiers' conduct during

---

58 Smallwood, David, 'Augustine's Law Revisited', *Sound and Vibration* (March 2012), http://www.sandv.com/downloads/1203smal.pdf. See also: Chapter 4 (Deployment), specifically: 4.6 (Swarming models). Also: Hambling, *Swarm Troopers*, p. 96. Written in 1984 by former under-secretary to the US Army Ralph Norman Augustine as a tongue-in-cheek set of business aphorisms, Augustine's law XVI states that 'in the year 2054, the entire defense budget will purchase just one aircraft. This aircraft will have to be shared by the Air Force and Navy three-and-one-half days each per week except for leap year, when it will be made available to the Marines for the extra day'.
59 Hunter, Stoll, 'Air Defence Shapes War-fighting in Ukraine, RAND Blog (24 February 2024), https://www.rand.org/pubs/commentary/2024/02/air-defense-shapes-warfighting-in-ukraine.html.
60 Vanderelst, Dieter and Alan Winfield, 'An Architecture for Ethical Robots Inspired by the Simulation Theory of Cognition', *Cognitive Systems Research*, 48 (May 2018), pp. 56-65.
61 Arkin, Ronald, *Governing Lethal Behaviour: Embedding Ethics in a Hybrid Deliberate/Reactive Robot Architecture* (Georgia Institute of Technology, 2007), generally. See also: US Surgeon General's Office, 'Mental Health Advisory Team (MHAT) IV Operation Iraqi Freedom 05-07', Final Report (November 2006), http://www.combatreform.org/MHAT_IV_Report_17NOV06.pdf. Human Rights Watch's Mary Wareham notes that Arkin's work dates from 2006 and that

operations Iraqi Freedom and Enduring Freedom was often 'questionable' with some ten per cent of soldiers reporting that they had mistreated non-combatants or damaged civilian property and only forty-seven per cent of soldiers agreeing that non-combatants should be treated with dignity and respect.[62] The case here is that machines might be able to do a better job in moments of stress and that there is a role for algorithms in reducing behavioural lapses in battlefield conduct.

F-35s are certainly too expensive to lose in great numbers, because of their cost and also because of their very long replacement cycle

Battlefield surveys, of course, suffer from bias and sample challenges as well as arguments about rigour (in what, after all, is a stressed universe and one that is unusually prone to noise). Sampling combatants raises questions about accountability, anonymity, independence as well as difficulties introduced by cross-cultural issues and accounting for context.[63] Using these data sets is therefore fraught with difficulty but, for the narrow purposes of a book reviewing humans' role on the battlefield, it serves at least to highlight possible fault lines. To this end, the report found that more than thirty per cent of soldiers agreed with the contention that torture should be allowed to save the life of a fellow soldier. Nearly half of participating soldiers also reported that they would not report a colleague if he or she had killed or injured an innocent non-combatant. Similarly, less than half of the soldiers polled said that they would report a team member for unethical behaviour. Empirically, this may not be unexpected given the prominent role of trust in military ethics. Neuroscientists have also found that human circuits responsible for conscious self-control are particularly vulnerable to stress. When these circuits shut down, primal impulses go unchecked.

Setting aside difficulties regarding these primary sources, evidence does therefore exist upon which Arkin can base his argument for an ethical driver: properly crafted algorithms may, in theory, produce a more

---

he has published little on the subject since that date: 'The attention that Arkin receives is more to do with the counterpoint his research provides to the AWS debate and the theory behind his construct' (in conversation with the author, June 2014).

62 US Surgeon General's Office, 'Mental Health Advisory Team', generally.

63 Statistical issues include sampling bias, under-coverage and social desirability in the answering of the survey's questions, non-response bias, the issue of leading questions in the absence of any control group, difficulties with apportioning causation and dealing with dependent variables. See: Metcalf, Jacob, 'Ethics Codes: History, Context and Challenges', *Council for Big Data, Ethics and Society* (9 November 2014), https://bdes.datasociety.net/council-output/ethics-codes-history-context-and-challenges/.

consistent, more compliant engagement outcome than has been achievable in a fraught combat situation where, time and time again, human discipline and supervision have been proved inadequate. Indeed, ethical arguments in favour of AWS deployment appear on first reading to have surprising heft. Weapons autonomy, after all, would reduce soldiers' physical participation on the battlefield (even if there remains a considerable tail in support of these technologies) and this has several long-recognised advantages. As far back as 1946, evidence suggested that after sixty days of continuous combat ninety-eight per cent of surviving soldiers suffer psychiatric trauma.[64]

Another driver, moreover, is how to create soldiers that are 'fit for purpose' in the first place. Again parking contextual and statistical concerns, Dave Grossman's Second World War study suggests that 'most men simply did not kill'.[65] A further US Army study, this time by Samuel Marshall (and since discredited given a sample size of just four hundred men), suggested that only fifteen per cent of the infantry soldiers

'The last 10% of performance generates one third of the cost and two-thirds of the problems'

'interviewed' had actually fired at enemy positions on any occasion despite eighty per cent of the sample having the opportunity to do so.[66] Korean War studies then indicate that fifty per cent of the F-86 pilots never fired their weapons and, ignoring area bombing, only ten per cent of those had actually hit a target.[67] Finally to this point, evidence also exists to suggest that less than one per cent of Second World War pilots accounted for thirty to forty per cent of all downed enemy aircraft.[68] While much of this early (and predominantly US) material is now judged unrepresentative for wars of survival rather than

---

64 No comparable European study exists, hence use here of another US study: Swank, R, *Combat Neuroses: Development of Combat Exhaustion,* Vol. 55 (Archives of Neurology and Psychology, 1946), pp. 236-47.
65 Grossman study, 1995, cit. Arkin, 'The Case for Ethical Autonomy in Unmanned Systems', *Journal of Military Ethics* 9, 4 (2010), https://www.cc.gatech.edu/ai/robot-lab/online-publications/Arkin_ethical_autonomous_systems_final.pdf, p. 9.
66 Marshall, Samuel, *Men against Fire: The Problem of Battle Command in Future War* (William Morrow Publishing, 1947), generally.
67 Sparks and Neiss Study, 1956, cit. Arkin, 'The Case for Ethical Autonomy in Unmanned Systems', p. 9. The study is also discussed by Arkin in 'Human Failings on the Battlefield', cit. Braden Allenby, *The Applied Ethics of Emerging Military and Security Technologies* (Routledge, December 2016), ch. 12.
68 Grossman, Dave, *On Killing: The Psychological Cost of Learning to Kill in War and Society* (Black Bay Books, 1996), generally. Despite controversy over Marshall's World War Two survey methodology, Grossman too uses Marshall's data to evidence soldiers' reluctance to kill their opponents.

recent wars of choice and expedition, a driver emerges here that humans provide a (surprisingly) low baseline if we are to make comparisons with machines undertaking these same tasks.

Regardless of training, it is unrealistic to expect human beings to adhere unerringly to the Law of Armed Conflict when confronted by challenges of the battlefield, an observation that is borne out by the continuing debate on ethics education across military training establishments. The argument is that these same soldiers, as a general sample and when compared to a machine agent, may at best be 'a variable tool in the waging of war', as Barbara Ehrenreich writes.[69] Arkin's suggestion is that a computational implementation of an ethical code (together, his 'artificial conscience'), when embedded into AWS control sequences, may in time provide enforceable limits on machine actions in engagements.[70] The construct may be weakened by its underweighting of battlefield intangibles such as a soldier's values, leadership and other behavioural traits, but the suggestion still provides a plausible driver to machine (rather than human) control in lethal engagements.

> *Neuroscientists have found that human circuits responsible for conscious self-control are particularly vulnerable to stress. When these circuits shut down, primal impulses go unchecked*

Moreover, although Arkin often acknowledged that it was 'too early to determine whether this software device is practicable'[71], this is not necessarily the point. The real issue is whether his construct might one day allow better compliance with IHL than is currently capable of being exercised by a human.[72] Can machines be additive? His argument more represents a prototype and, as such, 'a preliminary version of a device from which other forms may be developed'. Robots, after all, already process more information and do this faster than humans. They are likely to remain uninfluenced

---

69 See, generally: Ehrenreich, Barbara, 'Do Humans Have a Role in the Robot Wars of the Future?', *Guardian* (11 July 2011), https://www.theguardian.com/commentisfree/2011/jul/11/human-role-robot-war-future.
70 The test here is that such limits should be better, or at least as good, as the limits achieved by human soldiers on the battlefield. Arkin's actionable list is comprehensive and includes, *inter alia*, acceptance of surrender and humane treatment of prisoners, avoiding unnecessary suffering and damage and non-use of certain weapons.
71 Arkin, 'Governing Lethal Behaviour', p. 211.
72 The goal of Arkin's robotic controller design is to ensure that unethical responses are prohibited through an 'ethical governor' and, through an 'ethical adaptor', to 'prevent or reduce the likelihood of [an unethical action] via an after-action reflective review or an artificial affective function (guilt, remorse, grief)'. See: Arkin, 'Governing Lethal Behaviour, p. 20.

by fear or anger while at the same time better able to monitor the ethical behaviour of 'colleague' humans on the battlefield. Although Arkin's argument may appear frustratingly circular (the framework is theoretical without having to account for the inconvenience of context, relying much upon what the construct leaves out rather than what is included), it is still based on components which, individually sound, are difficult to refute.

The notion, however, is fraught with difficulty, not least its central hypothesis that lines of code will, in time and without error, be able to undertake the complex, interactive tasks required to control a lethal engagement.[73] Reviewing the feasibility of Arkin's construct (or ones that are generally adjacent to it) is therefore a key line of inquiry for this book, and one that occupies much of its subsequent chapters, in particular challenges relating to value and goal setting, the construct's required utility function, and whether these features can dynamically be managed. The coding complexity of ethical precepts is particularly challenging given that such programming must incorporate rules from all of the human rights conventions in order to be compliant. Arkin's framework, after all, requires that several building blocks must work seamlessly and in tandem.

In addition to his 'ethical governor' and 'artificial conscience', Arkin also posits a 'responsibility advisor' to make explicit 'just where *responsibility* vests should: (1) an unethical action… be undertaken by the autonomous robot…, or (2) the robot performs an unintended unethical act due to some representational deficiency'. Two enduring constraints arise. The first is the cumulative technical feasibility required for Arkin's mix. In discussing the challenges regarding 'action selection' in machines, the secretary of the Society for the Study of Artificial Intelligence and Simulation of Behaviour usefully observed in 2017: 'I don't mean that it's too difficult like "man will never fly" or "man will never land on the moon". I'm saying it's hopelessly misguided like "man will never dig a tunnel to the moon"'.[74]

---

73 Arkin 'Governing Lethal Behaviour', p. 4. Arkin justifies his framework as follows: 'It is a working assumption, perhaps naïve, that the autonomous agent ultimately will be provided with an amount of battlefield information equal or greater than a human soldier is capable of managing. This seems a reasonable assumption with the advent of network-centric warfare and the emergence of the Global Information Grid'. This book's Chapters Six (Wetware), Seven (Firmware) and Eight (Software) challenge these assumptions and the technical feasibility necessary to realise Arkin's framework.
74 Martin, Andrew Owen, senior technical analyst at the Tungsten Network, cited in Ben Sullivan, 'Elite Scientists Have Told the Pentagon that AI Won't Threaten Humanity', *Motherboard magazine* (19 January 2017).

Given humankind's manifest battlefield failings, AWS proponents also point to a moral component to these otherwise ethical drivers.[75] The context, after all, is that 'armies, armed groups, political and religious movements have been killing civilians since time immemorial'[76] and shortcomings in *jus in bello* often characterise military (and other) histories. Indeed, Arkin's supporters highlight exactly such battlefield lapses to justify AWS deployment. Ukraine has demonstrated that the current human-in-the-loop bombing practices still create unacceptably high civilian collateral damage, the inference being that human-out-of-the-loop machines might be able to perform better than (or certainly as well as) current in-loop and on-loop systems. That argument, however, is not new. Used in the nineteenth century for rapid-firing artillery, it has since been rolled out for developments in tank, aircraft and naval assets, and, in Ukraine, for drone warfare, sensor coverage and ubiquitous surveillance, precision fires and the ramifications of open-source intelligence.

> An 'artificial conscience', when embedded into AWS control sequences, may provide enforceable limits on machine actions in engagements

But Arkin's proposals actually create an interesting dilemma. Ideating the prospect of more humane armed conflict (and, with it, the saving of lives), putting in place legal means that restrict battlefield autonomy 'could [in itself] amount to not properly protecting life'.[77] Notwithstanding that such constraints would presumably vanish in the event of major war, the embracing of autonomy might thus be construed as a legal imperative. Indeed, still other constituents contribute to this ethical driver. AWS, for instance, might be able to limit their intervention to an 'appropriate' amount of force in a lethal engagement. Autonomous technologies might be capable of employing 'creative alternatives to lethality'[78] such as autonomous precision, the possibility of non-lethal immobilisation of humans as well as the disarming of targets that might otherwise be destroyed.

---

75 Bourke, J, *An Intimate History of Killing* (Basic Books, 1999), generally. See also: Walker, *Killer Robots*, pp. 59-60. A corollary here might be that such 'human failure' is an immutable state of the nature of war (not just the character of war). The theoretical significance of Arkin's system is therefore that it changes not just battle's character but also its nature.

76 Slim, H, *Killing Civilians: Methods, Madness and Morality in War* (Columbia University, 2008), p. 3.

77 Human Rights Council, 'Report of the Special Rapporteur', p. 6. See also: Mitchell, Paul, '"Three Laws Safe?" Autonomous Robots and Warfare', Laurier Centre for Military Strategic and Disarmament Studies (15 October 2012), paras. 8-12.

78 Human Rights Council, 'Report of the Special Rapporteur', p. 10.

Similarly, it is (again theoretically) possible that independent weapons might be programmed specifically to leave a 'digital trail' which could allow better *post facto* scrutiny of their actions and thus enhance accountability. From an ethical perspective, it might also be that AWS' modus operandi will be more 'hide and seek' in nature rather than formal force-on-force affairs, with consequently less collateral damage than one might expect from attritional warfare. In

'I don't mean that it's too difficult like "man will never fly" or "man will never land on the moon". I'm saying it's hopelessly misguided like "man will never dig a tunnel to the moon"'

this vein, deployment of self-directing machines might also be geospatially constrained, the ethical driver being (however far-fetched and prone to abuse it may be) that, just as soldiers are given defined rules of engagement, AWS may be disabled once in a specified no-kill zone.

Ethical arguments can also be based upon AWS' ability to improve battlefield practices. Economic considerations apart, uncrewed systems will not, after all, need to protect themselves. It is not a given that self-preservation must be an attribute in their decision-making. This has two ramifications. It affects the longstanding design equation across weapon-types that seeks to balance that armament's protection (now less necessary?) with its firepower and its mobility. There is also a theory that the AWS can be programmed to act *conservatively* with built-in buffer and delay

There is a theory that the AWS can be programmed to act conservatively with built-in buffer and delay mechanisms

mechanisms. Appropriately constrained, the idea is that AWS could be operationally superior to human soldiers in programming out the human heuristic of 'scenario fulfilment' whereby systemic expectation on the part of the soldier about how a set of circumstances will unfold can lead directly to bias in humans and human decision-making. All of this would be obviated by reducing the role of the human in battlecraft in favour of autonomous systems.

A final point (again rooted in high theory rather than the empirics of what might be technically feasible) relates to robots being built without the faulty psychological disposition of the human, a disposition that might otherwise lead humans towards immoral actions. Several advantages are implied from this. First, AWS may be able to disseminate information under fire to colleague machines and command structures without exaggeration, distortion or contradiction. This, in turn, might remove decision-makers from the front line. Others promote AWS deployment as part of a longstanding military aspiration of 'victimless' warfare that is based instead upon remote, pilotless and autonomous technology delivering victory through disruption rather than destruction. Finally, removing human supervision in weapons may lessen certain political pitfalls: in the case of a downed Reaper, for instance, there is no high value pilot to kill or to take hostage.

3.5 Operational drivers

Autonomous technologies also promise 'operational initiative'. Ukraine has shown the value of such asymmetry, whether across target recognition-to-engagement cycles or in the avalanche of information available to decision processes (examples being Ukraine's Kropyva and Delta systems and its Accelerator programme for integrating new means into legacy systems). Parties considering their options might study, for instance, Ukraine's Project FURY (First Ukraine Robotic Navy) and its unanticipated success in besting Russia's Black Sea fleet, and where autonomous and semi-autonomous means have provided a gamechanger to a defending nation without its own standing navy.[79] Peer parties, of course, have their own initiatives in place to preserve areas of pre-eminence, each setting up another idiosyncratic set of drivers. In mapping its own future procurement trends, the US DoD's set of roadmaps first appeared in the mid-2010s and was quick to assign autonomy a broad-ranging role across upcoming weapon platforms. The programme envisaged from its outset unsupervised assets working in seamless groups, benefitting from integrated communications and shared targeting data while all the time integrating more information from more sources in order to perfect its systems. The purpose of this chapter's final section is therefore to consider these operational catalysts.

Readers first need to note combat's increasing pace and how this acts as an accelerator. Just as Ukraine has demonstrated the pivotal role of satellite and surveillance technology, the availability of sensed real-time

---

79 Samus, Mykhailo, 'New Technologies of War: Ukraine's Asymmetric Advantage', *New Geopolitics* (10 February 2024), https://www.newgeopolitics.org/2024/02/10/new-technologies-of-war-ukraines-asymmetric-advantage/.

data has turned the battlefield upside down.[80] There is no longer sanctuary to be found in rear areas. The idea of safe replenishment behind the lines has disappeared. Just as decision-making has had to adjust to this new world, it has also had to capitalise upon this new speed of war, especially as more and more aspects of conflict not only leave the realm of human senses but also cross outside the limits of human reaction times. Indeed, there is little disagreement that weapon autonomy will accelerate the speed of war. Statistics illustrate the underlying issues. The human fighter pilot needs some 0.3 seconds to respond to a simple stimulus and more than twice as long to make a choice between multiple possible responses.[81] A robotic system faced with the same decision may need less than a millionth of one second to make that same action selection. The challenge, of course, is that quickening response times (here, the speed between an event and the autonomous weapon's response) do not necessarily equate to improvements in the quality of that response.

Humans already find it ever more difficult to participate in engagement sequences where machines already out-speed humans. This shortened decision-cycle also adds new risk to operations. The OODA Loop (observe, orient, decide, act) refers to a techno-strategic concept first developed by US fighter pilot and strategist John Boyd during his time as a Pentagon consultant in the 1990s. Boyd's insight was that 'advantage lies with the fighter whose OODA Loop is faster and more accurate than his opponent's, and who is able to throw his opponent's OODA Loop out of sync'.[82] Just as aircraft will have to manoeuvre too quickly for a pilot to control, so must its weapons be deployed at the same speed in order to match the beyond-human speed of the aircraft's own systems.[83] The operational driver is that speed becomes the imperative, the more so if friendly forces are to defeat an enemy's similarly autonomous counter-systems. Notwithstanding the dangers of 'claim inflation' when assessing new means, the direction here is that military systems will soon be too fast, too small, too numerous and too complex for humans to exert meaningful control.

---

80 Morris, Christopher, 'Ukrainian War: Offensive Use of Satellite Technology [Is] a Sign of How Conflict Is Increasingly Moving into Space', The Conversation (15 June 2023), https:// theconversation.com/ukraine-war-offensive-use-of-satellite-tech-a-sign-of-how-conflict-is-increasingly-moving-into-space-207641.
81 Singer, Paul, *Wired for War: The Robotics Revolution and Conflict in the Twenty-First Century* (Penguin Publishing, 27 January 2011), p. 127.
82 Marra, William and Sonia McNeil, 'Automation and Autonomy in Advanced Machines: Understanding and Regulating Complex Systems', *Warfare Research Paper Series*, 1-2012 (April 2012), p. 9: 'The fastest OODA Loop of the future combat plane will be an automated one and automated in both flight and weapons' functions'.
83 Anderson and Waxman, p. 5.

Hardware drivers also accelerate the adoption of autonomous means. As at early 2024, more than two hundred home-grown companies are involved in producing drones and uncrewed support systems for Ukraine's war effort.[84] The country's 2023 domestic drone production, moreover, was up one hundred times since Russia first invaded, and its government's 2024 target of some one million drone units demonstrates the rate of embrace of uncrewed technologies. Progress can be seen across the whole ecosystem with UAV power management a case in point. Having serviceable energy enables machine independence. Here, lithium-air batteries now provide a viable alternative to lithium-ion and, safety issues notwithstanding, units based on this technology already generate a similar energy density to that of gasoline. Other power advances show what might soon be possible. Within the first five years of the 2020s, researchers at George Washington University created a molten electrolyte battery using vanadium boride leading to battery packs that produce thirty times more energy than lithium ion. Battery disruption (and, through this, a driver to uncrewed weapon platforms) is also evident from lithium-sulphur chemistry, previously hamstrung by the unwelcome by-product of lithium sulphide and the degrading of cell capacity even after very few cycles.

The challenge, of course, is that quickening response times do not necessarily equate to improving the quality of that response

There have been similarly disruptive developments in fuel cells. Lockheed Martin has similarly demonstrated that its Stalker XE 240 drone could stay airborne for up to eight hours using propane-driven cells. The US Navy's Ion Tiger project uses a Protonex hydrogen fuel cell that can now fly for forty-eight hours using a cryogenic storage system. Commercial companies are also working with scavenging technology whereby uncrewed weapons might perch on power lines and recharge autonomously, enabling missions that continue more or less indefinitely with UAS able to undertake almost perpetual roles that would seem to bring feasible AWS deployment ever closer.[85]

---

84 Bielieskov, Mykola, 'Outgunned Ukraine Bets on Drones as Russian Invasion Enters Third Year', Atlantic Council (26 February 2024), https://www.atlanticcouncil.org/blogs/ukrainealert/outgunned-ukraine-bets-on-drones-as-russian-invasion-enters-third-year/.
85 Similar advances continue to be made in drone solar technology towards solving the key conundrum of 'high weight but low efficiency' power generation in uncrewed vehicles. Zephyr, for instance, has long been part of DARPA's Vulture project and has a seventy-foot wingspan which can fly non-stop for more than three hundred hours, almost two weeks, based on solar power alone. Developments in solar power, furthermore, will be particularly suited for smaller swarming autonomous drones given favourable wing area ratios; an aircraft that is half the

Hardware developments enabling machines to loiter have also proved a compelling driver. Israel's HAROP platform can reportedly stay in flight for nine hours. The weapon type is an attack system designed to locate and precisely engage targets. Currently controlled via data links for full human-in-the-loop operation, the current iteration of loitering munitions are consumable, single-use weapons that combine characteristics of a missile and a UAV, the point of such munitions being mission execution without relying upon other external systems for target and mission intelligence. As such, loitering munitions appear to fulfil a material step towards fully autonomous lethal engagement, already designed to fly without supervision to designated holding areas. The model is then for an operator (currently) to direct the selected munition towards the general target area before using video image to select and then attack that target.

Finally to this point, advances in miniturisation have also opened new possibilities across the range of AWS componentry. A laser target designator that weighed more than forty pounds two decades ago now weighs less than a golf ball. These smaller target designators have also disrupted autonomous operation by facilitating low altitude manoeuvre beneath cloud cover. Laser-based radar is a further key component in the weapon type (bouncing its laser off thousands of points, calculating the distance to each point and then building up a three-dimensional map of the weapon's surroundings). Units can now purport to work in darkness, fog and smoke are available with no moving parts and cost a few thousand dollars, a fraction of the price of recent similar systems.

What then is the effect of these developments on the thesis of independent UAS? The past two decades certainly demonstrate the technology's trajectory, from November 2001 when a Predator drone was first fired at suspected al Qaeda leaders in Afghanistan to the posting in September 2020 by the Azerbaijan military of high-definition footage of multiple, precise vehicle kills, taken by a Turkish-made Bayraktar TB2 drone.[86] Moreover,

---

size has just one quarter of the wing area and hence only carries a quarter as many solar cells. It also has just one-eighth the weight to support. Solar cell efficiency at the time of writing has also increased to more than thirty per cent using, for instance, gallium arsenide rather than traditional silicon, more than twice as efficient as earlier solar cell technology and at a fraction of the weight. Other commercial drivers include new wing materials that allow drones actively to optimise thermals as well as software advances transforming UAV efficiency. Examples here include dynamic soaring by optimising wind shear as well as software that allows UAVs to capitalise upon local air turbulence in urban environments.

86 Atherton, Kelsey, 'Mass Market Military Drones Have Changed the Way Wars Are Fought', *MIT Technology Review* (30 January 2023), https://www.technologyreview.com/2023/01/30/1067348/mass-market-military-drones-have-changed-the-way-wars-are-fought/.

dissecting the componentry on this second UAS would have revealed its very international provenance, further evidence of drone production's evolution over the period: a GPS device made by Trimble, an airborne modem and transceiver produced by Viasat and a GNC 255 navigation radio manufactured by Garmin. This same TB2 can communicate at a range of more than one hundred and fifty miles from its ground station and do so while travelling at nearly one hundred and fifty miles per hour. Even at these speeds, the platform can stay aloft for more than twenty-four hours at altitudes approaching twenty-five thousand feet. From this orbit it can share video links in order to coordinate attacks and direct friendly movement. It can also release laser-guided bonds onto people, vehicles or buildings.

The direction here is that military systems will soon be too fast, too small, too numerous and too complex for humans to exert meaningful control

Technology repurposed from the ubiquitous smartphone continues to be another key driver accelerating autonomy across military platforms. In 2007, twenty million smartphones were sold. By 2015, more than two billion smartphone units were in operation.[87] Smartphone adoption mirrors the same S-curve model of technology implementation discussed above. The context here for AWS deployment might be that, while it took landlines forty-five years to reach half of US households, smartphones achieved the same penetration in just seven years. Apple and Samsung each spent some fourteen billion dollars on research and development in 2016 alone. By way (again) of context, Apple had spent just one billion dollars in 2003. Moreover, the focus of these groups' research and development has evolved over time to include recent innovations in AI, camera technology, 5G connectivity and sustainable materials, all adjacent and precursor technologies to AWS deployment. According to the agencies of the United Nations, more than half of the world's population was already using the internet by 2019 with more than four billion consumers of online information and services able to communicate digitally with their fellow citizens. According to Internet Live Stats, this figure had risen by nearly a third in the intervening to years to more than five billion users.

This trajectory is extraordinary. As of 2020, ninety-six per cent of the world's population had access to a mobile-cellular telephone network. Almost more instructive as a driver is to look at the hardware enabling this

---

87 Hambling, *Swarm Troopers*, p. 161.

explosion in connectivity. Statista estimates that by 2021 there were 6.3 billion smartphone users in the world, a seventy per cent increase from five years previously. More than four million smartphones are reported to be sold every day. Connectivity is nowadays all about handheld devices with less than twelve per cent of the world's population now connecting to the internet by fixed-line telecoms.[88] The nexus between the number of connected users and the smartphone as a social and information tool has upended how war is chronicled, understood and amplified. This exceptional scale of adoption (both the number of handsets but, more importantly, the services, interventions and scope of capabilities that the underlying technology now enables) points to the systemic changes in ecosystems.

The revolution here is twofold; on the one hand, it is about the immediate availability and processing of information but, on the other, it also concerns that information's reliability and veracity. It is about trustworthiness but also new angles of attack that arise to compromise the underlying data that drives the whole discontinuity. A norm change here is that it is increasingly difficult to be a bystander in war. Just as people use their devices to record their experiences and surroundings, they may unwittingly be transmitting data points

Just as aircraft will have to manoeuvre too quickly for a pilot to control, so must its weapons be deployed at that same speed in order to match the 'beyond-human speed of the aircraft's own systems'

that are useful, for instance, to generating battlefield targets. This participation has collapsed the boundary between those observing war and those engaging in it, but it also arises from an invariably autonomous series of processes.

Why is this important to the subject in hand? Those same advances throughout the smartphone ecosystem have also accelerated algorithm construction, a central component to the data and probabilistic processes that will underpin unsupervised weapons. An example is useful. The replacement of the H264 data compression standard by HEVC (high efficiency video coding) means that four times as much data can be transmitted over an identical bandwidth than was the case twenty years ago. Similarly, a ten-fold improvement has taken place in the rate achieved by the Fourier Transform, a pivotal mathematical process converting signals from digital to analog and back and another prerequisite in AWS' technical spine. Advances

---

[88] Ford, Matthew and Andrew Hoskins, 'Radical War: Data, Attention and Control in the 21st Century' (Hearst, 2022), p. xix and generally.

in smartphone technologies have also contributed to that deluge of video output discussed above, outstripping humans' ability to analyse resulting data. In 2004, just seventy-one hours of video from UAVs was produced for analysis. By 2011 this had increased to more than three hundred thousand hours of output. Today the amount is reportedly one thousand times that figure. Data on this metric has become ever more meaningless as drones have entered the mainstream but, by way of reference, one minute of 4k video requires about one gigabyte of current memory.

The driver here for autonomous practice is the broadly held presumption that less than ten per cent of such footage is able to be analysed. The ubiquity of sensor fusion was laid bare a decade ago by the Pentagon's Gorgon Stare and Argus programmes, which then incorporated some sixty-five camera eyes on each uncrewed platform, clearly requiring sophisticated visual intelligence software to scour the output for potentially interesting sites. Its recent iteration covers an area of just over six square miles while the recent incorporation of the ARGUS-IS expands each unit's coverage area to thirty-nine square miles. The current system has three hundred and sixty-eight cameras capable of capturing some five million pixels each to create an image of some two billion pixels, exacerbating these units' processing problem exponentially with AI certainly now required to aid humans in what to look for in those datasets.

Autonomy is also accelerated by *fragility* across current military hardware. Ukraine is not alone in current conflict zones to have its battlefield defined by the effects, for instance, of electro-magnetic jamming and the ease with which communications can be compromised between weapon and operator. Electronic warfare, notes the *New York Times*, has affected the fighting in that conflict zone as much as weather and terrain.

A consequence has been to move decision-making ever further onto the weapon platform and so limit any requirement for remote and third-party communication

It was also the capture by Iran in 2011 of an American RQ 170 Sentinel drone (after adversarial compromise of its GPS) that first demonstrated the susceptibility of UAVs to hostile action. A consequence has therefore been to move decision-making ever further onto the weapon platform and so limit any requirement for remote and third-party communication. Parties' aims for AWS are that autonomous platforms will largely ignore radio transmissions and, importantly, send out few of their own, the effect being that these platforms increasingly require autonomy in order just to survive, with decision-making being built into what *de facto* becomes an increasingly autonomous platform.

AWS deployment is also accelerated by the attraction of deskilling operations, notwithstanding the complexity of this as a driver. After all, integration uncertainties mean that predicting the effect of AWS adoption on human resources remains a challenge. On the one hand, SIPRI notes that 'combat aircraft pilots must fly in real conditions to be properly trained and to fly between ten and twenty hours a month to maintain their skill set'. Autonomous assets, on the other hand, can sit on a shelf for extended periods of time without losing their operational capability and without involving human hours of training and familiarisation. A third consideration (and one which has long been embraced by the US Military's Future Combat System project) arises from the promise that autonomy enables the soldier on the ground to become a much more powerful nexus in the middle of what will be a cohort of autonomous and hybrid assets. Ukrainian reconnaissance units deploying uncrewed and largely autonomous aerial systems illustrate this point. Autonomous operations, after all, are not yet mounted from a remote headquarters with the press of buttons.

> *Parties' aims for AWS are that autonomous platforms will largely ignore radio transmissions and, importantly, send out few of their own*

To field autonomous means, the current reality is that users must spend 'time and treasure' deploying a convoy of vehicles with personnel threading their way into no-man's land, establishing a communications array and then holding that position for several hours.[89] Empirically, the introduction of robotic and autonomous systems is actually increasing both the number of people as well as the skills necessary within that force. The underlying technology may be sophisticated but that does not stop it from being labour and skills intensive, each mission requiring an operator, technician and communications specialist as well as force protection to keep them alive while they are doing their job. Nor is the personnel requirement limited to people directly in the field as provision must be made for planning flight paths, electromagnetic surveys and the exploitation of captured imagery.

The reality of emerging technology, notes RUSI's Jack Watling, is that it also requires people and, if the number of soldiers is reduced in one area, they

---

89 Watling, Jack, 'Automation Does Not Lead to Leaner Land Forces', War on the Rocks (7 February 2024).

are only then displaced to other parts of the battlefield. An enduring challenge, therefore, is that there may be *irreducible* minimums in warfare of how many people are needed to perform basic tasks. Watling uses the example of an infantry squad with a small autonomous tracked vehicle carrying a machine gun and sensor payload as well as a small drone for surveillance. Without these additional tools, an infantry squad usually numbers twelve, divided into three fire teams. The retooled squad still requires two such fire teams, the tracked vehicle, after all, cannot assault a trench, climb into a building or process prisoners. It also needs its own fire protection. The remaining team then manages the new systems. Oversight is still needed to stop these systems from shooting already engaged targets or to disengage themselves from the simplest of obstacles. There is also the maintenance burden of these systems, which now require much more than simple assistance

> *There may, therefore, be irreducible minimums in warfare of how many people are needed to perform basic tasks*

of a battalion armourer and their tools. Suddenly, there is a fleet of complex vehicles to be managed, all comprising mechanical drive trains and power packs, sophisticated electronics and their sensors as well as the weapons that make up the platform's inventory. The squad must manage the software and communications equipment that makes the platform functional with the whole arrangement requiring recovery assets in case the platform is damaged or malfunctions. Nor, notes Watling, is it reasonable to presume that one engineer can attend to both the platform's track, its power pack as well as the reprogramming, updating and debugging of the system's software.

Any notion, therefore, that autonomy allows that single soldier to do the job of what previously had taken several soldiers is not without challenge. Nevertheless, Lanchester's Square Law remains a longstanding but newly relevant driver to AWS deployment, providing a heuristic rule of thumb for the advantage of quantity versus quality in military engagements.[90] The law has been given new status given the attritional nature of the Ukraine war and states, all things being equal, that having twice as many units in a fight translates into a fourfold increase in combat power for units with aimed-fire weapons. The force-multiplication basis for the rule has been played out in Avdiivka, Bakhmut and Krynky.[91] Here, numerically superior forces can double up on attacking enemy units while the numerically inferior force

---

90 Johnson, Ronald, 'Lanchester's Square Law in Theory and Practice', School of Advanced Military Studies, Fort Leavenworth (1990).
91 Adams, Paul, 'Up to 30,000 Russian Casualties Claimed in Bakhmut', BBC: War in Ukraine (7 March 2023), https://www.bbc.co.uk/news/world-europe-64880268.

can only attack half of the opposing force at any one time. The chief value of mass (and the attraction, then, of low-cost, swarming UAS) is that it can be used to impose costs on adversaries because it forces them to encounter a large number of systems. Examples abound for the multiplication effect of uncrewed weapons from very-high-altitude, ultra-endurance, loitering units to blimps to and then to very small UAVs. More germane might be the example of Apache helicopter pilots using Lockheed Martin's VU-ITs to collaborate with partner UAVs in the role of being a remote sensor to investigate areas too hazardous to fly the helicopter.

Analysis of drivers must also account for *commercial* catalysts. A key notion is that the current drive towards a 'Robotic Age' is no longer being led by yesteryear's military-industrial complex. Instead, it is the result of a push strategy from a myriad of public/private companies (and, in the West, not by state enterprises) that are producing consumer goods and business-to-business services (several, presumably, with possible dual use to military applications). Their activities obviously embrace advanced computing, big data, AI, miniaturisation, additive manufacturing and small but high-density power systems. It is interesting that this landscape represents a departure from earlier procurement precedents such that the technologies that will enable AWS are now much more likely to come out of a thriving commercial sector. But this also masks several

All things being equal, having twice as many units in a fight translates into a fourfold increase in combat power

versos. First, the commercial sector has its own set of idiosyncratic frictions, from proliferation concerns, these entities' profit imperative and, in the case of AWS, opportunities that arise from COTS (commercial off the shelf) hardware where, in an age of 3d printing and bitcoin, uncrewed systems can be assembled with disturbing anonymity. Both attack and reconnaissance drones, as evidenced in Sudan, all of a sudden appear really cheap. Even the Iran-built Shahed drones, which have been used extensively in Ukraine and in the attack against Israel in April 2024, can deliver large explosive charges (and do this with precision some two thousand miles distant) reportedly cost between twenty and fifty thousand dollars.[92]

Second, commercial drivers are accelerated by the proven lobbying powers of the defence and drone industry, particularly in the US. In this

---

92 Kemp, Richard, 'Suicide Drones Threaten to Bankrupt Western Militaries', *The Telegraph* (16 April 2024), https://www.telegraph.co.uk/news/2024/04/16/suicide-drones-threaten-to-bankrupt-western-militaries/.

vein, the sheer volume of experimental military autonomy projects which are already in process and contribute towards a cumulative expectation for these uncrewed technologies. An example is DARPA's December 2023 tender which requested parties to solve for drones flying assigned missions when operator connectivity is lost or disrupted.[93] This again is the new context for these systems. In this vein, the Pentagon's Joint Robotics Programme is currently developing some twenty-two different prototype 'intelligent ground vehicles' on its own, ranging in size from small eight pound units to an autonomous seven hundred ton robotic dump truck that can move more than two hundred tons of earth in each scoop. A further driver towards battlefield robotics then becomes version improvement to existing platforms in order for these systems to take on much wider (and increasingly autonomous) battlefield roles.

> Numerically superior forces can double up on attacking enemy units while the numerically inferior force can only attack half of the opposing force at any one time

It is also the scale of resources being deployed that is a clear driver to systems becoming autonomous. Given overlap across programmes, accounting obfuscations and time lags in reporting, numbers for this phenomenon are difficult to abstract. Over the five years to 2022, the US budget[94] for uncrewed systems was expected to total more than twenty-four billion dollars.[95]

Context is again useful as this effort is no blip. In the half decade to 2010, flight hours for UAVs increased more than fivefold to more than five hundred and fifty thousand hours and the inventory of systems from less than three thousand to nearly seven thousand. UAV programmes might have begun in the mid-1990s, but it took sixteen years for the cumulative number of these assets' flying hours to reach one million. Doubling that to two million then took just over two years and was achieved in late 2013. Up-to-date analysis is increasingly meaningless given the explosion in platform type and, in Ukraine, a commitment to manufacture one million uncrewed

---

93 DARPA, 'DARPA Seeks Technical Solutions to Create Autonomous Capabilities for Commercial Drones'(9 December 2023), https://www.darpa.mil/news-events/2023-09-12.
94 Exact figures for spend appear to differ depending upon authors' definitions of particular programmes; see: US Department of Defence, 'Unmanned System Roadmap 2007-2032', p. 4.
95 Ninety-one per cent of this will be allocated to aerial UAS, eight per cent to maritime unmanned systems with the small balance being taken up by ground systems.

units in 2024.⁹⁶ The recent embedding of UAVs (and, indeed, personnel to run these assets) into the very centre of combat structures has only reinforced this trajectory. Every time a Ukrainian platoon is tasked, an exactly similar number of drone operators is in attendance. And there is clearly considerable room to expand further these uncrewed and eventually autonomous efforts. After all, US spending of two billion dollars per year on UAVs a decade ago was only one tenth of the amount that that government allocated just on its space capabilities.⁹⁷ Dollars here literally equate to fire power and, by 2021, the US was spending more than seven billion dollars on uncrewed assets, some ten per cent of its overall seven hundred billion dollar defence budget for the year.

This all has contextual consequences. First, militaries' growing experience around UAVs creates general confidence in the asset class. By 2010, the US Air Force was already training more uncrewed pilots than traditional pilots for the first time.⁹⁸ In 2011, the University of North Dakota chartered its first four-year degree programme in UAV piloting. Second, technical and tactical advances usually have particular resonance when earned in hot combat zones. The scale, for instance, of Ukraine's drone efforts in early 2024 is best illustrated by the fact that the country has been losing more than ten thousand drone units each month.⁹⁹ The battlefield, in many ways, has become a war of engineers where drone use means that fewer compatriots are killed and where targets are also engaged more effectively. More than two hundred domestic drone manufacturers have sprung up in Ukraine, often using parts cut out of lightweight wood and, using 3D printers, from rubber-like compounds. The paradox here is that drones may be launched by catapult but then use AI to run the platform and track those same targets. It is these quickfire advances that add self-fulfilling ambition to an application, feeding into that notion of the 'revolution in expectation' and, in the case of AWS, reinforcing the suggestion that weapon autonomy is close to being

---

96 Burunov, Oleg, 'From Raven to Coyote: How Many Military Drones Does the US Have?', Sputnik International (15 March 2023), https://sputnikglobe.com/20230315/from-raven-to-coyote-how-many-military-drones-does-the-us-have-1108417947.html.
97 Miller, Sergio, 'Creating a British "Army of Drones"', Wavell Room (15 December 2023), https://wavellroom.com/2023/12/15/creating-a-british-army-of-drones/. See also: Sabin, Philip, 'The Strategic Impact of Unmanned Aerial Vehicles', cit. Owen Barnes, *Air Power: UAVs: The Wider Context* (Royal Air Force Centre for Air Power Studies, 2009), p. 101.
98 As reported to the Committee on Oversight and Government Reform, Hearing on 'The Rise of Drones; Unmanned Systems and the Future of War', Committee on Oversight and Government Reform, Congressional Research Service (March 2010), generally http://digitalcommons.wcl.american.edu/cgi/viewcontent.cgi?article=1002&context=pub_disc_cong.
99 Schifrin, Nick and others, 'How Drone Warfare Has Transformed the Battle between Ukraine and Russia', PBS News Hour (13 December 2024), https://www.pbs.org/newshour/show/how-drone-warfare-has-transformed-the-battle-between-ukraine-and-russia.

feasible.[100] Drivers, therefore, are an important component to the subject's contextual framework against which the remainder of this book and its focus on deployment's challenges should be judged.

---

100 Hirsh, Michael, 'How AI Will Revolutionise Warfare', *Foreign Policy* (11 April 2024), https://foreignpolicy.com/2023/04/11/ai-arms-race-artificial-intelligence-chatgpt-military-technology/.

4
Deployment
Models for the removal of weapon supervision

Military technology has been automating for decades and the purpose of this chapter is to consider a range of likely deployment models for integrating unsupervised means into battlefield practices. The starting point here is states' initiatives to enhance their military capacities through pairing soldiers with technology. Although this may take on many forms (not all of which presage lethality), these deployment models generally reduce direct human involvement across battle's processes. This chapter is therefore an important scene-setter in this book's overall consideration of the feasibility autonomous weapon systems (AWS).[1] The chapter generally discusses deployment models based upon weapon capabilities. Friction exists, of course, between deployment's 'how' (the fielding and embedding of AWS) to its 'under what circumstances' (here, the context and appropriateness of AWS). All of a sudden, the drone operator more resembles a technologist than a traditionally uniformed combatant and, in this vein, context is more akin to the 1999 predictions of Chinese colonels Liang and Xiangsui that 'tomorrow's soldiers will increasingly be computer hackers, financiers, smugglers and agents of private corporations rather than members of a military'.[2] This all points to future armed conflict being unlikely to resemble a battlefield in its traditional sense.

---

[1] In so doing, the purpose of Chapter Four (Deployment) is not to question AWS' feasibility (the purpose of subsequent chapters) but instead to assume that appropriate progress is achieved in the capabilities that will be required to realise operationally and legally acceptable AWS (here, *inter alia*, progress across computer vision, natural language processing, machine learning, search and planning, logical and symbolic reasoning, human-machine interaction, physical manipulation means, power issues and locomotion, collaborative intelligence and issues around verification, testing and version control).

[2] Latiff, Major General Robert, *Future War* (Alfred Knopf Publishing, 2017), p. 4.

The common narrative, of course, is that it will be machines that fight battles on humans' behalf, that watch for humans and think for humans, that 'fight fast' and that are not limited temporally. Created in 2017, the dataset of the Stockholm International Peace Research Institute (SIPRI)[3] provides a useful starting point to unpick this phenomenon, identifying three hundred and eighty-one different such systems featuring degrees of autonomy in their critical functions.[4] That number, however, is meaningless without understanding that dataset's context. Indeed, it is more telling what the dataset does not reveal. It cannot, for example, account for task complexity. Given also that autonomous systems must model battlefield tasks mathematically, a deployment challenge must be the difficulty presented by each component task relative to established human standards and established human aptitude. Nor does the dataset quantify the degree of precision that is required by each of these tasks in respect to the whole endeavour. In ignoring permissible leeway or margin of error, it also overlooks coding factors whereby the more ill-defined that task's specification, the more challenging becomes its syntactic formulation. The dataset therefore underplays task tangibility (can each expected outcome of the independent weapon be qualified?), dimensionality (can the battlefield task be carried out in a single action or does it require sequential decisions and actions?) and interaction (are additional assets required in order to complete the task and, in so doing, is coordination required with human or other autonomous agents?). Weapon actions must generally be competitive or collaborative and, in the case of AWS, hard coded into the platform's initial representation (the moment when the weapon is first fielded). It should also be assumed that the SIPRI dataset cannot factor, moreover, for the dynamic state of the AWS' operating environment, whether

> *'Tomorrow's soldiers will increasingly be computer hackers, financiers, smugglers and agents of private corporations rather than the members of a military'*

---

3 This book uses as a basis Boulanin, Vincent and Maaike Verbruggen, 'Mapping the Development of Autonomy in Weapon Systems', Stockholm International Peace Research Institute (November 2017), https://www.sipri.org/sites/default/files/2017-11/siprireport_mapping_the_development_of_autonomy_in_weapon_systems_1117_0.pdf, generally. See also: Marta, Bo and Vincent Boulanin, 'Retaining Human Responsibility in the Development and Use of Autonomous Weapon Systems, SIPRI, October 2023, https://www.sipri.org/publications/2022/policy-reports/retaining-human-responsibility-development-and-use-autonomous-weapon-systems-accountability.

4 Boulanin and Verbruggen, p. 19. The dataset focuses on weapon systems rather than individual munitions.

at any moment it is observable, cluttered, adversarial, structured or stochastic (do AWS actions always produce the same effect?). The conclusion is that each such variable actually multiplies task complexity, because a whole set of relationships cannot be reflected in such datasets.

Acknowledging these limitations, what is the current picture for the deployment of weapon systems without human supervision?[5] For this, it is useful to understand how the SIPRI dataset is divided. Its two principal segments comprise weapons with autonomous mobility (forty-eight per cent of the dataset[6]) and autonomous targeting (twenty-three per cent of the dataset[7]). These two groups made up seventy-one per cent of the dataset, fully two hundred and seventy weapon systems. Excluding rounding, the balancing quarter of the weapon platforms also exhibited autonomous characteristics and related to intelligence systems (ten per cent), 'interoperability' systems (ten per cent) and health management systems (six per cent).

Nor does the dataset quantify the degree of precision that is required by these tasks in respect to the whole endeavour

Irrespective of issues related to data classification, SIPRI's breakdown provides a useful starting point given the clear heuristic that emerges. Just as full autonomy is generally (and unhelpfully) lumped together at one end of a capability continuum, the dynamic of these datasets shows that there must be several intermediate and transitional deployment models that will make up the rest of that continuum. This is best understood by drilling into the autonomous capabilities that drive SIPRI's dataset. Mobility-related functions, for instance, might comprise unsupervised homing/follow-me capabilities, autonomous navigation and, in time, takeoff/landing

---

5 In SIPRI's 2017 dataset, 195 (51%) of the military systems that were incorporating autonomy were unarmed, 175 (46%) systems were armed and 11 systems were classified as 'unknown'. Of the 381 weapon systems with autonomous capability, 58% were air systems, 24% ground systems and 18% maritime. Of those 381 systems, 225 had completed their development and 131 were still undergoing development. The development status of 14 systems was unknown and 11 systems had been cancelled. See: Boulanin and Verbruggen, p. 20. By way of background, the dataset was classified into three categories as follows: uncrewed weapon systems that featured some autonomy in their critical functions (that is, they could autonomously search for, detect, select, track or attack targets); uncrewed weapon systems that did not have autonomy in their critical functions but featured autonomous functions in any of the other capability areas covered by the study (here, mobility and intelligence, interoperability and operational management); and uncrewed and unarmed military systems (involved in intelligence, surveillance, reconnaissance and logistics missions).

6 A total of 277 systems of which 30% of that segment comprised armed weapon systems.

7 A total of 153 weapon systems of which a significant 85% related to armed weapon systems.

competences. Autonomous homing then underpins the finding and tracking of targets while follow-me capabilities allow AWS to shadow colleague system whether they be human or machine. While all these processes require substantial collaboration, they also point to more general requirements such as automatic sense-and-avoid routines to prevent collision when operated in cluttered environments.

Pell-mell developments across weapon classes (from Türkiye's deployment of Kargu-2 munitions in Libya to Russian loitering armaments in Ukraine) also mean that analysis here only provides a snapshot in time. For the reader, assumptions and norms have never been more fleeting. Some challenges remain enduring, but technical progress means that even the most long-dated practices are losing their relevance. Take, for example, the computational challenge for independent navigation. A purpose here has been to resolve variances thrown up by obstacles, by adversarial activities and from shortcomings in vision-based guidance. But unsupervised navigation is just as likely to be compromised by enemy spoofing and resulting loss of GPS as by general frictions arising from passage in a contested landscape (in which, moreover, an autonomous unit still needs to ensure best routing). This remains true today notwithstanding extraordinary advances in capability that have been made on the back of driverless vehicles. And autonomous navigation must still work first time, every time if that the weapon is to decide upon its route without supervision.[8]

> *Objects deform. Subjects are not necessarily solid but change shape. They may be partially obscured and difficult to abstract in a cluttered or textured background*

The SIPRI dataset also masks how battlefield characteristics influence deployment models. Just as new capabilities spur new use cases, so too may specific combat tasks benefit from tailored autonomous solutions. Examples include the unsupervised detection of perimeter intrusion, the pinpointing of artillery coordinates and remote processing of objects in an intelligence, surveillance and reconnaissance mission.[9] This is not to suggest

---

8 Several systems rely on waypoint navigation and, as such, may not be truly autonomous. Similarly, Northrop Grumman's MQ-4C Triton uncrewed aircraft system can autonomously plan a route but still relies on a human operator to set speed, altitude and mission objectives.
9 Examples of autonomy in intelligence roles include Israel's Counter-IED and Mine Suite (CIMS) developed by IAI, General Dynamics' uncrewed ground system called Mobile Detection Assessment and Response System (MDARS, developed for the US Army), Endeavour Robotics' RedOWL artillery targeting system, Boeing's ScanEagle system that can autonomously monitor objects of interest on the sea surface and, on a research basis, the US Office for Naval Research's

that the Delivery Cohort's technical difficulties are always solvable through technology. Regardless of improvements in object recognition, for instance, it is still an enduring challenge for a system to understand that an object is the same object when viewed from different angles. Objects deform. Subjects are not necessarily solid but change shape. They may be partially obscured (problems of occlusion) and difficult to abstract in a cluttered or textured background. Speed of movement, variety and illumination all create difficulties in machine routines that humans are able to navigate quite easily. It also ignores the practical challenges of having to disturb already entrenched routines with new practices.

In separating out deployment models, the architecture of each weapon proves a determining factor. It is also the degree of autonomy that has been designed *across* a weapon's whole componentry that dictates where agency will really sit in the engagement sequence. This, of course, gets ever more complicated given collaborative or distributed instances of autonomy (including, for instance, swarming), forcing the commander to identify where meaningful control really rests throughout a sequence. All of this then needs to be considered within a general test of reasonableness for each deployment model.[10]

> *It is still an enduring challenge for a system to understand that an object is the same object when viewed from different angles*

Analysis must also focus on concepts rather than specific weapon systems in order to remain current and not immediately out of date. After all, how weapons are used on the battlefield is all about control. This ranges along a continuum from humans engaging and selecting targets prior to initiating an attack (level one) to algorithms suggesting targets that humans then choose which to attack (level two). More technically demanding are models where algorithms first select targets that humans must approve before an attack is initiated (level three) to algorithms selecting targets that humans then have a restricted time to veto (level four). It is the continuum's

---

collaboration that seeks to infer intentions and threats in surveillance imagery through its Automated Image Understanding Thrust (AIUT) system.

10 MacCormick, Neil, 'Reasonableness and Objectivity', *Notre Dame Law Review*, 74, Issue 5, Article 6 (1999), https://scholarship.law.nd.edu/cgi/viewcontent.cgi?article=1648&context=ndlr. See also: Sharkey, Noel, 'Staying in the Loop: Human Supervisory Control of Weapons', cit. Nehal Bhuta and others, *Autonomous Weapons Systems: Law, Ethics, Policy* (Cambridge University Press, 2016), p. 26. Global consequences of AWS deployment (plausible deniability in first use of AWS, proliferation, ethics and non-compliance) are generally ignored in any consequentialist analysis but form a key basis of arguments in Chapter Five (Obstacles).

level five classification that concerns this book, with unsupervised programs selecting targets and initiating attacks without human involvement.

Although these bands appear neat and ordered, any such framework is obviously never that clearcut, and while it might be convenient that classification boundaries remain loose and dynamic, having weapons toggle between different deployment models on that continuum compromises accountability and is anyway technically very difficult. The later a human leaves the control loop (or the more limited that loop is in its operational context), the more systems can be argued still to be under human control. But because the selecting and engaging of targets involves a complicating variety of subtasks, this quickly becomes an intractably complicated relationship.[11]

> *The later a human leaves the control loop, the more systems can be argued still to be under human control*

No single set of deployment rules can ever be appropriate. Fielding, for instance, a self-directing weapon in a complex urban setting will always be different from that same lethal platform being wheeled out in a remote and controlled battle space with well-understood, minimal civilian density. Other more conceptual aspects of tasking also shape these deployment models as human supervision is removed. This includes the breadth of a weapon's competence (the ability to perform all the required actions necessary to achieve its task) as well as its epistemic competence (how can the machine extract all necessary knowledge in order to achieve that task while still accounting for every possible context?). Finally, to this point, deployment models must be shaped by the efficacy of how each weapon selects its actions (its ability to design and execute actions based upon that weapon's goals) and to do this in line with passing norms.[12] Together, these criteria comprise a starting set of deployment guidelines.

> *After all, a consequentialist view might be that fewer friendly casualties equates to more incentives to initiate violence and start wars*

---

[11] Sensory data must first be acquired and processed. Targets must be then identified from that output and thereafter tracked, selected and prioritised before, on the basis of engagement rules, an engagement decision and resulting application of force can be undertaken. Technical ramifications are discussed in Chapters Seven (Firmware) and Eight (Software).

[12] Walker, Paddy and Peter Roberts, *War's Changed Landscape? A Primer on Conflict's Forms and Norms* (Howgate Publishing, 2023). See also: Sartor, Giovanni and Andrea Omicini, 'The Autonomy of Technological Systems and Responsibilities for Their Use', cit. Nehal Bhuta and others, *Autonomous Weapon Systems: Law, Ethics, Policy* (Cambridge University Press, 2016), p. 63.

4.1 Capabilities, roles and use cases

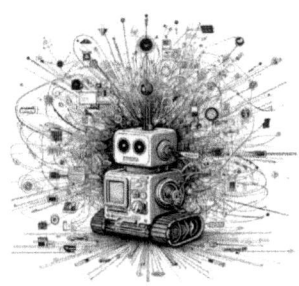

AWS deployment should be framed by the degree of fit between that weapon's capability and its expected role. Consider, for instance, the defence. What level of safeguard is required to deploy a defensive (albeit unsupervised) weapon that is narrowly programmed to engage specific enemy hardware in a specific battlefield area? Does restricting that weapon's tasking (such as the incursion-specific SGR-AI discussed in Chapter Two) or the imposition of narrow defensive purpose influence the degree of meaningful human control applicable to that use case? The AWS that is securing a perimeter can perhaps be portrayed as being defensive and perhaps therefore a more 'tolerable' application of autonomy as opposed to hunter-killer scenarios where unsupervised weapons are fielded in a manner that is spatially and temporally unchecked. But the difficulty here is that today's weapons rarely lend themselves to this neat division and persons under attack from an AWS will definitely be ambivalent whether the nature of that engagement casts them as a defender or an attacker; after all, every combat engagement, whether 'great or small, is defensive if we leave the initiative to the enemy and wait for his appearance on our front door'.[13]

Every combat engagement, whether 'great or small, is defensive if we leave the initiative to the enemy and wait for his appearance on our front door'

In considering deployment, three denominators are important to the Cohort; AWS' economic denominator (how many AWS can we afford?), its logistic denominator (have we got enough and appropriate AWS in theatre?) and, third, its capability denominator (how mobile, flexible and lethal is our AWS?). Moreover, the aggregate impact of each weapon system (and its role within each party's available arsenal) should be the determinant across deployment model, according with this book's emphasis on AWS' cumulative effects. This also dovetails with what is likely to be an incremental process of AWS implementation and the piecemeal removal of human oversight. Deployment costs, moreover, come in all manner of guises, including

---

13 Boggs, M, 'Attempts to Define and Limit "Aggressive" Armaments in Diplomacy and Strategy,' *University of Missouri's Studies*, XVI, Number 1 (1941), p. 68.

research and development, capital outlay and a portfolio of tooth-and-nail costs (maintenance, configuration, testing, validation, updating). They also arise from a set of intangibles, be that loss of legitimacy (should civilians suffer disproportionate harm) or the perils of public opinion reversal (in cases of manifest error or breakdown). Costs come from the sheer scale of effort that will be required to meet AWS' legal, technical and operational conditions precedent ahead of their inaugural deployment. States must also factor for a portfolio of macro inputs such as wider strategic interest, issues of reciprocity, the passing mix of one's own military means as well as the costs calculus around whether one's enemies may resort to similar means and actions.

Deployment models are therefore subject to a portfolio of exigencies. First, in no circumstances can the AWS make legal determination. Second, deployment of AWS is not in itself a preordained outcome but, rather, the deliberate decision of a group of human decision-makers, namely the Delivery Cohort.[14] Again we have that paradox. Removing human oversight is itself an inherently human endeavour. In determining the rules of each deployment arrangement, the Cohort is responsible (morally, legally, operationally and procedurally) for taking all reasonable steps to ensure that its deployment decisions comply with the several obligations detailed in this analysis. This constraint is certainly the case with Western powers. But although AWS deployment is always an inescapably human matter, the assumption that all states and participating parties will adhere to these norms is a stretch. The uncomfortable question remains the degree to which treaties and arrangements are just a Western construct, a set of aspirations driven by those with vested interests in democratic values but with little practical means of enforcement in the face of quite regular breaches. Empirically, after all, that notion of Western 'order' is not one that is adhered to by much of the global population.

> *In no circumstances can the AWS make legal determination*

---

14 Although set out above, it is useful to remind ourselves of the term 'Delivery Cohort', which is amplified in Chapter Six (Wetware, specifically 'The Delivery Cohort') to describe this decision group. It likely encompasses, *inter alia*, the following constituents: neurophysiologists to coordinate AWS networks; psychologists to coordinate learning and cognition; biologists for adaption strategies; engineers for control routines; logisticians; roboticists; electrical specialists; data engineers and data scientists; behaviorists; politicians; representatives from non-governmental organizations; sociologists; lawyers; company directors; weaponists; military tacticians; manufacturers; professionals involved in miniaturisation, simulation, configuration, coding, power supply and modularity; specialists in sensors, and in distributed and decentralised routines; ethicists; and specialists in tooling and calibration.

The Cohort will be influenced by AWS' more granular traits – the drivers of speed, agility and persistence, and the unsupervised weapon's reach and coordination. These five parameters provide a common-sense basis upon which models can properly be assessed. Two further categories are worth noting. First, weapon systems either incorporate autonomy at rest (operating virtually, in software, and including expert advisory systems) or autonomy in motion (stand-alone machines that have a presence in the physical world including robotics, autonomous vehicles and tangible weapon systems). Systems at rest might include, for instance, those monitoring and evaluation routines that enable autonomous assessment to help the commander, an increasingly important verification tool for swift *in loco* decision-making. Second, there still exists an important distinction between commander use and commander reliance on these autonomous means.

Although it is not the purpose of this book to rehearse specific weapon systems (the discussion is straightaway out of date), it is useful to consider some current examples to illustrate the passing ecosystem. Static systems, for instance, are generally fixed in location and have traditionally covered assets such as automated sentry guns and missile defence systems. Deployment examples include the Phalanx CIWS (Close-In Weapon System), long fielded by the US Navy for point defence against incoming threats.[15] Israel's Iron Dome is similarly designed to intercept aerial means of attack. The capabilities of these systems already pose considerable disruption. The THeMIS (an acronym for Tracked Hybrid Modular Infantry System and developed by Milrim Robotics), for instance, is touted as a multi-mission, uncrewed ground vehicle that can be equipped with several payloads including weapon systems. The platform's heft is demonstrated by it being procured by sixteen countries, eight of them members of the North Atlantic Treaty Organization (NATO). The weapon's stated tasking already includes reconnaissance, observation, target acquisition, communications relay, logistic support and medical

> *THeMIS's autonomous capabilities include real-time obstacle avoidance, autonomous resupply and unsupervised patrol capabilities. This is the emerging context from these machines*

---

15 Examples include the US Navy Phalanx system, which can autonomously search, detect and engage targets. Britain's 'fire and forget' Brimstone missiles can distinguish between armoured vehicles and civilian transport without human assistance and can hunt targets autonomously in predesignated areas. Israel's Harpy missiles can detect and autonomously destroy opponents' radars.

evacuation. But a glimpse of its intended use case is best provided by its recent pairing with Hunter 2-S Tactical swarming drones. These use machine-learning (ML) techniques to enable the platform first to identify, then monitor, then neutralise targets, even (it is reported) in demanding battlefield environments. The colleague drone can track a flight path towards a target area, collaborate autonomously with other drones in a swarm and then linger in the platform's area of interest. THeMIS's autonomous capabilities include real-time obstacle avoidance, autonomous resupply and unsupervised patrol capabilities. This is the emerging context from these machines.[16]

Three further model categorisations exist for AWS' likely deployment models. First, weapon systems are either single-use or recoverable. Single-use systems currently include the Switchblade drone (designed for one-time deployment). The tasking of more capable, expensive and multiple-use cousins cover reconnaissance and patrol missions after which the unit returns for maintenance and reuse.[17] Recoverable systems include the RQ-4 Global Hawk, a high-altitude, long-endurance uncrewed aerial vehicle (UAV) used for surveillance and reconnaissance. Introduced in 2001 and due for retirement

---

[16] Use-case models for autonomous defensive systems primarily fall into the category of providing protection and security and include counter-drone systems or autonomous interceptors for missile defence. In this vein, DARPA's Sea Hunter platform has been designed for anti-submarine warfare and developments in the category seems to make monthly headlines. Offensive systems are then used for attack missions and include autonomous aerial platforms or missile systems. Here, the X-47B is an uncrewed carrier-based combat platform developed by Northrop Grumman (with category examples first introduced in the 2010s) and has undergone considerable development since its inaugural flights. A further platform within this model type is the Harpy, an Israeli-made loitering munition designed to target radar systems. See: D'Urso, Stefano, 'Let's Talk about the Israel Air Industries Loitering Munitions and What They Are Capable of', *The Aviationist*, 7 January 2022, https://theaviationist.com/2022/01/07/iai-loitering-munitions/. It is also useful to triangulate these developments across the adversarial divide. The Uran-9 is roughly its Russian equivalent, designed primarily for reconnaissance but allegedly programmable to provide unsupervised fire support from its thirty-millimetre cannon, machine guns and anti-tank missiles payload. Russia has also deployed a range of aerial drones including its Orlan-10 platform, a tactical uncrewed drone vehicle intended for reconnaissance and electronic warfare. Operated remotely, the system features autonomous capabilities in flight management and data collection, the technologies' importance being reflected in the extraordinary growth in its deployment. By early 2024, Russia could reportedly manufacture (or otherwise procure) some ten thousand drones each month, a considerable stretch from its first recorded use of the Shahed-136 drone platform in September 2022, single units attacking military targets in Ukraine's Kharkiv region to the east of the country. Both sides in the conflict have employed loitering munitions which, in their deployment model, are a crossover between a drone and guided missile asset. These systems, such as the Ukrainian-made RAM-II and the Russian Lancet, can autonomously maintain holding patterns over a precise area before engaging targets based on predefined parameters.

[17] See underlying articles highlighted by Cole, Chris, 'Small Drones, Big Problem; Two New Reports Examine the Rise and Rise of "One-Way Attack" Drones', *Drone Wars UK* (21 June 2023), https://dronewars.net/tag/loitering-munitions/.

in 2027, several variants are now to be found in NATO arsenals. Second, deployment models are also defined by their domain of operation whether this be aerial (in this case, autonomous systems built into Lockheed Martin's F-35 Lightning II and, generally, the Kratos XQ-58A Valkyrie), maritime (the uncrewed Orca, a large autonomous submarine developed by Boeing and the Knifefish for the US Navy as a mine countermeasure platform[18]) or ground-based (the Mission Master, an uncrewed ground vehicle first developed by Rheinmetall and Israel's Guardium for Border Patrol and related tasks). A third category depends upon the level of autonomy of those platforms within each model. The Taranis, developed by BAE Systems, is a quite long-dated British demonstrator program for uncrewed combat aerial vehicle technology and also envisaged as a fully autonomous platform. In this vein, the crewless Mojave, with its fifty-five foot wingspan, has subsequently taken off from the UK's HMS *Prince of Wales* in November 2023. Other platforms such as the MQ-9B SkyGuardian and South Korean SGR-A1 Sentry Gun have onboard autonomous functions but seemingly still require human confirmation before engaging their targets.

4.2 Planning tools

AWS must know if it is to decide, and then when and what to decide. To this end, nearly one third of the systems identified in the 2017 SIPRI dataset already use variations of autonomous target recognition (ATR) as an unsupervised 'decision aid' for human operators. ATR is central capability if the controlling agent (here, the commander or, eventually, the machine) is to translate aims and goals into a sequence of actionable tasks that can be undertaken autonomously by a colleague machine. As discussed in later chapters, the programming challenge is that battlecraft and its decision-making are partly science but also, in large part, art. Nevertheless, a driver here has been the several digital initiatives attempting to automate and then make autonomous these planning routines. The US Department of Defense's (DoD) Project Maven was started in earnest in 2017. Intended to improve autonomous object recognition, the initiative was initially trained using drone footage recorded by US Navy Seal units in Somalia.[19] The early program performed disappointingly. The source images were often too

---

18 Harper, Jon, 'Navy Receives First Orca Unmanned Submarine from Boeing', *Defense Scoop* (20 December 2023).
19 Manson, Katrina, 'AI Warfare Is Already Here', *Bloomberg Business Week, The Big Take* (28 February 2024).

hazy or shot from angles that confused the algorithms. They were often so poorly labelled so that model training failed. They were also inconsistent; one iteration might (or might not) flag an object simply as a tank while another would detail the vehicle precisely as a Soviet-designed T-72 variant.

Performance was even worse outside the lab where systems struggled to handle poor network connections and different computer configurations. In the intervening half decade, however, progress has purportedly been steady, improving the project's artificial intelligence (AI) through many incremental changes to training techniques. In addition to video imagery, Maven's autonomous target management now unites data from radar systems (which can see through clouds, darkness and rain), audio devices as well as from heat-detecting infrared sensors, reputedly enabling the software to search for objects of quite broad interest such as engines or weapon factories. The DoD's program also analyses non-visual information by cross-referencing geo-location tags drawn from electronic surveillance and social media feeds. It has also benefited from wide testing, including (again purportedly) in Ukraine. Indeed, during the first ten months of that war, Maven's offering is alleged to have undergone more than fifty separate cycles of improvement. Again, this is the emerging context for these machines and their capabilities.

> The programming challenge is that battlecraft and its decision-making are partly science but also, in large part, art

Recently reclassified from a developmental project to a 'program of record', Maven's functioning still very much depends upon the environment in which the program is running. Algorithms, for instance, that are trained on desert conditions remain quite poor at identifying more generalised objects. Nevertheless, while human analysts may get their descriptors correct eighty-five percent of the time, that figure for Maven is already around sixty per cent. Tanks may empirically be the easiest to spot but the program's accuracy rate can fall below thirty per cent with objects such as anti-aircraft artillery or when snow or other conditions make images harder to parse. And, while benefit also comes from the speed of its algorithms, using AI-targeting in a crowded terrestrial battlefield (the more so with civilians in the vicinity) remains an enduringly unproven use case.

The new context here is that the US military reportedly has more than eight hundred separate programs in play using AI to solve for military tasks, from processing weapons-sensor data and planning resupply routes for ammunition to traditional targeting across the battle space. But although some aspects of military operations (for instance, movement rates, fuel consumption, weapons effects) may be quantifiable and reflect what is the science of war, most battlefield variables (for instance, the ambiguities and nuances of leadership, the complexity of adversarial operations, serendipity, the weather...) belong instead to the art of war. These comprise the subjective, the ambiguous and the contextual, all of which (as we will see in later chapters) complicate the matter-in-hand's coding. Thus, while the starting point to a deployment model may be the removal of human supervision, its efficacy still depends upon human-centric management of intangibles as well, of course, as the integration of that system into the user's wider portfolio for combat options.

Autonomy is already deployed across war's planning processes, a complex set of tasks long considered byzantine, cumbersome and inflexible. Without the lift of technology, staff teams would be condemned to work with slow, poorly tailored information (and all of this under the time constraints of a manual process) in order to arrive at what are very consequential battlefield decisions. Two such decision aids have been deployed in the Israel-Gaza conflict since October 2023.[20] The Israeli Defense Forces' (IDF) Gospel system identifies buildings and structures that their intelligence suggests has relationships with militant representatives. A colleague product, termed Lavender, marks people and then prioritises their targeting.[21] The systems' relevance here is twofold. First is the emerging sophistication of being able to knit previously heterogenous information feeds that cumulatively now enable militaries to create thousands of targets. Second is the empirics of what happens when human oversight is relaxed in these processes, the loosening of previous 'zero error' policies and the treatment of mistakes as purely a statistical matter. At its peak, notes a source quoted by NGO +972, these systems generated thirty-seven thousand people as potential human targets.[22] Human verification of these engagements appears

> Although some aspects of military operations may be quantifiable and reflect what is the science of war, most battlefield belong instead to the art of war

---

20 McKernan, Bethan and Harry Davies, 'The Machine Did It Coldly: Israel Used AI to Identify 37,000 Hamas Targets' (3 April 2024), *Guardian*, https://www.theguardian.com/world/2024/apr/03/israel-gaza-ai-database-hamas-airstrikes.
21 Abraham, Yuval, '"Lavender": The AI Machine Directing Israel's Bombing Spree in Gaza', +972 (3 April 2024), https://www.972mag.com/lavender-ai-israeli-army-gaza/.
22 Abraham, '"Lavender"'.

to have been dialled down to just a few seconds, notwithstanding that some twelve hundred new targets were being added to the IDF's tracking system each day.

Such uncrewed systems have also been deployed despite having demonstrated material shortcomings, most of which are directly relevant to the introduction of AWS.[23] By removing human oversight, there appears to have been poor correlation between those living in their home when added to these systems at the start of that conflict and those who were listed as living there prior to the war. In delegating these decisions to machines, this particular Delivery Cohort appears to have prioritised speed over accuracy, the assessment of collateral damage being statistical and calculated without appropriate human confirmation. As highlighted by United Nations secretary-general António Guterres, the results have been 'pretty catastrophic'.[24]

> *These [decision] systems managed to generate thirty-seven thousand people as potential human targets. Human verification of those targets appears to have been dialled down to just a few seconds*

Although not a use case involving direct lethality, these planning systems are a relevant precursor to full deployment of AWS with, according to Paul Scharre, at least thirty countries now operating similar 'defence systems' that have autonomous modes.[25] The destruction of Gaza and the purported killing of more than thirty thousand Palestinians certainly posits a disturbing vision of war sustained in large part by machine systems tasked to recommend targets. To put that into perspective, prior to Gospel and Lavender's deployment, Israel's targeting efforts would produce fifty pre-screened targets in Gaza per year. The updated program now produces one hundred targets each day, with half of this universe subsequently being engaged.[26] According to figures released by the IDF in November 2023, more than fifteen thousand targets were attacked in Gaza's densely populated coastal territory in less than one

---

23 Ibrahim, Mohammed, 'Use of Lavender Data Processing System in Gaza', Stop Killer Robots Campaign (4 April 2024), https://www.stopkillerrobots.org/news/use-of-lavender-data-processing-system-in-gaza/.
24 Galloni, Alessandra, 'UN Chief Says Gaza Deaths Show Something Is 'Wrong' with Israel's Tactics', Reuters (8 November 2023), https://www.reuters.com/world/middle-east/reuters-next-un-chief-says-gaza-deaths-show-something-wrong-with-israel-2023-11-08/.
25 Scharre, Paul, 'The Perilous Coming of AI Warfare', *Foreign Affairs* (29 February 2024), https://www.foreignaffairs.com/ukraine/perilous-coming-age-ai-warfare.
26 Davies, Harry and others, '"The Gospel": How Israel Uses AI to Select Bombing Targets in Gaza' (1 December 2023), *Guardian*, https://www.theguardian.com/world/2023/dec/01/the-gospel-how-israel-uses-ai-to-select-bombing-targets.

month. By comparison, between five thousand and six thousand targets were struck during the whole fifty-one day 2014 war. The inputs for these decision systems, moreover, are now being generated in real time and immediately centralised into multi-domain databases which are updated moment by moment, providing the IDF with targets that can be struck by a mix of artillery and drones in what is in effect a super-fast kill chain.

> *It is this acceleration in a party's ability to sort data, establish patterns, prioritise and then act on this accumulated information in seconds, rather than hours, that has disrupted practices and confirmed AI as an indispensable tool of modern warfare*

It is this acceleration in a party's ability to sort data, establish patterns, prioritise and then act on this accumulated information in seconds, rather than hours, that has disrupted practices and confirmed AI as an indispensable tool of modern warfare. It also provides what is now a tested lens into autonomous weapons' likely trajectory. The discontinuity here is that the IDF examples underwrite a thesis that military systems be ever more deeply connected into AI-enabled networks to collect information and anchor decisions from many thousands of data points in order to provide commanders with an almost limitless (but essentially unverified) set of actionable options. Once again, this is the new context for AWS deployment.

Two doctrinally defined products, the course of action (COA) sketch and statement, provide a further litmus test for this analysis. The premise here is that a machine (and not just the commander) must develop multiple courses of action before making its decision. The process is involved. For our purposes, it must

> *Most combat scenarios have no end-state, and it quickly becomes clear that the basis of all such deployment models is intractably anchored in human behaviour, heuristics and biases*

incorporate the relevance, configuration and then application of available assets, at the same time as factoring for that commander's primary and subsidiary aims. But it is here that the subjective and tricky-to-abstract stages of the COA must rely on the tenets of experience, interpretation and an understanding of the art of war. Indeed, until this pivot point, it has almost been plausible to suggest that autonomous treatment might conceivably improve outcomes. But deployment challenges arise exactly at this point of the process, the balance of this book being an effort to demonstrate why human involvement is required to achieve repeatable outcomes that maintain trust between commanders, staff and the deployed weapon. The parallel here is doubly useful; even autonomous decision aids, while not directly initiating violence, may be just as capable of state-defining actions as the lethal AWS.

The planning component within each engagement sequence is too often understated, comprising a discrete set of routines that is disproportionately complex: identifying options, ascertaining feasibility (and drawing upon all relevant data sources necessary for this phase) establishing probabilities and then identifying all possible executable actions (and points of synchronisation) across these processes.

Two challenges arise. First, planning is both resource consuming and prone to error. It remains dependent upon

Even autonomous decision aids, while not directly initiating violence, may be just as capable of state-defining actions as the lethal AWS

experiential (and thus subjective) inputs. Regardless of an individual's training, the enduring norm is that command and action selection is a direct product of experience which has been built up over long careers on the battlefield. Second, most combat scenarios have no end-state and it quickly becomes clear that the basis of all such deployment models is intractably anchored in human behaviour, heuristics and biases (and the coding challenges that they likely involve) rather than any set of rules-based instructions.

4.3 Machine and human teaming models

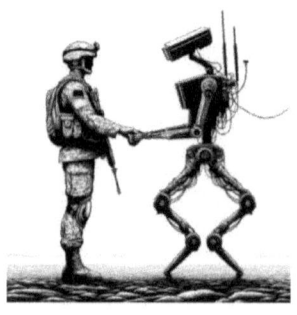

Force multiplication is much touted as an advantage of low-cost, uncrewed and autonomous weapons. Unsurprisingly, however, the driver is not clearcut. First, it is unrealistic to assume one-for-one substitution. Swapping out battlefield assets generally requires experimentation with all of the delays that trials and checks usually entail. Wholesale exchange of weaponry involves updating doctrine and, in the case of AWS, an attendant challenge of capturing that same experience that was previously in the realm of the human commander. Nevertheless, it is here where models for machine-human deployment emerge. The clear notion is that machines can be tasked with carrying out tedious, complicated, repetitive or dangerous tasks to improve efficiencies. Machines can better withstand extreme physical conditions and operate in challenging environments whereby the proven effectiveness of the trained soldier can presumably be leveraged by contiguous, accompanying hardware that is designed as 'companion technology'.

Current war zones would seem to signal a coming-of-age for battlefield robotics, yet the Hamas attack of October 2023 also demonstrates that it is not possible for parties to rely solely on technology, either for their homeland security or to replace crewed assets and means.[27] 'History and experience', noted David Mindell in 2015 (erstwhile professor of aeronautics and astronautics at MIT), 'show that the most difficult problem is not for autonomy but instead for the mixing of human and machine'. The overarching issue, he further commented, remains 'the optimal amount of automation to offer trusted, transparent collaboration'.[28] Indeed, the conundrum across teaming models is the putting in place of trusted mechanisms to toggle command between human and machine and to do this in moments of high stress where decision-making is compromised by partial information in a fast-moving environment. Here, even the simplest articulation of this challenge throws up a whole list of conditions that must be in place for these models, from a feasible human-machine ratio (here, the teaming ratio as it relates to pairing human soldiers with autonomous assets) to operating in the highly contextual nature of each battlefield (and where all inputs must be expressible by lines of code).

Moreover, this ratio (between human operator and colleague system) is not obvious. In uncrewed aerial systems, this ratio is currently many to one (that is, many human operators to just one UAV) and, for ground systems, sometimes two to one. The aim here for deployment models is to move these ratios either to one to many or zero to many. The challenge, however, is that very little of this can be set in stone but always remains conditional upon a string of intangibles: the type and number of tasks to be executed, the nature and complexity of the team's operating environment, the sophistication of participating systems, and the cognitive workload of the human operator at any moment whether in or out of hot contact.

Teaming models, then, can take many different forms. At one end of this continuum, we have seen that autonomous systems are already serving

---

27 Cohen, Sagi, 'Gaza Becomes Israel's Testing Ground for Military Robots', *Haaretz* (3 March 2024), https://archive.is/P6mAQ.
28 Mindell, David, 'Driverless Cars and the Myth of Autonomy', *Huffington Post* (14 October 2015), para. 12 of 14 https://www.huffingtonpost.com/david-a-mindell/driverless-cars-and-the-myths-of-autonomy_b_8287230.html.

as advanced decision-support tools, sifting through and analysing huge datasets to identify threats and suggest courses of action. Here, therefore, a basic expression of teaming might be information-sharing between platforms and human operators, the human operator in command of a fleet of uncrewed aerial vehicles for surveillance or early warning. In these modes, humans are making critical decisions (target engagement and its means) but based upon intelligence gathered independently by teaming machines. But even this most straightforward arrangement requires thought about access and interoperability. Without reliable, secure and real-time communication between humans and those systems, all such models break down. After all, this is quite unlike human-to-human partnerships, where initiative, enterprise and training take over when contact is lost.

> *The conundrum is putting in place trusted mechanisms to toggle command between human and machine and to do this in moments of high stress where decision-making is compromised*

Further along this continuum is mixed-initiative cooperation involving humans and autonomous systems collaborating on actions that leverage each party's respective situational awareness and capabilities. This represents a leap in intricacy, involving distinction between defence and attack, single or multiple use, actions that are event driven or deterring, specialist or generic. Teaming models here are characterised by the machines' potential for learning and adaptation and also by their potential deadliness (the degree to which they can initiate lethality). The more elaborate the tasking, the more complicated the partnership model. Is the team's intended operation solo or swarm? Is it stealth, covert or open? Does it depend on domain or duration? Its operation will also be framed by the shape and degree of each team's agency depending upon that model's expected level of human supervision. Collaborations can then be differentiated by the team's range of engagement; in this way, the US Phalanx System is defending a ship by point defence while a missile defence system such as Iron Dome is protecting a more substantial geographic block. Overlapping or duplicatory tasks add further complexity, with even the simplest models requiring a trusted framework of verification and permission. These must be sufficiently sophisticated to

> *'History and experience show that the most difficult problem is not for autonomy but instead for the mixing of human and machine'*

manage a whole portfolio of technologies, standards and use cases whether aerial, maritime or surface, a sovereign arrangement or a swarm of systems. Moreover, each team must also reform without consequence if one component of that team becomes unworkable or otherwise compromised.

Cooperation models further split between human machine collaboration and human-machine teaming (HMT). The former concerns the optimising of cognitive tasks while HMT relates to the efficient execution of complex tasks.[29] Teaming is currently understood in the context of humans interacting with anywhere from one to several hundred autonomous and uncrewed systems. But it is the recent and runaway progress of AI and robotics that has drawn attention to the new capabilities, efficiencies and advantages that these technologies can enable.[30] Although SIPRI's 2017 database found little evidence of sophisticated uncrewed systems which could act as so-called loyal wingman, this equation has changed considerably over the intervening half decade. Weapon systems that incorporate human-machine autonomy include, *inter alia*, air defence systems, certain active protection systems and robotic sentry systems as well as certain guided munitions and loitering weapons. In this vein, the 'loyal wing' man concept posits a human pilot controlling the tasking and operations of a handful of relatively inexpensive, modular and attritable autonomous platforms. Wingman platforms will likely operate forward of the crewed system (whether aerial, maritime or land based), carrying out an increasing number of missions including electronic attack or defence, intelligence and surveillance. Already capable of autonomous reconnaissance, these assets can also act as decoys to attract fire or, in the case of UAVs, otherwise 'light up' enemy air defences and to achieve this at a fraction of the cost of an F-35 airframe (which, even considering inflation, costs almost two hundred times more than a single P-51 Mustang cost in 1945).[31] Indeed, simply developing

After all, this is quite unlike human-to-human partnerships, where initiative, enterprise and training take over

---

29 Kaushal, Sidharth and others, 'Leveraging Human-Machine Teaming' (18 January 2024), Royal United Services Institute, https://www.rusi.org/explore-our-research/publications/special-resources/leveraging-human-machine-teaming.
30 Nurkin, Tate and Julia Siegel, 'How Modern Militaries Are Leveraging AI', Atlantic Council (14 August 2023), https://www.atlanticcouncil.org/in-depth-research-reports/report/how-modern-militaries-are-leveraging-ai/.
31 Venckunas, Valius, 'Loyal Wingmen: The Cyberpunk Future of Aerial Warfare', *AeroTime Hub* (30 March 2023), https://www.aerotime.aero/articles/25825-loyal-wingmen-the-cyberpunk-future-of-aerial-warfare.

uncrewed versions of current combat jets might solve for the issue of pilot fragility but fails to address the rising cost of these platforms.

The wingman drone is also intended to operate with quicker-than-human reaction times and with the ability to execute complex procedures. Examples are useful to demonstrate the direction of travel. First in the public eye in 2014 (with Dassault's nEUROn stealth drone undertaking a formation flight with a Rafale fighter), the promise of these technologies continues to be another driver to lessening human supervision. In 2018, Airbus announced that it had successfully tested an arrangement with an airborne commander controlling five target drones in an effective system-of-systems where different components of the overall team pooled their strengths to perform better than the sum of their parts, namely the development of purpose-built disposable drones that act as a colleague swarm to an airborne controller. DARPA's LongShot program builds on this arrangement to field systems with narrowly focused attack missions but with the carrying of air-to-air missiles extending the whole team's range.[32] More recent iterations point to the deployment of small disposable drones carried by existing fighters and bombers which, launched at long range, are intended to close the distance to the enemy before either launching missiles or acting as a booster stage for other guided weapons.[33]

> *These must be sufficiently sophisticated to manage a whole portfolio of technologies, standards and use cases whether aerial, maritime or surface, a sovereign arrangement or a swarm of systems*

Regardless of the nomenclature used for these systems (manufacturers tend to emphasise their wares' 'teaming capabilities', preferring the term 'collaborative combat aircraft' to that of wingman), the development by Kratos Security of its XQ-58 Valkyrie points to the progress made in this colleague weapon category. Capable of high-subsonic speeds with decent maneuverability and stealth, the platform can carry up to two hundred and fifty kilograms of ordnance in internal weapon bays. Costing less than three million dollars, the platform competes head on with current anti-air missiles. Kratos' non-stealth variant, its UTAP-22, costs even less and is currently being

---

32 Mehta, Aaron, 'General Atomics LongShot Drone for DARPA to Start Flight Tests in December', *Breaking Defense* (7 September 2023), https://breakingdefense.com/2023/09/darpa-longshot-general-atomics-drone-flight-test-first-look/.
33 Franks, Johnny, 'DARPA's Armed LongShot Drone Takes Flight', *Warrior Maven* (10 January 2024), https://warriormaven.com/sea/darpas-armed-longshot-drone-takes-flight.

repurposed from being an aerial target drone into a CCA where 'AI systems' will purportedly allow it to act as co-pilot to crewed aircraft. It is these precursor experiments that are informing the introduction of broad-tasking AWS. Indeed, this embrace of autonomy is evidenced by the development of Skyborg's Autonomy Control System and the X-62A Vista (Variable Stability In-Flight Simulator Test Aircraft), both offering algorithmic solutions to control uncrewed aircraft. The sophistication of HMT's tasking is therefore increasing quickly but also incrementally, progress also being tied to the speed with which behavioural constraints can be ironed out. These include the extent to which parties are prepared to delegate decisions to algorithms as well as the expanding target types that commanders will accept without human supervision.

Precursor systems also include machine-machine collaboration and where the human is ever further from the decision loop. The trend is already evident and includes, for instance, adoption of active protection systems (APS). These operate in a similar manner to air defence systems (combining a sensor system, a tracking, evaluation and classification system, and a fire control system) but in this case to protect assets against incoming munitions. Similar to AWS, the triggers for APS rely entirely upon data received from onboard sensors that then collaborate with colleague hardware to initiate an appropriate kinetic response. At that point, APS' engagement is without oversight and without veto. Although its current application may be defensive (and there is also no 'learning component' within their function[34]), APS only requires a small operational stretch before its deployment more mimics the likes of robotic sentries and gun turrets that independently detect, track and then engage broader targets.[35] After all, capabilities such as a 'readjustment capacity' (allowing, for instance,

> *The overwhelming majority of these interactions involve algorithms that possess no physical form*

---

[34] Deployed examples include Israel's Trophy APS on its Merkava tanks. SIPRI's dataset lists only nine countries developing APS technology and seventeen hard-kill APS products, all to degrees capable of autonomous target detection, identification, prioritisation and target engagement. As a category, APS reaction time is pivotal: just three hundred milliseconds to intercept an anti-tank missile launched from four hundred metres.
[35] SIPRI's dataset identifies just three such century weapons (Samsung's SGR-A1, Raphael's Israeli Sentry Tech and South Korea's DODAAM Super aEgis II). Guided missiles are excluded from this analysis as the subgroup is generally assigned targets in advance by human operators and only uses autonomy to track, navigate or engage what is therefore already a preassigned (and therefore not autonomously selected) target.

flight path adjustment mid-mission) or autonomous target designation that is either go-onto-location (a particular geographic location) or go-onto-target (target engagement that is based on signature or other predefined designation) are merely matters of the Cohort's instruction and coding.

4.4 Developing models for autonomous weapons

An extention to this category then involves autonomous loitering weapons. The class operates from a simple premise: what if a missile could become more accurate by slowing down? Indeed, a useful way to understand a loitering weapon is to think of it as an airborne munition where developments in communications, processing and miniaturisation mean that this weapon can now undertake a variety of functions that were once the preserve of crewed aircraft or artillery. The utility of the weapon category is derived from its generally offensive tasking and its ability to engage over a geographical area. The munition can be deployed singly or in mass and from a range of launch platforms and locations. Capable of mid-course correction, the system increasingly integrates intelligence, surveillance and reconnaissance to engage targets. It combines the precision of missiles with the versatility of drones, able to loiter or hover over targets until an optimal window presents itself to engage. The weapon class is not new and first appeared during the 1973 Yom Kippur War, the IDF using uncrewed drones to provide real-time targeting links to its warplanes, particularly to neutralise air defences. Loitering munitions, principally the Israeli Harob and Harpy, notably enabled Azerbaijani forces to overwhelm Armenian air defences during the Second Nagorno-Karabakh war in autumn 2020.[36]

Notwithstanding these systems' unsupervised operation, their compliant use still remains the responsibility of the local commander. This adds complexity. Few other weapons have combined such high reliance on sensed data coupled with a wide window of time during which the system

---

[36] Hinz, Fabian and Tom Waldwyn, 'Europe Comes Full Circle on Loitering Munitions', International Institute for Strategic Studies (2 February 2024), https://www.iiss.org/online-analysis/military-balance/2024/01/europe-comes-full-circle-on-loitering-munitions/. See also: Shaikh, Shaan, 'The Air and Missile War in Nagorno-Karabakh: Lessons for the Future of Strike and Defence', Center for Strategic and International Studies (8 December 2020), https://www.csis.org/analysis/air-and-missile-war-nagorno-karabakh-lessons-future-strike-and-defense.

is looking for targets. The likelihood of error compounds given the duration of each AWS deployment. Moreover, the commander is *de facto* absent in its processes and the platform must derive its 'awareness' through its camera bank regardless of inclement weather or adversarial measures. Notwithstanding that these models (and the new use cases upon which they are built) may be upending commanders' dispositions, the Cohort must act within the United Nations Institute for Disarmament Research's warning that there remains more volatility in the real-world data of the battlefield than those limited data samples upon which autonomous systems are trained. All teaming collaborations, after all, must factor for target variety, aborting or suspending any action that fails international humanitarian law tests.[37] It also follows that teams must be able to assess the incidental effects of each attack to learn from each outcome and so inform both their own and their colleagues' subsequent actions.

> *Few other weapons have combined such high reliance on sensed data coupled with a wide window of time during which the system is looking for targets*

Common goals must underpin these models. Here, the human component of the team is performing three simultaneous roles. He or she is the operator such that the weapon cannot complete engagements without that operator's direct participation. The human is also the moral agent whose dynamic, value-based judgement decides whether the use of force is appropriate (and, indeed, whether local rules of engagement have been met). Third, the human is the attack's fail-safe should technologies disappoint or circumstances change such that the parameters of the original engagement are no longer appropriate. All of this provides practical and legal reasons to retain humans in these weapons' decision loop. These three roles, moreover, sit uncomfortably with autonomously initiated lethal means. It is solely humans who can make moral and legal judgement and solely humans who are responsible and accountable for those weapons' outcomes. In any deployment model, HMT is subject to the same legal, ethical and practice constraints that are in place for stand-alone soldiery. This conflation is important. As currently drafted and in the event of degraded or lost communications, US DoD Directive 3000.09 requires that weapon

---

37 Sharkey, 'Staying in the Loop', p. 28. Paul Scharre has highlighted, however, that Sharkey's list could eliminate a number of conventional weapons already in use. The point here might be to look forward to technical developments and 'upgrade our sensibility to civilian harm as a result of [those] developments'.

systems neither choose nor engage targets that have not previously been selected by an authorised human operator.[38]

How then might collaboration models work in practice? In the first instance, HMT should be based on a narrow-tasked partnership. Weapons within a team should then draw on several data sources for their operation, the purpose being to integrate goals and actions with the human's knowledge, experience and situational awareness. Diversifying data sources brings resilience and better context. This might sound obvious, but what happens when the teaming weapon encounters ambiguity or disagreement in its collaboration with its human colleague? To what degree do disrupted communications or poor data warrant terminating the arrangement? How does the participating human wrest back control from the now-not invigilated colleague asset, the more so given its mutability and extent of overlap between asset, tasking and practical agency? In this vein, the convention for compromised teaming might be that the weapon immediately stops its activity (failsafe mode). It may also 'fail dangerous' whereby the machine continues to engage targets that have been preauthorised by human controllers. A further convention might then be for a fail-deadly mode whereby the isolated weapon continues to engage 'emergent targets of opportunity' that have not specifically been approved by human operators. Fail-deadly mode might also permit lethal force in defensive engagements.

There remains more volatility in the real-world data of the battlefield than those limited data samples upon which autonomous systems are trained

4.5 Flexible autonomy

Instead, a deployment model based on flexible autonomy and dynamic task allocation might appear more feasible whereby colleague control of battlefield tasks as well as the means to kill is passed, in theory, back and forth between soldier and weapon system depending upon changing circumstances. The narrative remains that machines are providing force-multiplication

---

[38] US DoD, Directive 3000.09, Autonomy in Weapon Systems (25 January 2023), https://media.defense.gov/2023/Jan/25/2003149928/-1/-1/0/DOD-DIRECTIVE-3000.09-AUTONOMY-IN-WEAPON-SYSTEMS.PDF.

and expanding the battlefield, still removing humans from harm's way, confusing the adversary through less predictable actions and, through advances in natural language models, providing more efficiently set priorities and better communicated tasks. Teaming models may also increase overall task robustness, resulting, in part, from permitted redundancy with teams comprising either homogenous (multiple similar systems requiring less technical in their coordination) or heterogeneous assets (with quite different and non-interchangeable weapon 'members').

Unsurprisingly, the model is a priority for interested parties. The vision is for a portfolio of frontline functions that can be undertaken with varying levels of autonomy, moving operations from a relationship that is based upon fully manual, human-in-the-loop supervisory control to one that is fully autonomous. In circumstances when stress is highest and disaster imminent, control might pass from the *in situ* soldier to the teaming weapon. The promise is that the commander should make informed

Practices must correctly apportion workload levels while ensuring ease of interaction between colleague weapon and human party

choices about where and when to invoke autonomy and do so based on trust in the teaming machine. There should be a condition present, namely the ability to verify colleague machine behaviour and performance, factoring for risk the opportunity to mitigate that risk, balancing between the passing environment and the capabilities of each such partnering model.

Certain operational characteristics are common across these flexible deployment models. First, these models involve multiple assets. Protocols need to work across two, three or many teaming parties, raising issues of operator capability, the sharing of awareness and the creation of trust in so diverse a relationship. Second, practices must correctly apportion workload levels while ensuring ease of interaction between colleague weapon and human party. The teaming system, moreover, may or may not be directly lethal. Different levels of autonomy will also be appropriate at different times, within different teams and, presumably, across different componentry, terrain, weather and environments. It is unsurprising, therefore, that flexible autonomy contains several enduring challenges, not least how autonomy be wrested back from machine to human if that machine is incommunicado or working autonomously. How might such toggling work given the many different scenarios and trigger points that will face these collaborating assets? The challenge is only deepened by the traits of current ML models, including inappropriate suppression of doubt, disregard of ambiguity, and poor inference of causes and intentions.

These shortcomings are well documented. Unsupervised decision aids have previously been dogged by biases, from automation bias (the uncritical acceptance of a proposed outcome as suggested above) to confirmation bias (seeking and then accommodating specific information to confirm a prior belief). A further systemic fault line is that machine reasoning's suggested COAs are empirically unable to incorporate contextual information in its processes. The condition here is referred to as WYSIATI (what you see is all there is), an acronym that usefully captures what is a general flaw. A further constraint is worth highlighting: the more reliable that the human operator judges the weapon system, the less rigourous might be the monitoring of that system once deployed, generally complicating the calibrating of these models if that weapon is to be both compliant and reliable as a battlefield asset. This has always been a human challenge, but the battlefield context for coders remains that twenty-three hours and fifty-nine minutes of boredom are typically followed by one minute of frenetic and disorientating commotion.

Flexible autonomy requires that weapon control should always be partially manual as it is the human operator who must choreograph the task's overall performance while the weapon carries out specific tasks. Given this refinement, additional components are needed in the model if the operator is to decide where control sits between weapon and human at every juncture. At the very least, the weapon should be providing potential options and perhaps be ranking these opportunities, whether as a recommended target list or a simple COA assessment. The human operator must still set goals and boundaries before agency can be delegated to that colleague machine. Two challenges arise. First, those levels of autonomy will change very rapidly in lockstep with battlefield developments, and the model will put unreasonable responsibility upon the human (who must also decide how conservative to be with that authority). Second, the model still relies upon the human operator being able to interact, understand and, crucially, predict this partnership with the colleague machine.

> *The challenge is only deepened by the traits of current machine-learning models, including inappropriate suppression of doubt, disregard of ambiguity, and poor inference of causes and intentions*

The matter of prediction is addressed in subsequent chapters and depends on several associations, most of which are difficult to abstract.

An adjunct constraint arises from routines becoming circular, whereby the weapon's output (the projections arising from its prediction routine) in turn corrupts the issue that the weapon is trying to forecast (the notion of the 'undeclared consumer'). Resolution must take place notwithstanding that one or more of the weapon components (and, moreover, the soldier) may be operating outside expected design assumptions. Complexity also increases exponentially as the number of assets and relevant data points increase. While the actions of a single robot may depend upon sensors and effectors, multiple robots must then factor for partner robot states, partner actions and intentions, and how these matters are distributed.

Flexible autonomy is therefore an important consideration in why we need humans on the battlefield. Human-machine teams, whether loosely or tightly coupled, generate idiosyncratic complexity. Partnership characteristics are inherently unstable, each relationship's evolution dependent upon each team's sensed data points and then the inferences derived from these often exogenous events. One human will certainly operate quite differently from his or her peer. Partnerships can only develop sequentially with each new battlefield task or, more complicatedly, in task depending on the path and progress experienced within that individual task. Teaming weapons must either be deployed within an understood plan (itself a complication in what is likely a rapidly changing combat environment) or must be entered *ad hoc* into each task without the guardrails of a plan. Task fluidity similarly complicates the deployment model given that the human operator (who is more than likely engaged elsewhere) must unexpectedly take back control from the autonomous agent that has faltered. In what is termed the out-of-the-loop control problem, humans prove particularly poor having to intervene unexpectedly to diagnose a complex problem in the face of partial or contradictory information.

> *The condition here is referred to as WYSIATI (what you see is all there is). The more reliable that the human operator judges the weapon system, the less rigourous will be the monitoring of that system once deployed*

Two further challenges arise from the coupled nature of this relationship. First, these models are based upon seamless communication, and its layering (which weapon receives what, which weapon received what and the degree of priority attached) creates a control conundrum which does not lend itself to machine mediation, the more so in an environment where the adversary is disrupting transmissions through spoofing, decoy, cyber or other electronic intervention. As broadcasting degrades, messages or parts of messages likely drop off or experience corruption and, given that data sharing is a fundamental precept across these models, predictable outcomes can no longer be assumed. In coordinating teaming tasks, it is then the Cohort that alone must control task delegation.

Second, various sub-models emerge, each with its own vulnerabilities. In the first instance, centralised control requires precise management whereby sensed information is collected, filtered, cleansed, checked and then prioritised before being distributed to all partnering entities. This is a complicated obligation given the unacceptable bottleneck created by the model. A workaround might be to enable some decentralised processing of that information, each deployed weapon then using (or becoming) its own controller to decide autonomously its own course of action. Theoretically, agency will then be unaffected as a team grows or changes in size. But distributed battlefield control requires prior definition of collective behaviours unless this is to become a new bottleneck and additional source of variance.

Predicting a group of robotic weapons is complicated by what is mathematically an exponentially larger set of data points

A final vulnerability arises from the need to reconcile weapon actions and effects. Given the model's inherent non-linearity, impact will be enduringly difficult to attribute. This is exacerbated by the generally small number of individual components that comprise each teaming group and where the universe of collected data will be statistically irrelevant and prone to noise. Just as toggling control between parties compromises the predictability of the whole, moving from local rules (one machine) to global behaviour (a swarm of machines) creates friction. The inverse problem (when a number of units must toggle to precise local rules) is harder still. Indeed, the takeaway is that any grouping should be role-specific, narrow-mission and incremental in tasking.

Partnerships can only develop sequentially with each new battlefield task or, more complicatedly, in task depending on the path and progress experienced within that individual task

4.6 Swarming model for AWS deployment

One such application is the automation of small weaponised units into a swarm of self-directed armed drones. The scenario dominates tabloids the world over with commentators fanning another revolution in expectation.[39] The promise of a weaponised swarm is that while individual elements may not themselves be threatening, a swarm's deployment in numbers is likely to be difficult to defeat.[40] This is supported by its characteristics of group robustness, of low-cost and rapid evolution. For the purposes of this section, swarming is best understood as a convergent attack from many directions. Swarming is seemingly amorphous, but it is a deliberately structured, coordinated and strategic way to strike from every angle with the intention of mounting a sustainable pulsing of force and fire power.

Several countries are pursuing swarm technology. China's *Global Fortune Forum* in December 2018 provides useful opening context for the phenomenon, the hosts setting a record for the largest drone swarm then deployed, with more than one thousand drones maneuvering as a group, dancing and flashing lights in coordination as part of an aerial show. Singer reported soon afterwards that each drone cost a disruptively affordable $1,500, a figure that included the data links and software used to operate it.[41] Able to manoeuvre within a flight deviation of just two centimetres horizontally and one centimetre vertically, an individual drone would automatically land rather than compromise the swarm's overall integrity if unable to reach its programmed position.[42] In the intervening half decade, parties' vision for military swarms has developed considerably. China, in

---

39 Vernadakis, George, 'US Pentagon Is Developing a New Weapon of Mass Destruction', *Daily Mail* (10 February 2023), https://www.dailymail.co.uk/sciencetech/article-11737323/US-Pentagon-developing-new-weapon-mass-destruction-includes-THOUSANDS-drones.html.
40 Lagrone, Sam and others, 'Pentagon Puts Out Call for Swarming Attack Drones that could Blunt a Taiwan Invasion', USNI Naval Institute (30 January 2024), https://news.usni.org/2024/01/30/pentagon-puts-out-call-for-swarming-attack-drones-that-could-blunt-a-taiwan-invasion.
41 Singer, PW and Jeffrey Lin, 'China Is Making 1,000 UAV Drone Swarms Now', *The War Zone*, 13 July, 2021, https://www.popsci.com/china-drone-swarms/.
42 Krepinevich, Andrew, Origins of Victory, *Yale*, 2023, p.94.

particular, has articulated plans for highly complex, heterogeneous swarms that integrate scouting, command-and-control and strike elements to be delivered by a mix of stealth drones, hypersonic vehicles and high-altitude air ships and using near-space to thwart an adversary's defences.

Nor are swarm operations limited to the land domain; here, the notion is also for mobile smart mines, deployed to work together as a swarm, able to position themselves as a 'dynamic' minefield that can react to enemy sweeping by detecting gaps in their area of operation and closing them on the basis of which geographic location (such as a maritime chokepoint) has coverage priority. What is notable is that even quite *simple* swarm assets can attack highly valuable targets notwithstanding their quite modest payload; even the most complex platforms are dependent on soft components such as radars and stealth coatings, all of which are vulnerable to a small delivery of explosives. Ukraine has shown that infrastructure, logistics, supply and command assets are similarly exposed to swarm attack, either by presenting more targets than defenders can engage simultaneously or by exhausting an adversary's counter-swarm weapon stocks.

How then might swarming look as a deployment model? Its theoretical advantages, after all, appear plausible, attractive and already supported by advances in technology. Swarming is generally ascribed to a population of agents using decentralised and self-organising behaviours in their decision-making to adapt to environmental changes and solve problems. The model here is that multiple weapon systems, operating as an autonomous pack, could return 'mass' to the battlefield by augmenting crewed combat systems with a number of cheap uncrewed systems that expand materially the number of sensors and shooters in a fight. This in itself is an important driver given ever-rising large-platform costs. As precision-guided munitions become commonplace amongst adversaries (both state and non-state), a shrinking number of combat assets can become a strategic liability, adversaries concentrating ever-more elaborate weapons on what is then an ever-smaller number of principal ships, bases and other high-value battlefield assets.

Swarming may seem an appealingly new paradigm, but it is of course a naturally occurring phenomenon that for centuries has enhanced the chances of survival across different animal species. Swarming can readily be seen in flocks of birds, schools of fish, colonies of ants and clouds of bees, with 'swarm intelligent algorithms' being driven by scientific interest in studying and replicating these behaviours. As a deployment model, it hypothetically combines the highly decentralised nature of combat with mobility of manoeuvre and a high degree of organisation. Swarming should enable the fielding of forces that are qualitatively sufficient but able to operate in a much-dispersed environment and marshalling a greater number of entities.

Swarms, moreover, can theoretically take more battlefield risk, balancing survivability against mission factors, all accelerated by pell-mell advances across drone technology. The concept has two parts. By engineering agent behaviour, the promise is that swarm weapons will remain dispersed while searching for targets but then rapidly concentrate depending upon the swarm's mission and environment. Second, greater numbers of incoming ordnance complicate an adversary's targeting priorities. Finally to this point, disaggregation of combat power into less complex, less costly and less exquisite systems will also enable a 'family-of-systems' approach

Swarming hypothetically combines the highly decentralised nature of combat with mobility of manoeuvre and a high degree of organisation

(in procurement terms), thereby increasing system diversity, reducing technology risk and mitigating commanders' reluctance to risk losing high-cost, difficult-to-procure assets.

Swarming shares many of the characteristics with other deployment models, able to deliver lethality without supervision but also positing disruptive advantages. A swarm can absorb casualties, a general heuristic being that battlefield morale cracks when casualties reach a tipping point. This is a complex subject: David Hambling contentiously cites a rate of thirty per cent, such that when some third of a party's troops are killed or incapacitated, an attack falters or defenders retreat.[43] Within a swarm, however, the notion is that 'asset casualties' are less consequential. Similarly, while a single broken component can end a Reaper's mission, an autonomous swarm continues to operate almost regardless of its loss ratio.

Three general concepts of operations now become relevant, all cemented by swarming: the imposition of costs, the denying of terrain and the buying of time. Should, for instance, the People's Liberation Army try to cross the Taiwan Strait, its progress will likely be checked by deployment of small, uncrewed and swarming speedboats intent on damaging billion-dollar troop transports and destroyers. Air-launched drones will then create clutter and, by attacking radars, blind those assets. The Chinese military has to either assume more risk or divert additional resources to protect its force,

---

43 Hambling, David, Swarm Troopers: How Small Drones Will Conquer the World (Archangel Ink, 2015), p. 182.

slowing the operation.[44] Swarm operations therefore change anticipated costs relative to anticipated benefits, and also point to swarms being as important as a deterrent as for warfighting.

This deterrence piece is interesting. Consider, for instance, Ukraine's success in keeping open its trading routes in the Black Sea. The asymmetry here has been created by the combination of large numbers of uncrewed surface vessels, but also deployed in concert with raids, missile strikes, intelligence, deception and electronic warfare keeping Russia's navy constantly guessing and wondering what will hit it next. Rather than outright defeat, it has been the risk of suffering further losses (and, by the way, without a clear counterattack option) that has required Russia's fleet to retreat to safer waters further east. Unsurprisingly, swarming has again reinvigorated proponents of mass. Moreover, imposing asymmetric costs has a long history in competitive strategy and, in this vein, cheap and attritable swarms may prove critical if they can alter an adversary's decision-making.

> *The disaggregation of combat power into less complex, less costly and less exquisite systems also enables a 'family-of-systems' approach, thereby increasing system diversity, reducing technology risk*

By way of context, it is useful to consider technical requirements for fielding swarm systems. Swarms must be networked, dynamically re-taskable and capable of several different missions. First, the swarming agent must be able autonomously to search for relevant targets. In this vein, radio frequency is a long-dated technology, first embedded within the US AGM-88 High-Speed Anti-Radiation Missile (HARM) platforms in the late 1980s. As discussed earlier in this chapter, several parties have embedded this means into autonomous drones designed to loiter while searching the battlefield for targets. Other technologies now meet this requirement, including robot-vision algorithms programmed to recognise patterns within an image and thence to find, classify and match objects of interest. The swarm's communication capabilities also decide the swarm's range and resilience, its scalability and adaptability. They determine its command and control. Early centralised models relied upon unsupervised agents staying in range of a base station, the challenge being the limited range of the model and the vulnerability of such a critical node. Newer models already resemble

---

44 Jensen, Benjamin, 'Bringing the Swarm to Life: Roles, Missions and Campaigns for the Replicator Initiative' (13 February 2024), War on the Rocks, pp. 1-2.

the often-pictured starling flock whereby hub-and-spur variants extend the swarm's range, decrease required bandwidth and allow the swarm dynamically to scale in size.

A further envisaged characteristic is that the swarm will sense and adapt to emerging opportunities faster than its target, and it is here that the mathematics of swarming becomes particularly compelling.[45] Swarming assets are dispersed, forcing the adversary to expend more munitions on more incoming targets. Second, platform survivability is replaced instead with a concept of swarm resiliency. The swarming model relies largely upon overwhelming enemy defences such that 'leakers' get through and take out the target. An adversary's guns, after all, generally shoot in just one direction at any one time and still require periods of processing time to deal with each incoming asset that poses a threat.

The power of the model, however, lies in more than just the mathematics of greater numbers and here it becomes useful to consider the swarm's coordination. Several models exist, usually based on maintaining separation from nearest neighbours, steering with regards to the locus of the neighbours (while attempting to move towards the average position of those neighbours) and thus keeping the swarming weapon flock together. The benefit here arises from the swarm's self-healing networks.[46] None of its members

> *Swarm operations change anticipated costs relative to anticipated benefits, and also point to swarms being as important as a deterrent as for warfighting*

needs be in overall charge (with control therefore decentralised), resulting, in theory, in no loss of cohesion during an engagement: half a swarm is still a swarm and still capable of the same actions. There are also range benefits to swarming models, the potential distances achievable by AWS deployed in swarm formation increasing by as much as the square root of the number of flyers in the formation. In this vein, the US Air Force's Air Vehicles Directorate has concluded that swarming units can achieve an eighty per cent increase in range compared to the distance they can fly alone. This is likely disruptive so

---

45 Naval Postgraduate School in Monterey, 'UAV Swarm Attack Protection System: Alternatives for Destroyers', Monterey Publishing (2012), http://calhoun.nps.edu/handle/10945/28669. Although undertaken a decade ago, the modelled outcome of just ten incoming autonomous attack drones indicated that the defences of the destroyer would be overwhelmed by at least one significant hit.

46 Frei, Regina and others, "Self-Healing and Self-Repairing Technologies', *International Journal of Advanced Manufacturing Technologies* (29 November 2012), http://cui.unige.ch/~dimarzo/papers/JAMT.pdf.

long as the mathematics is not then undone by the challenges of deploying these platforms.

The model's elasticity also suggests that swarms benefit both attack and defence. The US Navy's low-cost swarming UAV LOCUST program can already manoeuvre multiple 'autonomous' units without those machines having to be individually controlled. The organisation has similarly demonstrated control architectures to manage swarming uncrewed surface vessels: equipped with CARACaS software (Control Architecture for Robotic Agent Command and Sensing), robotic boats respond as a swarm when approached by a potentially threatening enemy ship. A recent requirement by the US Pentagon (for attack drones to deter Chinese aggression over Taiwan) demonstrates that multiple parties are aggressively developing swarm models to blunt specific threats.[47]

Whether machine autonomy favours offensive or defensive military operations is an important determinant for the Cohort. If developments here favour the offence, then two consequences arise. The first is the dichotomy whereby parties find themselves increasingly able to use force but also increasingly aware of having that same force used against them; both conditions are catalysts for arms race and first mover conflict. Second, while qualitative changes in military capability tend towards the introduction of new

> *Platform survivability is replaced instead with a concept of swarm resiliency. The swarming model relies largely upon overwhelming enemy defences such that 'leakers' get through and take out the target*

forms of force, merely quantitative change in capability often leads to those parties expanding their existing forms of force.[48] This should inform the Cohort, notwithstanding that qualitative developments are difficult to anticipate given adoption of new tactics and other attribution difficulties. In the case, however, of swarm technologies, numbers would seem to matter, the more so in domains where there are multiple vulnerable points that a defender hopes to protect, swarms of individually expendable drones likely overwhelming platform defences by attacking either from more directions or

---

[47] Lagrone, Sam and others, 'Pentagon Puts Out Call for Swarming Attack Drones That Could Blunt a Taiwan Invasion', *USNI News* (30 January 2024), https://news.usni.org/2024/01/30/pentagon-puts-out-call-for-swarming-attack-drones-that-could-blunt-a-taiwan-invasion.

[48] Garfinkel, Ben and Allan Defoe, 'Artificial Intelligence, Foresight, and the Offense-Defense Balance', *War on the Rocks*, 19 December 2019, https://warontherocks.com/2019/12/artificial-intelligence-foresight-and-the-offense-defense-balance/.

in more waves then that platform's defences are capable of managing. Should this method of attack prove viable, the Cohort might embrace ever larger swarms that initially benefit those doing the attacking with AWS focussing upon less well defended targets and, importantly, parts of targets.

Uncrewed technology is of course not limited to the land domain: as above, Ukraine's success in the Black Sea has highlighted its potential across maritime drones, with boats already remotely controlled for as much as one thousand nautical miles. They can do this regardless of sea state, with a payload of some one thousand pounds, and they can loiter in position for several days. Ukraine's boat swarms can purportedly manage occasional speeds of more than thirty-five knots, incorporate sensing and avoidance technologies and operate in low-visibility conditions where global navigation satellite systems are already denied. Indeed, recent successes using uncrewed maritime platforms (by both the Houthis and Ukraine) have only added to their attraction and renewed focus upon autonomous versions of seaborne lethal systems.[49]

Half a swarm is still a swarm and still capable of all the same actions

Such analysis, however, ignores technical and behavioural challenges to swarming deployment. First, swarm coordination is fundamentally complicated. Swarms require considerable intelligence and logistics. Swarm deployment also makes it difficult for the Cohort to attribute responsibility – a team of weapons might be composed of individually understood units but together, as a swarm, may deliver unanticipated outcomes (or simply fail). The model's behavioural challenge also underplays the cognitive demands that will be placed on the Cohort (including, of course, the ceiling number of units that can be controlled without general autonomy). Command-and-control is also prone to third-party interference given the rise in vulnerable attack surfaces created by fielding multiple assets, the more so given the Cohort's need to match levels of swarm intelligence to the complexity of the underlying task.

To be compliant under the Law of Armed Conflict (LOAC), this must also be organised in advance of each mission. The process must then account for any state changes encountered during the mission and the speed of reaction required by the swarm, as well as the extent to which cooperation

---

49 Knickerbocker, Mike, 'Written in Black and Red: Asymmetric Threats and Affordable Unmanned Surface Vessels' (3 January 2024), War on the Rocks, https://warontherocks.com/2024/01/written-in-black-and-red-asymmetric-threats-and-affordable-unmanned-surface-vessels/.

among swarm components is required in order to complete each task. Any deployment model must account for the degree of risk (both in terms of probability and consequence) of task failure. The added complication is that traditional tasks must now be defined and allocated in wholly new ways. A further element to this equation is that militaries the world over are already heavily invested, financially, doctrinally and bureaucratically, in their current in-the-loop equipment and methods of fighting. An enduring problem is also that the enemy still gets a vote.

4.7 Operations and causes of failure in AWS models

Real-world empirics complicate deployment models. More than ever, AWS will comprise a weapon system that is based upon an intricate mix of motherboards, central and graphic processing units, transistors and resistors, and where efficacy depends upon voltages, bytes and electrical charges made across fragile silicon. The underlying technologies are frail to an extent not previously seen in arsenals and there will be little ability to bootstrap repairs when an AWS breaks down. A broken uncrewed aerial system is a broken uncrewed aerial system. As detailed in later chapters, the Delivery Cohort must also factor for these systems' other in-field challenges, particularly the configuration, testing and validating of the software required to underpin these assets' deployment, especially given that failure may occur at multiple points in all these models. Here, there is the risk of practical failure (not accomplishing a mission, wasting resources and requiring reallocation of assets), technical failure (a subset, whereby the AWS fails to behave as envisaged, requiring rework and reappraisal) and legal failure (whereby LOAC and protection of civilians falls short).[50]

Understanding low-probability, high-consequence risk is critical to understanding AWS deployment models. Militaries' track records in managing risks of this type have traditionally been quite poor. Indeed, NASA's 1986 Challenger accident provides a useful starting point. Physicist

---

50 Ferrell, Cynthia, 'Failure Recognition and Fault Tolerance of an Autonomous Robot', *MIT Adaptive Behaviour*, 24 (1994), p. 3 and pp. 4-5 <http://web.media.mit.edu/~cynthiab/Papers/Breazeal-AB94.pdf>.

Richard Feynman first looked into dispersion within in-house views when calculating the probability of an accident in that program. Estimates, he found, ranged from one in one hundred (the position of engineering departments) to one in one hundred thousand (as articulated by management departments).[51] The upshot here is that it is really hard to quantify this type of exogenous risk. In considering AWS deployment, calculating exposure to ensuing accidents actually depends on the risk in any given instance multiplied by the number of exposures to that risk over a given period. Thus, a one in ten thousand chance of fratricide might convert into an impressive 99.9% safety rate, which, once verified by testing (which may or may not be the correct type of testing), might lead policymakers to conclude that such a weapon without supervision is appropriately safe. If, however, the number of potential weapon interactions with friendly forces in a combat environment is sufficiently large (the case with the US Patriot system and certainly with roving AWS), this would translate into an actual number of fratricides in the hundreds, thousands or larger, enough to have operational impact and, presumably, to affect popular embrace of weapon autonomy.

This is both a deployment and operational point. Even a very low probability of failure can result in an unacceptably high number of fratricides if the number of possible interactions with friendly systems is high. Akin to the sale of lottery tickets, there exists a deployment

Swarming assets are dispersed, forcing the adversary to expend more munitions on more incoming targets

paradox whereby even very low probability events can become effectively inevitable given enough exposure; this then makes unlikely accidents 'normal' in complex (and, here, autonomous) systems. As one commentator explains of AWS, 'these systems are currently too complex and tightly coupled to prevent accidents that have catastrophic potentials. We must live and die with their risks, shut them down or radically redesign them'.[52]

This coupling phenomenon is pivotal to understanding deployment risk within AWS models. Ever-more complex weapons are vulnerable to system failure simply due to their components interacting in unexpected ways, whether within the system itself, involving human operators or within the weapon's wider environment. Indeed, risk 'non-linearity' remains an

---

51 Feynman, Richard, 'Volume 2: Appendix F – Personal Observations on Reliability of the Shuttle', p. 49.
52 Perrow, Charles, *Normal Accidents: Living with High Risk Technologies* (Princeton University Press, 1999), p. 354.

enduring challenge for AWS and its colleague assets. The tightly coupled nature of AWS technology removes, for instance, any time slack in routines, limiting exercise of external judgement and any bending of rules that might otherwise alter AWS behaviour. On its own, this represents a discontinuity.

System failures within AWS will likely cascade rapidly from one silo to the next with no opportunity for external agents either to react, reboot or otherwise intervene. In tightly coupled systems, 'normal accident' theory suggests that accidents are certain given the time and use horizons which are posited for these systems. Research similarly indicates that accident rates will be understated in the early adoption phase of these technologies, just as these new models are being bedded in. An adjunct point here is that hidden failure modes generally lurk undetected, the more so given the number of outwardly independent components that now interact to compound that initial failure.

Calculating the risk of ensuing accidents actually depends on the risk multiplied by the number of exposures to that risk over a given period

Systemic vulnerability in complex systems is not simply conjectural. Even by 2006, seventy-seven robot-related incidents had been reported in the UK in that year.[53] Given the breadth of tasking, the number is no longer collected. A worked example is helpful. The USS *Vincennes* is a Ticonderoga Class Cruiser equipped with the Aegis Special Weapon System intended to counter multiple air, surface and subsurface targets. The system operates with variable autonomy in its prioritising of these targets. In July 1988, its systems wrongly identified an inbound civilian aircraft carrying nearly three hundred passengers as an Iranian F-14 fighter plane. The systems then engaged the aircraft and there were no survivors. Subsequent investigations found that procedural errors had failed to reset the Aegis computer, which was still displaying data relating to an earlier grounded F-14. Moreover, this event did not represent 'peak complexity' as AWS will operate in ever more layered environments characterised by incomplete information, quicker interactions, unanticipated connections between component systems and adversarial meddling.

The *Vincennes* tragedy has been a marker for what might go wrong with weaponry without supervision which, as well as the 2010 Flash Crash

---

53 Singer, Paul, *Wired for War: The Robotics Revolution and Conflict in the Twenty First Century* (Penguin, 2011), p. 195.

in the US equity markets, should remind us that unsupervised interactions between individually accredited components can often produce unwelcome effects.[54] Systemic risk, moreover, particularly increases as new elements are introduced or the size of the product set is expanded. While personal computer software used to use just one megabyte of RAM memory, the recommendation today is for eight gigabytes of RAM for casual computer usage and internet browsing, sixteen gigabytes for spreadsheets and other word-processing programs, and at least thirty-two gigabytes for gamers and multimedia creators. Windows 95 required just four megabytes of RAM at launch, but this had increased eightfold for the introduction of Windows 2000. Today, Windows 10 requires a processor running at two gigahertz or faster, RAM storage of more than eight gigabytes and general storage of more than one hundred and twenty-eight gigabytes, all twice the requirement suggested for Windows 8 simply to turn on the computer.

Malfunction risk is just one deployment consideration. It is really failure in all its forms that should prompt concern as human oversight is removed. First, the Delivery Cohort must factor for adversarial meddling of these assets, compromising system integrity and enabling unintentional system availability. Hacking might occasion misfires, missed opportunities, inappropriate targeting, return-to-base sequences and other unsuitable actions. Intervening with system inputs might also 'mis-train' that weapon's systems. It might compromise system confidentiality (the likes of command intent, mission orders, rules of engagement and target parameters), its effects then exacerbated by AWS' modular nature. The weapons' broad manufacturing provenance is also unhelpful, whereby components are likely to have been procured from a wide set of commercial suppliers whose rewards may often be determined by designing in capabilities rather than designing out vulnerabilities.

> *The tightly coupled nature of AWS technology removes any time slack in routines, limiting exercise of external judgement and any bending of rules that might otherwise alter AWS behaviour*

How important to deployment models is adversarial action? Consider again simple hacking. A convention in April 2024 invited the general public

---

54 See: Treanor, Jill, 'The 'Flash Crash' of 2010: How It Unfolded', *Guardian* (22 April 2015), https://www.theguardian.com/business/2015/apr/22/2010-flash-crash-new-york-stock-exchange-unfolded.

to goad eight leading AI chat box programs into generating a range of problematic responses.[55] The categories included political misinformation, demographic biases, cybersecurity breaches and claims stated in a AI sentence. Three quarters of the attempts were judged successful. Chat bots, moreover, proved to be poor guardians of sensitive information. The context here is that the price of waging spectrum warfare becomes cheaper each year as retail electronics and telephony increase in sophistication. Without human oversight, the challenge for the uninvigilated system is to distinguish between adversarial and friendly information resources. How is the AWS to differentiate between malicious denial of service and other plausible sources of disruption from malfunction, system changes or other explainable phenomena? The verso here is that uncrewed platforms also remain discoverable from their own electronic emissions. Encryption still remains rare across military assets, a fault line that is compounded by the limited spectrum available for these assets' communication. Even by 2016, research was suggesting that some eighty per cent of military transmissions still travelled on vulnerable commercial satellite communications channels and, as late as 2018, only one per cent of defence communication was protected against even modest jamming.

In tightly coupled systems, 'normal accident' theory suggests that accidents are certain given the time and use horizons that are posited for these systems

Other challenges exist for the Delivery Cohort in this space. Signal fratricide occurs when friendly antennas unintentionally overpower other friendly communications assets. Spectrum management is complex, contextual and requires moment-by-moment intelligence to map an enemy's electromagnetic activity that currently requires substantial human involvement. It is also a thoroughly dynamic process. 'Lost link' incidents, triggered when a satellite moves out of range or a machine otherwise drops such signal, are common and require immediate (human) ingenuity to sort. A further deployment challenge then becomes that spectrum control is always fragile and is even difficult to establish whether this state been achieved. Equally, a reeling opponent can rebound very quickly in this space, the more

---

55 Oremus, Will, 'Hackers Compete to Find AI Harms; Here's What We Learned', *Washington Post* (4 April 2024), https://www.washingtonpost.com/politics/2024/04/04/hackers-competed-find-ai-harms-heres-what-they-found/.

so given that electromagnetic counterattack requires technical (and, again, invariably human) intervention rather than high-cost hardware.

Lastly, all deployment models must factor for operational and in-field challenges of supporting independent assets in the field. This covers basic resupply, the replenishment of ammunition and engine materiel, inventory updates and patches as well as these assets' general servicing. Given that these machines are sensor- and software-intensive, this logistics tail must include maintenance of innumerable subsystems and, in order to maintain military advantage, the facility for beta-testing (with appropriate safeguards) new capabilities. All of this requires that the correct number of the correct spares are correctly positioned. It also requires that these be available at the correct time and with the correct support at hand to carry out all necessary work. In the case of long-duration assets, this becomes doubly difficult if the autonomous unit has not been in recent communication. Given the high-regret nature of AWS failure, this must all take place without impacting performance.

> *Hidden failure modes generally lurk undetected, the more so given the number of outwardly independent components that now interact to compound failure*

The Cohort's invariable goal must be that autonomous weapons operate fault-free, with any 'operational conditioning' (otherwise termed trial and error) clearly being an inappropriate basis for their deployment. This is additionally complicated given that several subsystems may work satisfactorily in some but not all aspects of their operation. This is also evidenced by subsequent requests for remedial funding to rework these technically complex platforms; more than eight billion dollars of ancillary funding was required to upgrade the Reaper programme between 2010 and 2015, addressing early performance faults (a failure, for instance, to detect other aircraft while in flight and having to abort in high winds, snow and rain) and to ensure general conformance.

> *By 2016, some eighty per cent of military transmissions still travelled on vulnerable commercial satellite communications channels and, as late as 2018, only one per cent of defence communication was protected against even modest jamming*

5
Obstacles
General challenges to the removal of weapons supervision

This book's next five chapters now focus on fault lines impacting the deployment of autonomous weapon systems (AWS). The aim of this first chapter is to consider deployment impedimenta that are *not* rooted in technology. Its first two sections discuss existing legal frameworks into which AWS deployment must be shoehorned. In subsequent sections, the chapter looks at political, behavioural and contextual factors. It concludes by considering the frictions created by proliferation, by ethical constraints and, finally, by matter of accountability as human supervision is eroded. In order to isolate these behavioural hurdles, this chapter makes the assumption, later to be undone, that AWS deployment is technically feasible. The argument clearly has two quite opposing ends: at one end, commentators suggest that autonomous weapons may be as good as humans in adhering to the Law of Armed Combat (LOAC). At the other, machines are certainly not yet able to replicate the cognitive skills required to use force consistently and without oversight.[1]

The underlying context for this book is that of an ongoing transformation in the character of warfare with evermore control over violence

---

[1] Sartor, Giovanni and Andrea Omicini, 'The Autonomy of Technological Systems and Responsibilities for Their Use', cit. Nehal Bhuta et al., *Autonomous Weapons Systems: Law, Ethics, Policy* (Cambridge University Press, 2016), p. 66. They note that machines already excel in several cognitive skills relevant to the Laws of Armed Combat, including the calculation of positions and trajectories, recognising relevant patterns in large data sets as well as applying complex rules to a given situation. Conversely, the author also notes machine weakness in reading peoples' intentions and attitudes, anticipating behaviours, reacting creatively to unexpected circumstances, applying latitude and rule-deviation in exceptional circumstances, and then assessing significance in battlefield gains, losses and other attributes.

being delegated to computer systems.² This then prompts consideration of the several halfway houses that exist in this transformation, from task-specific autonomy taking place in discrete weapon sub-components to wholly unsupervised weapons operating independently. AWS deployment is neither linear nor obvious along this continuum and requires that this chapter isolate the factors that come in and out of play as the tasking of each individual weapon changes. This can also happen moment by moment and defining these non-technical challenges is complicated as the issues are difficult to abstract, the more so given the pace of innovation across componentry as well as the maze of deployment models for these independent weapons.

One overarching question anchors this chapter's analysis: what is the *legal* bar for an autonomous operation to be properly legitimate? A starting point is provided by the Geneva Academy: the AWS must be able 'to evaluate a person's membership in the state's armed forces (distinct from an armed police officer, for instance), his or her membership in an armed group (with or without a continuous combat function), whether or not he or she is directly participating in hostilities, and whether or not he or she is *hors de combat*... The weapon system would also need to be able, first, to recognize situations of doubt that would cause a human to hesitate before attacking and, second, refrain from attacking objects and persons in those circumstances'.³ While the Academy's narrative usefully identifies the debate's key constituents, our discussion must bring in several other factors. In the first place, there is no definitional difference between defensive weapons and those intended for hostile application. Second, autonomous targeting should be understood in terms of the

> *Autonomous targeting should be understood in terms of the weapon's determination between several presented objects and which, when and where to engage chosen targets and to do so with chosen force*

---

2 Models for this control shift were discussed throughout Chapter Four (Deployment), specifically: 4.3 (Machine and human teaming model) and 4.5 (Flexible autonomy). Here, the mechanics of human-machine autonomies, hybrid systems, joint-cognitive systems, co-agency and other interaction models severally complicate any deployment. See: Hollnagel, E and DD Woods, *Joint Cognitive Systems* (Basic Books, 2005), https://epdf.tips/joint-cognitive-systems-patterns-in-cognitive-systems-engineering.html. Hollnagel and Woods consider in detail the matter of agency and the importance of weapon configurations in determining relationships between operator and technology.
3 Weizmann, Nathalie, 'Autonomous weapon systems under international law, Academy briefing 8', Geneva Academy (2014), p. 14, https://www.geneva-academy.ch/joomlatools-files/docman-files/Publications/Academy%20Briefings/Autonomous%20Weapon%20Systems%20under%20International%20Law_Academy%20Briefing%20No%208.pdf.

weapon's determination between several presented objects and which, when and where to engage specified targets and to do so with chosen force. This book's purpose, after all, is to consider lethality taking place without human involvement, and this excludes merely remotely controlled weaponry where a human operator has undertaken prior determination for each separate engagement. Unsurprisingly, this is rarely clearcut and, for the purposes of this chapter, is taken to mean where the human operator has (or, in the case of pushing a button, has not) meaningful participation and meaningful control in the process. Similarly, many instances exist where autonomous technologies do not infringe on LOAC, such as autonomous refueling and autonomous navigation, and these are largely ignored by the analysis. The review also bypasses instances where a weapon may be activated without supervision but where this is undertaken in a very controlled environment with minimal civilian exposure or where the weapon can be deactivated should the situation suddenly change mid-engagement.

5.1 The Geneva Conventions and the Law of Armed Conflict

The St. Petersburg Declaration of 1868 was the first formal treaty prohibiting the use of certain weapons in war, in this case banning exploding bullets weighing less than four hundred grammes.[4] Since that date, the international community has built two principal structures to regulate new technologies in warfare. International humanitarian law (IHL) broadly consists of a body of rules that apply during armed conflict with the aim of protecting persons who do not, or who no longer, participate in hostilities. Such rules regulate the conduct of hostilities.[5] IHL sets limits on armed violence with the aim at least to reduce suffering. It is, notes the International Committee of the Red Cross, 'based on long-standing

---

4 The origin of the restriction was in Russia's 1863 invention of a bullet exploding on contact with soft substances. Given that such ordnance would have been 'an inhuman instrument of war', the Russian government suggested that its use be banned by international statute.
5 The published work of the International Committee of the Red Cross (ICRC) is a key reference for this study; in particular, see: ICRC, 'A Guide to the Legal Review of New Weapons, Means and Methods of Warfare: Measures to Implement Article 36 of Additional Protocol I of 1977', *International Review of the Red Cross* 88, 864 (December 2006), p. 932.

norms that are rooted in the tradition of all societies'.⁶ International human rights law (IHRL) is then the body of law designed to promote human rights on social, regional and domestic levels, setting down everyday rights such as the right to life, equality before the law, freedom of expression, and the right to work, social security and education.

Human rights law therefore tends to govern the conduct of a state towards its people in peace time and is traditionally distinct from IHL, which governs state conduct during armed conflict. The rules of IHL have been developed and codified over the past century in international treaties, notably the 1949 Geneva Conventions and their Additional Protocols of 1977. Widespread acceptance exists, moreover, that human rights protection does not simply cease in times of armed conflict and that IHL and IHRL apply concurrently. Certain of these rights (including, *inter alia*, the right to life) are specifically *not* subject to derogation, whatever the circumstances. Given this interplay between IHL and IHRL, the crux here is that the extent to which AWS can legally be deployed. The distinction is relevant. IHL tends to provide stronger protections than IRHL against lethal force and the destruction of civilian property.⁷

The purpose of this section is then to provide foundation to enable understanding of the legal challenges to AWS. It is not to do more than this, as the subject is complex and already well documented.⁸ Accordingly, this section focuses upon the principal frameworks as they relate to AWS deployment. First, it looks at obligations conferred by the Geneva Conventions.⁹ Second, it considers the nexus that exists between, on the one hand, specific legal requirements arising from IHL and IHRL and, on the other, relevant technical bottlenecks that complicate the fulfilment (in AWS operation) of those parallel constraints. In this vein, the Hague Conventions of 1899 and 1907 are the first formal international declarations of the laws of war and war crimes

---

6 ICRC, 'The Basics of International Humanitarian Law – December 2017', *International Review of the Red Cross* (27 January 2018), https://reliefweb.int/sites/reliefweb.int/files/resources/0850_002-IHL_web.pdf.
7 ICRC, 'The Basics of International Humanitarian Law – December 2017'.
8 For a useful overview, see: Petman, Jarna, 'Autonomous Weapon Systems and International Humanitarian Law: 'Out of the Loop'?', *Helsinki*, pp. 24-52, https://um.fi/documents/35732/48132/autonomous_weapon_systems_an_international_humanitarian_law__out_of_the. For a compendium of reports, see also: Human Rights Watch and Harvard Law School International Human Rights Clinic, 'The Need for New Law to Ban Fully Autonomous Weapons', Memorandum to Convention for Conventional Weapons Delegates (November 2013), pp. 2-3.
9 For a primer of the Geneva Conventions, see: ICRC, 'Geneva Conventions and Commentaries', undated, https://www.icrc.org/en/war-and-law/treaties-customary-law/geneva-conventions.

in the body of secular international law.[10] They define the qualifications of belligerents including what must comprise AWS characteristics such as acceptably proportionate methods of engaging the enemy and, tangentially, the prohibition of pillage within seized territory as a result of war. The Geneva Conventions then comprise four treaties and additional protocols that establish standards within the international transaction of war. Signed in 1949 by 195 countries, the documents define the basic, wartime rights of prisoners (Convention I), protections for the wounded (Convention II) as well as for civilians in a combat zone (Convention IV).

> Not only must these concepts be understood but it is only through their encoding that an autonomous machine can ensure adherence in its subsequent actions

The combined corpus is directly relevant to AWS deployment as a set of rules maintaining human dignity and protecting the vulnerable and defenceless during conflict. Sitting underneath these two conventions is the Law of War, a legal term of art that refers to that aspect of international law which determines acceptable justification to engage in war (*jus ad bellum*[11]) as well as the limits to acceptable conduct once war is being fought (*jus in bello* and IHL, detailed above). Not only must these concepts be understood but it is only through their encoding that an autonomous machine can ensure adherence in its subsequent actions. Assessing the technical feasibility of lethal machines meeting LOAC then comprises the basis of this book's following chapters.

Conforming to these principles presents the most deep-seated obstacle to fielding independent weapons. Can delegating the decision to kill to an algorithm ever be LOAC compliant? As noted above by the Geneva Academy,

---

10 By way of context, Convention II of the Hague's 1899 document specifies the treatment of prisoners of war, forbids the use of poisons and killing combatants who have surrendered as well as the attack of undefended towns. Convention IV codifies 'the prohibition of the discharge of projectiles and explosives from balloons or by other new analogous methods', 'the prohibition of the use of projectiles spreading asphyxiating poisonous gases' and 'the prohibition of bullets which can easily expand or change their form inside the human body'.

11 Comprising 'proper authority and public declaration', 'just cause', 'proportionality', 'last resort', 'reasonable probability of success' and 'right intention'; see also: Abney, Keith, 'Autonomous Robots and the Future of Just War Theory', cit. *Routledge Handbook of Ethics and War*, Fritz Allhoff et al. (eds) (Routledge, 2013), p. 340. The definition of *jus in bello* is best provided by the ICRC; see: Moussa, Jasmine, 'Can "Jus ad Bellum" override "Jus in Bello"? Reaffirming the Separation of the Two Bodies of Law', *International Review of the Red Cross* (31 December 2008), generally https://www.icrc.org/en/international-review/article/can-jus-ad-bellum-override-jus-bello-reaffirming-separation-two-bodies. *Jus in Bello's* position here is set out in the peer-reviewed Encyclopaedia of Philosophy: http://www.iep.utm.edu/justwar/.

self-directing weapons must be capable of distinguishing between combatants and civilians.[12] This extends to combatants who surrender and who must similarly be spared from harm. The Conventions then forbid methods (here, AWS action sequences) that inflict unnecessary human suffering or destruction of civilian property. Moreover, even if affirmative obligations are retained by humans, autonomous routines must still adhere to all passing obligations including, for instance, the provision of medical attention to wounded combatants and the sick.[13] Congruence also requires that that captured combatants and civilians are protected against acts of violence. Here and elsewhere, AWS' empirical implementation of these measures is complicated by their very lack of definition.

Three circumstances mean that the LOAC framework is usually ambiguous: first, its lengthy evolution; second, the fact that it has rarely been tested in courts of law; and, third (and as recognised in the US Army's field manual), the 'customary' nature of its law whereby much of the corpus's interpretation comes from the unwritten nature of its rules. Instead, these rules are based on traditional practices, values and customs which of them are considered lawful in that particular society, and these invariably differ across borders, cultures and communities.[14] These are thus material sources of friction, and other examples from that the Field Manual illustrate

> *The 'customary' nature of law whereby much of the corpus's interpretation comes from the unwritten nature of its rules based instead on traditional practices, values and customs*

the problem set. The unsupervised weapon must refrain, for instance, from employing any violence that is unnecessary for military purposes. Similarly, the Delivery Cohort should make its decisions with appropriate regard to 'humanity and chivalry'.

While LOAC does not specify that decision-making must be carried out by a human, it is clearly essential that the AWS undertake extensive steps that are complementary to human care before lethality can be deemed compliant by the Delivery Cohort. And this decision waterfall (the sequence of phases that comprises the weapon's decision) may be easy to commit to paper but

---

12 ICRC, 'Basic Rules of the Geneva Convention and Additional Protocols' ICRC publication 1988 ref. 0365, generally, https://www.icrc.org/eng/assets/files/other/icrc_002_0365.pdf.
13 Combatants (in this case, the AWS) must also be able to distinguish the universal Red Cross with attacks upon facilities and vehicles displaying this universal symbol being forbidden.
14 US Army, 'Field Manual, FM 27-10 as amended', *Department of the Army Field Manual* (6 April 1976), para. 1.

is thoroughly intricate to execute in practice. Indeed, simply listing some of these processes that comprise the legal engagement becomes a useful device just to frame this complexity: identifying and verifying targets; matching, authorising and prioritising targets prior to attack; establishing the means of that attack having first considered colleague assets, minimising civilian collateral damage and adhering to the command's broader goals; confirming that each attack is proportionate, necessary and accretive[15]; and, finally, incorporating assessment of the engagement after the event into the agent's subsequent behaviour (and how that attack should either be suspended or followed up).

This list, moreover, is by no means exhaustive and the challenge here is that most inputs for the decision process are sinuous and defy abstraction, the more so given that they invariably arise from the agent's own and uninvigilated sensors. This imprecision regarding the weapon's own environment is therefore matched by the wooliness of interpretation of legal frameworks. Either those tenets have not been tested (whether in a changing combat environment or in a court of law) or existing legal frameworks just do not cover emerging practices.

> *Different data applies at different phases of the engagement sequence and, in chasing exactness, where parameters must multiply quicker than the decision processes can manage*

Prescriptions that were written down half a century ago cannot be expected to factor for the technical advances required to realise autonomous machines. There is also the matter, covered in subsequent chapters, of how these machines manage and then interpret their data when factoring for LOAC.

Indeed, achieving precision in these systems depends entirely upon data, that precision being contingent upon both that data's relevance (issues of accuracy, processing and appropriate fit to the weapon's training set) and its sourcing (issues of depth, veracity and the number of usable parameters that it allows). The complication for LOAC is that different data apply at different phases of the engagement sequence and, in chasing exactness, where parameters must multiply quicker than decision processes can manage. Again covered in later chapters, none of this is facilitated by the variable degree of human presence that actually exists across this process. In this

---

15 For the ICRC's discussion on 'military necessity' and its sanction, see: ICRC, 'Military Necessity', ICRC Casebook, undated, https://casebook.icrc.org/glossary/military-necessity.

Obstacles 143

vein, efforts to secure a statutory ban on lethal autonomous weapon systems have been in play since 2013, primarily concentrated in the Convention on Certain Conventional Weapons (CCW) in the United Nations in Geneva.[16] The context here is that previous control regimes have had a poor reputation and, in earlier campaigns, ensuing legal restrictions have often proved weak, ineffective or illusory[17], the International Court of Justice's inconclusive 1996 ruling on nuclear weapons being an example of an institutional posture 'not to decide'.

5.2 Proportionality and distinction in AWS deployment

Although the rules of proportionality can readily be written down on a piece of paper[18], establishing whether an attack is actually proportional is profoundly contextual.[19] If lines of code are the only way of capturing this setting,

---

16 For a useful summary on these negotiations, see: United Nations, Office for Disarmament Affairs 'Convention on Certain Conventional Weapons – Group of Governmental Experts on Lethal Autonomous Weapons' (2023), https://meetings.unoda.org/ccw-/convention-on-certain-conventional-weapons-group-of-governmental-experts-on-lethal-autonomous-weapons-systems-2023.
17 For example, nuclear weapons are not subject to an explicit legal prohibition. The International Court of Justice (ICJ) in July 1996 reported as follows on the issue: 'The threat or use of nuclear weapons would generally be contrary to the rules of international law applicable in armed conflict, in particular the principles and rules of humanitarian law. However, the Court cannot conclude definitively whether the threat or use of nuclear weapons would be lawful or unlawful in an extreme circumstance of self-defence... so while the threat or use of nuclear weapons was generally held to be against international law, the judges could not determine that it would always be'. See: ICJ, 'Legality of the threat or use of nuclear weapons', *ICJ Reports* (1996), p. 266.
18 *Weapons Law Encyclopaedia*, 'Proportionality in attacks (under International Humanitarian Law)', http://www.weaponslaw.org/glossary/proportionality-in-attacks-ihl: 'The international humanitarian law of proportionality in attack holds that in the conduct of hostilities during and armed conflict parties to the conflict must not launch an attack against lawful military objectives if the attack may be expected to result in excessive civilian harm (death, injuries or damage to civilian objects, or a combination thereof) compared to the concrete and direct military advantage anticipated. If conducted intentionally, a disproportionate attack may constitute a war crime'.
19 See: Chapter Eight (Software), specifically 8.1 (Coding methodologies). PAX uses the equation of 'one low-level terrorist versus three children?' to illustrate this conundrum; moreover, a slight change in circumstances (is the soldier's hands up in surrender?) might fundamentally change the legally compliant response. See also: Walker, *Killer Robots?*, pp. 47-59: 'How, for instance, will a collection of autonomous machines, each self-learning based on individually set confidence limits and working without external interference, independently and identically assess the likely harm that may be caused to civilians and then decide on an engagement's proportionality in relation to any anticipated military advantage before undertaking that lethal action?'

then the process requires, *inter alia*, the attachment of values to targets, objects and even categories of human beings. Dynamically shifting these values is then essential if the machine is to make appropriately probabilistic assessments, and to do so while accounting for almost limitless contextual factors. The obligations of proportionality also pivot upon subjective assessment of tactical advantage. Indeed, simply trying to define such advantage reveals the subject's complexity[20], given its dependence upon probabilities and smart prediction in what are fundamentally matters of subjective judgement. It must also factor for policy and mission. Two questions arise. How, where and how often should these weightings and confidence thresholds be reset in AWS routines? And how can control mechanisms be implemented in real time in order to meet these legal requirements?

Proportionality is always a dynamic construct. Priorities and asset readiness change moment by moment, and battlefield situations remain forever context specific as both civilian risk and military advantage are situational, uncertain and rarely predictable. They are also holistic

> *Although the rules of proportionality can readily be written down on a piece of paper, establishing whether an attack is actually proportional is profoundly contextual*

because military advantage can only really be assessed in relation to larger strategic considerations as well, of course, to the actions of the enemy.[21] Here, for instance, the value and engagement risk of an opposition leader as a potential target in a crowded market square changes as a battle unfolds. Indeed, proportionality is hugely complex even for the human chain of command to decide given its dependence upon trusted, real-time inputs sourced from several agencies.[22] It is not by coincidence that the US Air Force

---

20 Brown, Bernard, 'The Proportionality Principle in the Humanitarian Law of Warfare: Recent Efforts at Codification', *Cornell International Law Journal*, 10, 1 (December 1976), Article 5, 140-142. See also: UK Army, 'Land Operations', *Land Warfare Development Centre*, Army doctrine publication AC 71940, https://www.gov.uk/government/uploads/system/uploads/attachment_data/file/605298/Army_Field_Manual__AFM__A5_Master_ADP_Interactive_Gov_Web.pdf, Chapters 5 and 8.
21 Kalmanovitz, Pablo, 'Judgement, Liability and the Risks of Riskless Warfare', cit. Nehal Bhuta and others, eds., *Autonomous Weapons Systems: Law, Ethics, Policy* (Cambridge University Press, 2016), pp. 150-151. This is corroborated by the ICRC's conclusion that 'an attack carried out in a concerted manner in numerous places can only be judged in its entirety'.
22 ICRC, 'Decision Making in Military Combat Operations', ICRC Publications (October 2013), pp. 13-14 ('Commander's direction and review') and p. 15 ('Evaluation of factors'), https://www.icrc.org/eng/assets/files/publications/icrc-002-4120.pdf. See, also, plot summary of Gavin Hood's *Eye in the Sky* movie, 2016, https://www.imdb.com/title/tt2057392/plotsummary.

Judge Advocate General focuses on the matter when assessing whether the use of violence is legitimate: as set out in *Air Force Operations and the Law*, proportionality is 'an inherently subjective determination'[23], the more so given that individual engagements are usually prompted by operational factors which, from the perspective of a machine tasked with assessing the battle, must often appear dislocated from wider strategic considerations.

The issue of proportionality is again one of empirics and reasonableness. It is also a technical issue (the design, for instance, of software that is capable of predicting civilian harm) as well as an ethical issue. After all, how many civilians need to be judged at risk in that market square before the autonomous weapon amends its action selection?[24] In either case, weightings must be attached to variables, and thresholds assigned to decision paths before specific weapon actions are triggered. Moreover, the constituents of those decisions are very broad. They must factor for the weapon's efficiency, risks relating to whether it will perform as expected or, for a host of unforeseen reasons (environmental, adversarial, simple malfunction), its overall predictability. It is, after all, never the weapon that is making such judgements but rather human-directed algorithms and choices on threshold values that alone frame weapon outcomes. It is an enduring matter of indifference to victims whether the threat in front of them comes from crewed or uncrewed, supervised or unsupervised weapons. In this vein, machine-generated predictions must minimally require that the weapon can navigate contradictory values and contradictory interests in what is a contested environment. Three issues arise. It is challenging to understand what is reasonable *ex ante* an event. Second, it is insufficient that the AWS' rules of engagement rely solely on the test of a 'reasonable attacker'. Finally, the Cohort's decisions must factor for the completeness and relevance of information that is processable for each such engagement sequence.

Tests for reasonableness, moreover, cannot be the sole litmus test of appropriate engagement by the AWS but rather the actions of the human

> *It is an enduring matter of indifference to victims whether the threat in front of them comes from crewed, uncrewed, supervised or unsupervised weapons*

---

23 Hagmaier, Tonya, *Air Force Operations and the Law: A Guide for Air and Space Forces* (Air Force Advocate General's School Press, first ed., 2002), generally.
24 See, for example: Human Rights Watch, 'A Wedding that Became a Funeral: US Drone Attack of Marriage Procession in Yemen' (2013), generally, https://www.hrw.org/report/2014/02/19/wedding-became-funeral/us-drone-attack-marriage-procession-yemen.

beings beyond the deployment of AWS. Distinction and proportionality assessments are probabilistic, the decision for the Cohort being how and how often these processes' confidence levels are re-based in each and every one of its uninvigilated weapons. Let us set aside for a moment the technical complexity of this requirement: it is also not straightforward on any *practical* level. Will thresholds be set differently by democratic states as opposed to non-democratic states? To what degree should popular opinions be factored into these thresholds? Will a UK AWS have exactly similar threshold levels to, say, a French AWS?[25] How are differences here to be understood and communicated? This conundrum is also aggravated by strategic considerations. Practitioners already use game theory to suggest that a state's legal obligation is to deploy AWS assets if such weapons confer military advantage to the side that first uses them. A norm then arises based on increasingly automated and increasingly persistent warfare, all brought about by the principle of necessity.

A second complexity then arises from human rights protection of both life and physical integrity; just as parties must refrain from arbitrary deprivation of life, AWS must at least consider positive steps to secure the right to life demanded by that state's jurisdiction.

> *Will thresholds be set differently by democratic states as opposed to non-democratic states? Will UK AWS have exactly similar thresholds to, say, French AWS?*

However this is framed, AWS deployment must at least adhere to states' duty to control the use of lethal force, including relevant assessment of surrounding circumstances, minimising recourse to legal force and incidental loss of life. Case law, moreover, extends this obligation to conducting investigation upon individuals' deaths in order to secure accountability. The hurdle for the Cohort is that failure to fulfil these positive obligations constitutes in itself a human rights violation. It also exposes the unacceptable gulf between machine and human practices, the more so given that legal and ethical assessment cannot simply be binary (is the system acceptable or unacceptable?), but instead must be contextual and specific to environments.

---

25 Clearly the application of force has much to do with the politics, national morale, historical tradition and cultures of warring states. See, for instance, the different stances adopted towards Germany by France and Great Britain between 1919 and 1939 (Professor Lloyd Clark, in conversation with the author, 13 September 2023).

A second hurdle for warring parties is that they must be able to distinguish between civilians and combatants.[26] While this may appear a straightforward process, compliance here is complicated on several levels. Illegal practices have long been adopted by parties seeking to frighten and coerce civilian populations. Conflict in urban settings and across parties' infrastructure assets have made distinction between legitimate targets and non-combatants ever more difficult, Ukraine being a current case in point.[27] The United Nations projects that by 2050 just under seventy per cent of the world's population will live in cities. Today, it is reckoned that five hundred and seventy-eight cities have more than one million residents and that thirty-two megacities each comprise more than ten million citizens. What then are the ramifications here for deploying AWS? In Gaza, Israel has faced a well-armed and deeply entrenched militant group nested inside a civilian population of a million city-dwelling residents, all under the intense glare of international media and within the apparent strictures of LOAC.[28] When was the last time a US infantry battalion cleared a hospital or skyscraper? Indeed, militaries can clearly no longer rely solely upon stand-off tactics and precision strikes during urban operations, consequent complications of using weapon systems without human oversight being almost unthinkable.

Combatants, moreover, in unconventional armed conflict are ever less likely to be wearing uniforms or insignia. They are tricky for humans to distinguish, and even trickier for machines to adjudicate whether an individual (or, more difficult still, an individual within a group) can reasonably be categorised as a legitimate target, the more so given the ease with which participants can game a weapon's training by, say, exhibiting unexpected behaviour, by concealing armaments or by exploiting the machine's sensing vulnerabilities. Unambiguous definition of this 'civilian' does not even exist within the legal principle of distinction, frustrating the coding between recognition algorithm and sensor. In considering civilian populations, the 1949 Geneva Conventions instead require the use of 'common sense' while

---

26 Rule 1, St. Petersburg Declaration, www.icrc.org. Now codified in Articles 48, 51 and 52 of Additional Protocol I of the Geneva Regulations (against which, interestingly, no objects or reservations have been made).
27 HRW, *Losing Humanity: The Case against Killer Robots*, p. 30. As above, it is beyond the scope of this paper to consider in detail hybrid warfare. See instead: Mattis, Lt Gen James and Lt Col Frank Hoffman, 'Future Warfare: The Rise of Hybrid Wars', *Naval Institute Proceedings* (November 2005), Vol 132/11/1,233, https://www.usni.org/magazines/proceedings/2005/november/future-warfare-rise-hybrid-wars.
28 Barno, David and Nora Bensahel, 'Learning from Real Wars: Gaza and Ukraine' (6 December 2023), War on the Rocks, https://warontherocks.com/2023/12/learning-from-real-wars-gaza-and-ukraine/.

the 1977 Protocol I defines (unhelpfully for AWS deployment) a civilian in the negative sense as someone who is not a combatant.

A further wrinkle arises from whether a combatant is wounded or otherwise incapacitated, a key tenet of battlefield law.[29] Under Article 41 of Protocol I, an individual is considered to be *hors de combat* if three tests are met: if he or she is in the power of an adverse party, has clearly expressed an intention to surrender, or has been incapacitated by wounds and is therefore incapable of defence (provided in all cases that that person abstains for hostile acts and refrains from escaping).[30] The test for unsupervised weapons, in order to be LOAC compliant, must also be that algorithmic targeting routines perform these tasks at least as well as a human soldier; after all, it cannot be assumed that everyone present in a war zone is a combatant.

This presents a host of challenges. First, AWS must be able to attribute intention to those in its immediate vicinity. Here, humans understand one another in ways that machines cannot; cues can be subtle and certainly subjective, and there is an infinite number of circumstances where lethal force is inappropriate.

> *Humans understand one another in ways that machines cannot; cues can be subtle and subjective, and there is an infinite number of circumstances where lethal force is inappropriate*

A second complication arises. If the purpose of military hostilities is not necessarily to kill combatants but to defeat the enemy (even if this requires the killing of combatants), this suggests AWS routines that must comprehend and factor for tactics, wider strategy and planning.

The principle of distinction is therefore another thoroughly subjective process. *Killer Robots*, the report published by Human Rights Watch (HRW), identifies several scenarios to illustrate such engagement dilemmas, epitomised by the frightened mother running after her children who are playing with toy guns near a soldier. Adjunct challenges arise. Unable to 'identify' with humans, AWS will presumably be incapable of showing compassion, a proven check on the willingness to kill. In this manner, although an AWS might self-authorise engagement of a child holding what its routines have identified as a gun, a human soldier in the same circumstances might

---

29 ICRC, 'Customary IHL: Rule 47, Attacks against Persons Hors de Combat', IHD Database, undated, https://ihl-databases.icrc.org/customary-ihl/eng/docs/v1_rul_rule47.
30 Protocol I, Article 41(2); the ICJ in its case 'Legality of the Threat or Use of Nuclear Weapons' affirmed the importance of the Martens Clause 'whose continuing existence and applicability is not to be doubted' (Advisory opinion, 8 July 1996, para 87). The judges also found that the Martens Clause represents customary international law: ibid., para 84.

remember his or her own children and hold fire, seek the child's capture or adopt courses of action which avoid that child. The issue's complexity may be well understood, but technical solutions to the hurdle are enduringly challenging. In this vein (and perhaps unsurprisingly), the ICRC took until 2009 to provide what is still convoluted guidance on how best to classify civilians engaging in hostilities: 'In order to avoid erroneous or arbitrary targeting of civilians, parties to a conflict must take all feasible precautions in determining whether a person is a civilian and, if that is the case, whether he or she is directly participating in hostilities. In case of doubt, the person in question must be presumed to be protected against direct attack.'[31] This is clearly a difficult call for an experienced human soldier and becomes a question of technical feasibility, explored in later chapters, for machine systems to perform the same.

If the purpose of military hostilities is not necessarily to kill combatants but to defeat the enemy (even if this requires the killing of combatants), this suggests AWS routines that must comprehend tactics, wider strategy and planning

Other obligations arise from the Geneva Conventions that cumulatively frustrate AWS deployment. The matter of civilian protection is covered by Article 57. The proscription here is unusually clear, stating that 'an attack which may be expected to cause incidental loss of civilian life, injury to civilians, damage to civilian objects, or a combination thereof, which would be excessive in relation to the concrete and direct military advantage anticipated is forbidden'.[32] Its legal basis is corroborated by subsequent precedent: under the Statute of the International Criminal Court, 'intentionally launching an attack in the knowledge that such an attack will cause incidental loss of life to civilians or damage to civilian objects… which would be clearly excessive in relation to concrete and direct overall military advantage anticipated… constitutes a war crime'.[33] State practices, furthermore, have established this ruling as a norm of customary international law, applicable in both international and non-international armed conflict. The

---

31 ICRC, 'Direct participation in hostilities: questions and answers' (2 June 2009), cit. Cole, Chris, 'Submission from Drone Wars UK to the Defence Select Committee Inquiry "Towards the Next Defence and Security Review" on the Use of Armed Unmanned Aerial Vehicles (UAVs)', Drone Wars (April 2013),, p. 6, https://dronewars.net/wp-content/uploads/2013/04/dwuk-submission-to-dsc-april-2013.pdf.
32 The principle of proportionality in attack is also contained in Protocol II and Amended Protocol II to the Convention of Certain Conventional Weapons.
33 ICC Statute, Article 8(2)(b)(iv), UNTAET Regulation 2000/15.

deployment challenge is best summarised by the ICRC's own definition that proportionality is fundamentally subjective and 'above all must be a question of common sense and good faith for military commanders',[34] requiring, notes HRW in *Losing Humanity*, that any such test requires 'more than a balancing of quantitative data'.[35]

The Convention's Article 57 also requires that constant (and specific) care shall be taken to spare the civilian population, civilians and civilian objects. In programming terms (as well as in semantics), this is a fundamentally abstruse definition. It is also further complicated by the obligations inferred by Article 57(2) whereby directing parties take all feasible precautions in the choice of means and methods of attack be taken in order either to avoid or to minimise loss of civilian life, injury and damage to civilian objects.[36] AWS must suspend its attack if it becomes apparent that its object is not a military one, or apparent (following due analysis) that the attack may be expected to cause incidental loss of civilian life, injury to civilians and damage to civilian objects that would be excessive in relation to the direct military advantage anticipated. Furthermore, Article 57(3) further stipulates that when a choice is possible between several military objectives, then the objective to be selected shall be one expected to cause the least danger to civilian life and civilian objects. Still other conditions are listed that would seem directly to hamstring AWS deployment: Article 54(b)(4) of Protocol I bans 'inherently indiscriminate weaponry', and Article 35(2) of Protocol I rules against weapons that cause unnecessary suffering or superfluous injury. Setting aside arguments on the constitution of a target, a weapon may be deemed indiscriminate simply if it cannot be aimed specifically at that target. This then is the legal context facing the Cohort as it considers removing supervision from lethal engagement.

5.3 Accountability in AWS deployment

These various legal obligations together comprise AWS' 'framework of responsibility'.[37]

---

34 ICRC, Rule 14, Proportionality in Attack and Attendant ICRC's Authoritative Commentary on the 1977 Additional Protocol, generally but specifically pp. 231-237 ('Identification') and 237-241 ('Neutrals').
35 Docherty, Bonnie, 'Losing Humanity: The Case against Killer Robots', Human Rights Watch (2012) http://www.hrw.org/reports/2012/11/19/losing-humanity-0.
36 ICRC IHL database, 'Rule 14 – Proportionality in Attack'.
37 Beard, Jack, 'Autonomous Weapons and Human Responsibility', *Georgetown Journal of International Law*, 617 (2014), 642.

In this vein, customary law attributes accountability as the key determinant in responsible weapon operation. Failure in AWS' function therefore points directly to causality: 'harm would not have resulted had the responsible component correctly exercised the function attributed to it'.[38] This is reinforced by the notion of 'blameworthiness' when harm occurs through a system fault (arising from substandard behaviour throughout the agent and therefore the responsibility of the AWS Delivery Cohort). Tort law similarly extends liability into all areas impacted by the removal of supervision. These include that same causality of harm (strict liability), ownership or custody of a weapon that causes the harm, being 'principal to the agent who caused the harm' (vicarious liability) as well as to general product, design, negligence and organisational liabilities. The challenge is that AWS' intent will be very difficult to determine given the human's increasing remoteness and, indeed, the likely inaccessibility of the weapon system. A consequence then becomes diffusion of agency and the blurring of attribution. *Animus belligerendi* (the intention of a party to fight against a chosen opponent) cannot practically be proved without real-time analysis, less and less likely in the case of an unsupervised platform.

Establishing direct connection between action and intention is even more far-fetched when these means of warfare are either 'spatially distributed' or 'temporally deferred'. It is similarly uncertain that mistaken use of AWS in a neighbouring territory constitutes a legal trigger for conflict in instances where the error was due to sloppy procedures, underlining that AWS deployment must be governed by IHRL standards on the use of force and not just by the law of hostilities. Furthermore, for the law of hostilities to be relevant, control over the weapon's actions must be demonstrable in order for that weapon

'Means of warfare' (and, by extension, AWS deployment) must be evaluated according to the 'principles of humanity' and within the 'dictates of public conscience'

to be classified as a targeted 'means of warfare' against another party. It will rarely be clearcut whether AWS (or parts thereof) are being used as that relevant 'means of warfare', the distinction becoming ever more blurred in instances of war-fighting, in-fighting, policing and other state uses of violence. When then does AWS deployment constitute a legal use of a means of warfare? Establishing in real time the appropriate legal framework remains a material impediment to any lawful deployment by AWS' Delivery Cohort.

---

38 Sartor and Omicini, p. 62.

In considering legal impediments, one further constraint requires particular consideration: the Martens Clause has formed part of LOAC since its appearance in the preamble to the 1899 Hague Convention. Protocol I Article 1(2) states that 'means of warfare' (and, by extension, AWS deployment) must be evaluated according to the 'principles of humanity' and within the 'dictates of public conscience'. Apart from being a wonderful use of language, there is actually no tested interpretation of the clause, which, at its most restricted, suggests that customary international law is over-arching and transcends all battlefield circumstances. A wider interpretation of the clause is that conduct in armed conflict should be judged according not only to treaties but also to the principles of international law as referred to by the clause. This second interpretation is borne out by the International Law Commission's 1999 statement that the Martens Clause provides that civilians and combatants remain under the protection and authority of the principles of international law as derived from established customs, from the principles of humanity and from the dictates of public conscience.[39] The Cohort must therefore decide upon the degree of adherence that its programmers should adopt; is the Clause simply a Western construct to be ignored by autocratic adversaries or are its routines to operate under problematic legal constraints that may cause it disadvantage.

5.4 Article 36 and LOAC-compliant weaponry

Other legal hurdles exist that complicate states' attempts to field compliant AWS. Article 36 of the Additional Protocol I to the Geneva Conventions confers the obligation on state signatories to evaluate new or modified weapons in order to ensure their compliance with the provisions of humanitarian law.[40] States' deployment of AWS must be managed within the prescriptions of these weapon reviews, which should also be a continuous process throughout states' procurement processes. Given that

---

39 UN Report of the International Law Commission, 'Work of Its 46th Session', UN Publications, GAOR A/49/10 (May-July 1994), p. 317. See, also: HRW, 'Heed the Call: A Moral and Legal Imperative to Ban Killer Robots', HRW Publications, (September 2018), https://www.hrw.org/report/2018/08/21/heed-call/moral-and-legal-imperative-ban-killer-robots, p. 3.

40 Protocol I Additional to the Geneva Conventions of 12 August 1949, and Relating to the Protection of Victims on International Armed Conflicts (Protocol I) adopted June 8, 1977, 1125 UNTS 3 (entered into force, 7 December 1978).

Article 36 forms part of the agreed Conventions, much of civil society's case (including that of the non-governmental organisation HRW[41]) rests on this assertion of best practice. By definition, removing weapon supervision complicates states' ability to undertake these formal assessments.

Few states, moreover, have appropriate review mechanisms in place to satisfy the obligation, given that the full implementation of these duties constrains how independent weapons might be used. Under the article, the 'study, development, acquisition or adoption of a new weapon, means or method of warfare' places any state party 'under an obligation to determine whether its employment would, in some or all circumstances, be prohibited', either by Protocol I or by 'any other rule of law applicable' to such party.[42] The scope of an Article 36 review is therefore very broad.[43] It covers weapons of all types and the way in which these weapons are to be used. It also captures any weapon in a state's arsenal.[44] The obligation requires that formal identification be made of where and under what legal limitations the use of that weapon is lawful. In so doing, states must also consider whether the use of these new technologies would be contrary to international law in some or all circumstances. Failure to do this renders a state internationally responsible for a breach of its obligations vis-à-vis the other parties to that Additional Protocol I. For those few states that are not signatories to Additional Protocol I, it is even arguable that these reviews should still be undertaken 'as a corollary to other international obligations' and as a matter of best practice.[45] Article 36 is deliberately catch-all and refers to any other rule of international law that applies to each deploying state including weaponry already subject to a ban.[46]

---

41 For a useful distillation of the arguments see: Garcia, Denise, 'The Case Against Killer Robots', *Foreign Affairs* (10 May 2014), http://www.foreignaffairs.com/articles/141407/denise-garcia/the-case-against-killer-robots. For the original report, see: Docherty, 'Losing Humanity'.

42 US Defense Department lawyers have rejected various proposed new weapons on this basis including blinding laser weapons in the 1990s and, reportedly, various cyber-technologies for use in cyber-conflict. See: 'Legal Reviews of Weapons and Cyber Capabilities', Air Force Instructions, number 51-402 (27 July 2011).

43 ICRC, 'A Guide to the Legal Review of New Weapons,' p. 938.

44 It is useful to provide evidence for Article 36's scope: 'All weapons to be acquired', 'further to research and development' or 'off-the-shelf', 'acquired for the first time without necessarily being new in a technical state', 'and existing weapon that is modified in a way that alters its function', 'and existing weapon where a State has joined a new treaty'; when in doubt, Article 36 stipulates that legal advice should be sought.

45 ICRC, 'A Guide to the Legal Review of New Weapons', p. 940.

46 In chronological order, the following specific weapons have been considered under international instruments: 1868 – explosive projectiles under 400g weight; 1899 – asphyxiating gases, 1899 – expanding or dum-dum ammunition; 1907 – poisoned weapons; 1907 – automatic submarine contact mines; 1907 – Martens Clause: weapons according to 'principles of humanity and the dictates of public conscience'; 1925 and 1972 – biological weapons; 1976 – environmental

As well as examining the legality of the weapon's design and characteristics, the instrument confers the obligation that reviewing authorities also look at how the weapon under review might be used. Consideration of health and environmental factors[47] are included in the language.[48]

Given there may be several ways for a weapon system to fail its Article 36 review[49], such testing appears time-consuming, thankless and expensive. It offers no concrete benefit to the commissioning party and it is consequently little wonder that it is poorly prioritised with States usually opting for expediency. The process also contains several practical difficulties, its stipulations running exactly counter to states' efforts to gain technical advantage over their adversaries. It does not help that the provisions have also been flouted since their inception with neither specific policing nor sanctions included in their definition. Their undertaking is also weakened by the obligation that 'all relevant scientific evidence pertaining to the foreseeable effects on humans has been gathered' including empirical (rather than desktop) evidence. This should include 'what is the expected field mortality' as well as consequences that might impact the victims' psychology

---

modification techniques; 1977 – means of warfare unable to be directed precisely at a specific military objective and might thus include civilians and civilian objectives without distinction; 1977 – bombardment which treats as a single military objective a number of clearly separated military objectives located in a town, city, village or other area containing a similar concentration of civilians or civilian objects; 1977 – attacks expected to cause excessive damage to civilians and civilian objects (excessive in relation to concrete and direct military advantage anticipated ('rule of proportionality'); 1980 – non-detectable fragments; 1980 – mines, booby-traps; 1980 – incendiary weapons; 1993 – chemical weapons; 1995 – blinding lasers; 1997 – anti-personnel mines; 1998 – International Criminal Court definitions of poisons, weapons causing superfluous injury or unnecessary suffering; 2003 – explosive remnants of war; 2008 – treaty to ban cluster munitions. Mary Wareham notes that such weapons in this list have not been prohibited. For instance, non-detectable fragments (CCW Protocol I) are not regarded as a weapon. Nor are such statutes watertight. CCW Protocol II and Amended Protocol II do not prohibit antipersonnel mines or anti-vehicle mines. CCW Protocol III, for instance, prohibits the use of air-delivered incendiary weapons in the civilian areas but not ground-incendiary weapons.

47 Action 2.5.2 of the Agenda for Humanitarian Action adopted by the 28th International Conference of the Red Cross and Red Crescent notes the importance of ensuring a multi-disciplinary approach to the review of weapons.

48 It is telling that the US Department of Defense's 2012 Directive on Autonomous Weapons contains a separate enclosure defining such a set of review guidelines for future AWS development. See: US Department of Defense Directive, 'Autonomy in Weapon Systems', Number 3000.09, Enclosure 3 (21 November 2012) p. 7. These specify policies on system capabilities, doctrines, training, appropriate levels of human judgement and care in deploying these systems, as well as security and testing of the systems.

49 Examples include predicted reliability of targeting mechanisms, evidence that the foreseeable effects of an unsupervised weapon can be limited to the target and can be controlled in time or space. Similarly, Article 36's prescriptions on precise injury and damage levels are complicating, especially in their treatment of mortality rates and 'whether the weapon would cause anatomical injury or anatomical disability or disfigurement'.

or physiology.⁵⁰ Notwithstanding that, three quarters of a century after its parent instrument was agreed, the framework might advantage those who ignore its obligation, it does remain a minimum legal requirement.

Indeed, several of these legal frameworks gradually lose fitness for purpose.⁵¹ An example is in the practical problems of definition. The iRobot Warrior was an early example of an uncrewed weapon to fall outside the UN Register's technical specifications⁵² of either 'Battle Tanks' or 'Armoured Combat Vehicles'. Revision and the contentious updating of legislative listings is clearly required just at a moment when little appetite exists to do so.⁵³ Furthermore, identifying a circumvention of the obligation is complicated by the pace of development in weapon systems as well as the confusingly modular nature of their procurement. Procurement of the US Raven uncrewed aerial vehicle (UAV) provides a case in point: characteristic of a spiral development process, its deployment has been driven by a portfolio of incremental developments, each shaped by rapidly changing end-user requirements and technology rather than any traditional Version 1 release being followed by a Version 2. Article 36 processes are unclear. Nor is there a replicable, transparent *modus operandi* that can conveniently be regulated by statute. Given multiple contractor relationships, no common build system or indexing procedure exists. Similarly, no repository for source code or any recognised (and continuous) testing infrastructure is in place to enable reliable reporting. States' depressing behaviours in other of the Conventions' control regimes means that there is even less incentive to comply with what appear to be the constraining requirements of Article 36.

> *It does not help that the provisions have also been flouted since their inception with neither specific policing nor sanctions included in their definition*

Legal ambiguity has also led to policy ambiguity. Directive Number 3000.09 was first published by the US Department of Defense (DoD) in

---

50 States generally agree that suffering which has no military purpose is a legal violation.
51 HRW and Harvard Law School International Human Rights Clinic, 'The Need for New Law to Ban Fully Autonomous Weapons', Memorandum to Convention for Conventional Weapons Delegates (November 2013), pp. 2-3.
52 It does not, for instance, have the 'high cross-country mobility' or 75mm bore cannon of the tank nor is it able to 'transport a squad of four or more infantrymen'. See: UN Office for Disarmament Affairs, 'Categories of Equipment and Their Definitions' (undated).
53 For instance, the Switchblade and Fire Shadow weapon systems currently fall into a grey area between 'munitions' (under the less tightly controlled Article 3 of the treaty) and a 'combat aircraft'.

156 War Without Oversight

November 2012 as the first significant policy announcement by a technically developed state on the development of semi-autonomous and autonomous weaponry. The document instigated a temporary ten-year moratorium during which the DoD should only develop autonomous systems that deliver non-lethal force. That suspension, however, was not (and, despite revision and restatement in June 2023, remains not) clear-cut. There is a carve-out, for instance, that allows a waiver 'in cases of urgent military

> *Given that the point of AWS trialing is precisely to determine the probabilistic range of action without which no appropriate limits can be defined ex ante, it is right to highlight the implausibility of any 'test-as-you-go' regime*

need'. The directive also defines wide testing requirements. Given that the point of AWS trialing is precisely to determine the probabilistic range of action without which no appropriate limits can be defined *ex ante*, any 'test-as-you-go' regime is fundamentally implausible. Any elasticity of standards only underpins the requirement for human judgement in what is the deployment of independent weapons without human judgement. Finally, the directive only relates to the US DoD and notably excludes other procuring parties such as the Central Intelligence Agency.

5.5 Behavioural constraints to AWS deployment

Legal challenges to AWS deployment are themselves framed by behavioural factors. Decades of experience confirm, after all, that fielding new methods of war is rarely straightforward, prompting (notes the ICRC) 'unprecedented issues that make the legality of an attack more complex'.[54] Indeed, measuring the effects of individual weapon innovation is always tricky, this missing correlation between

---

54 For discussion on 'new methods of warfare' see: ICRC, 'New Technologies and Warfare', *International Review of the Red Cross*, 886, 94 (2012) pp. 457-467, https://international-review.icrc.org/reviews/irrc-no-886-new-technologies-and-warfare. See also: Lorber, Azriel, *Misguided Weapons: Technological Failure and Surprise on the Battlefield* (Potomac Books, 2003), generally. An interesting narrative to this point is provided at: Dutton, Judy, 'Nine Bizarre Weapons That Failed Spectacularly', Mental Floss Blog (29 April 2014), https://www.mentalfloss.com/article/30669/9-bizarre-weapons-failed-spectacularly.

weapon system and outcome pointing to context being just as important as technical hurdles in the Cohort's deliberations. The battlefield's changing character is about people, not systems: the driver remains that 'armies, air forces, and navies function with people who use and employ machines and weapons'.[55]

States' militaries, moreover, have often failed to exploit technical innovation. The Mongols missed the Gunpowder revolution; the Chinese, Turks and Indians let pass the Industrial Revolution; and the Soviets missed the early phases of the Information Revolution.[56] The behavioural point here is that leveraging military disruption is rarely straightforward. Sources of inertia exist at all points of a weapon's deployment process exactly because past performance has so often proved no guarantee to future returns in defence planning, the more so in a landscape of fast-moving technologies. It is just as likely that a rival state (or even non-state grouping) introduces a new variant or countermeasure which quickly erodes the advantage of a particular technology. It may also be that a state devotes unexpected resource to creating temporary (although still destabilising) leadership in a new weapon class. The issue, of course, is always more nuanced. The French, for instance, had the main battle tank advantage in 1940 but perhaps squandered that benefit through poorly deploying that asset. As noted by Max Boot, 'the end can come with shocking suddenness even after a long streak of good fortune'.[57]

Institutional reserve in procurement practices has deep historic roots. Forces' personnel know that they have a job to do regardless of the kit with which they are provided, and integrating novel technology into organisations that have been traditionally suspicious of change makes that task harder. Over the span of a career, perhaps twenty-five years for those responsible on the parade ground for bedding in new means, technology comes and goes. Behaviourally, those practitioners are much more concerned about everyday tasks than betting all on new ideas. Embedding new equipment is too often characterised by remedying inappropriate design, creating workarounds, training colleagues, creating doctrine and devising new practices, and arranging testing and validation, as well, in time, as coordinating the

---

55 Tilford, Earl, *Reviewing the Future* (Parameters, 2002), p. 151. The point is made in a different manner by Toby Walsh: 'There isn't this separate part of the world called the battlefield that's signposted "Battles over here please". Battles are fought in cities right where we live'.
56 Boot, Max, *War Made New: Weapons, and the Making of the Modern World* (Gotham, 2006), generally.
57 Boot, Max, 'What the Past Teaches about the Future', *Joint Force Quarterly*, 44 (2007), p. 109, http://indianstrategicknowledgeonline.com/web/MIL%20HIS%20JFQ44%20boot.pdf.

withdrawal of superseded equipment and 'training out' their use.[58] This represents a long tail of often underappreciated tasks.

Other sources of inertia discourage efficient integration, whether they be issues regarding user motivation and user apprehension, risk aversion by commanders and operators alike, or cultural and other local calculi. Integration is also compromised by users' fear of failure, criticism and career impact. All of these traits are natural characteristics of service life but, as a norm, they combine to work against embrace of change. Integration, moreover, is often ambitious, requiring bold bets in face of uncertain outcomes, the more so if human oversight is being substituted for machine intelligence. In the case of AWS, the norm remains a willingness to persevere in face of setbacks, self-doubt and loss of control.

Planners and the Cohort alike, moreover, must understand that not all *'The end can come with shocking suddenness even after a long streak of good fortune'* technologies necessarily become fully fledged capabilities.[59] Instead, they come about through collaborative and well organised absorption which takes place throughout the entire organisation. It must be foundational in reach, matching available human capital to the assets being introduced. Indeed, once kit has been delivered, the embedding of new assets is fundamentally an exercise in behavioural science.

The Cohort must also factor that military acquisition programmes have long represented a demanding series of cliff faces. They are measured over decades and that norm is unlikely to change over the period considered by this book. There are also several degrees of separation between the personnel specifying new systems and those in the Cohort tasked with integrating and then deploying these assets. 'Not-invented-here' biases are difficult to remedy, the more so (in the case of AWS) when outputs defy explanation and where the weapon's initial configuration is so specialist that third-party 'outsiders' are assigned that task. Three issues reinforce the point. First, novel platforms often risk obsolescence at this very point of delivery given the long timelines required to deliver systems in order to iron out integration issues, to combat inertia and, paradoxically, to do this while factoring for the

---

58 Walker, Paddy, 'Agile Procurement? Norms and Challenges to the Integration of Novel Systems into Legacy Force Design', *Centre for Historical Analysis and Conflict Research Journal*, 2nd Quarter 2023, generally.
59 Breaking Defense Staff, 'Vehicle Platform Integration: Where Technologies Become Capabilities', *Breaking Defence* (14 October 2019), https://breakingdefense.com/2019/10/vehicle-platform-integration-where-technologies-become-capabilities/.

fast-changing character of war. Second, change across sectors, geographies and categories conspire to complicate the integration process. A third matter is again behavioural and is rooted in how forces have traditionally procured their capabilities; inter-service rivalry remains a factor in procurement. So does traditionally siloed thinking that still persists between navy, army and air force, the more so given that tactics and doctrine, themselves

> *Institutional reserve has deep historic roots and forces' personnel know that they have a job to do regardless of the kit with which they are provided*

additional sources of friction, are key precursors to the integration of new systems. The very ground rules for these systems must be agreed and articulated before these competing parties can even think about successful deployment across arms and services.

The Cohort's equation is also complicated by contradictions between legacy forces that are still held in states' arsenals and the arrival of new technologies, many of which have been procured to address specific requirements (and shortfalls) which date back to earlier generations and with only passing relevance to the matters in hand. The norm, after all, is that new platforms are rarely procured unencumbered. Furthermore, the erratic pace of innovation and the often random traction of new equipment (today's purported 'revolution in military affairs' is often tomorrow's old news) tends, as a generalisation,

> *Integration is also undone by users' fear of failure, criticism and career impact*

to frustrate development of a clear, unambiguous use case for these novel military solutions. This is exacerbated by the 'baggage' that often accompanies procurement projects (government interference, changes in personnel, muddled responsibilities, opaque commercial priorities), a verso being parties' fear of missing out on that new disruptive technology.

All of these frictions mean that battlefield outcomes rarely depend upon hardware procurement alone. Today's norm, moreover, is that there is less requirement simply to kill the greatest number of one's adversaries. Instead, victory is a more managed process which is shaped by getting the right narrative accepted by the right audiences. It is about directing all available means to the fight. It is still about breaking the adversaries' will but, in operations, by being more prepared and more durable than one's enemies. These are broad issues common to procurement but do not suggest that removing human oversight is another key determinant to the Cohort.

Procurement may define passing battlecraft (and, in so doing, that moment's character of war) but does not alter war's nature. After all, it is only through the efficient integration of new means (in our case, the doctrinal, behavioural and operational integration of AWS) that novel systems can even start to be additive in legacy force design.

One further point regarding the promised asymmetry of AWS is worth noting. Western procurement anxieties usually concern cost and long timelines, but less assiduous non-state and non-peer competitors now have the opportunity in AWS to source lethal means that may be inexpensive, irregular and innovative. The democratisation of technology means that suddenly an almost inverse relationship exists between complexity in new weapon systems and the 'low-tech', 'good enough' characteristics offered by simple off-the-shelf capabilities that can be pitched against them. It is unclear, for instance, whether America's Reaper or other similarly sophisticated uncrewed systems would have proved universally appropriate in Ukraine where low-cost, bootstrapped advances have appeared each week, each month, often requiring little change in delivery practices but causing material adjustment in how the intended target then operates. Change is also signaled by the emerging gulf between solutions that are still institutionally sourced and this plethora of commercially available technology. The latter is usually dual use, is ubiquitous and cheap to acquire, needs minimal bespoke integration, and can increasingly be repurposed for the battlefield.[60]

> *Other sources of inertia discourage efficient integration, whether they be issues regarding user motivation and user apprehension, risk aversion by commanders and operators alike*

> *Today's purported 'revolution in military affairs' is often tomorrow's old news*

A further heuristic dampening AWS deployment is that new battlefield technology rarely gives lasting advantage. Parties facing autonomous weaponry (and without the wherewithal to field similar) will likely respond with behaviour 'that works well enough, be it ever so inelegant and probably decidedly irregular' in order to defeat that new technology.[61] This, after all, is a repeated theme for those looking to deploy new means and some examples are useful to demonstrate the point. In the case of the air power theorists of

---

60 Walker, Paddy and Peter Roberts, *War's Changed Landscape? A Primer on Conflict's Forms and Norms* (Howgate, 2023), pp. 95-113.
61 Gray, Colin, *Another Bloody Century* (Phoenix, 2005), p. 52.

the 1930s, Giulio Douhet grossly underestimated the neutralising effect of a defending air force and anti-aircraft fire on attacking bombers.[62] Planners have similarly (and persistently) overestimated the effect of bombing. Defeat has also often been a very particular spur to innovation, leap-frogging and negating the original technical catalyst. The Israelis almost lost to Egyptian and Syrian anti-tank and anti-aircraft missiles in the 1973 Yom Kippur War, having beaten them just six years previously. In the same vein, technologies and concepts rarely remain the property of one party for a long time. France matched the needle gun less than four years after Konniggratz, Germany matched Britain's Dreadnoughts some three years after their first launch, and the USSR had its own atomic bomb four years after Hiroshima.

Just as Douhet's advocacy of pre-emptive aerial attack was undone by the arrival of radar, the argument must be that AWS will be compromised by the introduction of some other workaround or new technology. In the same way that trench systems were to offset the impacts of artillery and quick-firing machine guns (and improvements in radar negate the edge achieved by first-generation stealth aircraft), the assumption

Low-cost, bootstrapped advances have appeared each week, each month, often requiring little change in delivery practice

must be that combatants quickly muster alternative tactics to blunt uncrewed weapons. The response to drone warfare in Ukraine has been a case in point: a whole industry has quickly sprung up to counter uncrewed aircraft systems including directed radio wave guns, handheld jamming packs, smoke and camouflage kits, and an array of stand-off netting, slatted armour and hanging chain measures designed to defeat incoming munitions.[63] In the case of AWS, this will likely take on several guises, from straightforward wire obstacles to neutrino beams, electromagnetic pulsing, flux compression technologies, microwave and other high frequency interference, energy lasers and signal jamming. The norm for the Cohort is that competing parties will always seek asymmetrical and equalising tactics.

Those parties that can create advantage in a narrow field of combat are also likely to stall an adversary's more general technical superiority and, on this

---

62 MacIsaac, David, 'Voices from the Central Blue: The Air Power Theorists', cit. Paret, Peter, *Makers of Modern Strategy from Machiavelli to the Nuclear Age* (Oxford University Press, 1986), p. 634. See also: Biddle, Tammi David, *Rhetoric and Reality* (Princeton University Press, 10 January 2009), pp. 69-128.
63 Radio Free Europe, 'The Anti-Drone Technologies Evolving on Ukraine's Front Lines' (3 July 2024), https://www.rferl.org/a/anti-drone-evolution-ukraine-war-russia/33020303.html.

basis, the assumption that high tech can only be defeated by similar high tech is likely wrong. The phenomenon is often termed nullification: the first time an army faces disruptive battlefield technology (the needle gun at Konniggratz, the machine gun at Omdurman, the tank, the smart bomb in the Gulf War) it is likely to be caught off guard. The next time the other side is likely to be less impressed. The behavioural constraint for AWS deployment is that a series of quickly learned tactical innovations by an adversary will likely dull the effect of the innovation. High tech is expensive, complicated, in short supply, and not necessarily available when and where required. Serbia's use of UN hostages as human shields in Bosnia illustrates how a simple although asymmetrical act can quickly negate the advantages conferred by a party's investment to the wrong technical question. Indeed, the issue is best captured by Einstein's purported advice to President Truman: 'I know not what weapons World War Three will be fought but World War Four will be fought with sticks and stones'.[64]

> *The introduction of new battlefield technology rarely gives lasting advantage*

Further friction arises from the relationship between AWS' intended function and that weapon's transparency and accountability. It is impossible to punish a lethal robot for unlawful acts that it carries out. Civil society regularly highlights that AWS, free from human supervision and with no facility for emotion, will lack any limbic response which can condition that weapon's behaviour. Unless the weapon can 'understand' (and not merely respond to sensed data as a trigger to its actions) that it will be admonished for breaking IHL, then its decisions cannot be influenced by the potential peril of that accountability. Removal of human supervision, moreover, might paradoxically invest too much power concentrated in the hands of too few parties: binary deployment decisions then become the preserve of what may be a very small cohort. Indeed, transparency is an important behavioural consideration for the Cohort whereby no robot should have the expectation of 'privacy in a public place'. This points to a further contradiction. Consider first the irony that exists in the Cohort's wish that the robot should learn to undertake its mission in the best possible manner, the Cohort also limiting its machines' latitude and creativity.

The Cohort must also be able to predict the human workload required by each different AWS deployment model. As already noted, removing

---

64. Brooks, Rosa 'Why Sticks and Stones Will Beat Our Drones', *Foreign Policy* (4 April 2013), https://foreignpolicy.com/2013/04/04/why-sticks-and-stones-will-beat-our-drones/.

supervision does not translate on the ground to any absence of human engagement. But capacity planning varies over different parts of the same task and depends on that task's complexity, duration and contingencies. If AWS deployment does not deliver efficiencies, then why should a commander allocate resource or permissions to such assets? Commentators, moreover, point to a material increase in the time required by human parties to participate in decision-making where autonomy is deployed: outcomes must be reviewed, configurations tweaked and operators must now factor for additional unknowns in their processes.[65] An adjacent behavioural twist is sometimes called the automation conundrum, whereby the better a system's autonomy, the more (and quickly) out of practice a human operator becomes and, it follows, the more demanding the situation that he or she will have to face when intervention is eventually required.[66] The challenge is that automatic and then autonomous systems erode human skills by removing the need for practice even if human operators were once expert. This is best explained by the psychologist James Reason, author of *Human Error*: 'manual control is a highly skilled activity, and skills need to be practiced continuously in order to maintain them... yet an automatic control system that fails only rarely denies operators the opportunity for practicing these basic control skills. When manual intervention is necessary, something has usually gone wrong. Acknowledging automation bias, operators need to be more rather than less skilled in order to cope with atypical conditions'.[67]

High tech is expensive, complicated, in short supply, and not necessarily available when and where required

These same automatic systems often 'accommodate' prior incompetence by being easy to operate. They also keep in place facility for humans to rectify mistakes. Indeed, inept operators and mildly erroneous practices can remain in place for quite some time before an operator's incompetence really surfaces, doubly relevant if autonomous systems are to fail in unusual situations. In either case, an unrealistically skilled response will be required from the human who is intervening in face of likely bewildering evidence. Wiener's Laws of Aviation provide a useful generalisation, noting that digital

---

65 Deci, Edward and Richard Flaste, *Why We Do What We Do: The Dynamics of Personal Autonomy* (Putnam, 1995), generally.
66 Hartford, Tim, 'Crash: How Computers Are Setting Us Up for Disaster', *Guardian* (12 October 2016), p. 4.
67 Reason, James, *Human Error* (Cambridge University Press, 1990), generally and: Reason, James, 'Human Error: Models and Management', *British Medical Journal* 320/7237, (2000), 768-770.

devices tend to 'tune out small errors while creating opportunities for large errors'.[68] The context here is always the simple math that a computer (which, for argument's sake is broadly one hundred times more accurate than a human and one million times faster) statistically makes ten thousand times as many mistakes as the human.

Other behavioural constraints should concern the Cohort. First, the adoption of autonomy is likely to be accelerated by the phenomenon of threat inflation whereby national commentary represents risks as being much larger than they often are. Milliman's *Psychological Rationales for Threat Inflation* explains

'When manual intervention is necessary, something has usually gone wrong. Operators need to be more rather than less skilled in order to cope with atypical conditions'

the behaviour in terms of a pre-eminent state having no locally relevant enemies but nevertheless being 'suffused by an inherent insecurity'. It is the mere existence of a threat, in this case weapon autonomy, which cannot ever be eliminated, and which can prompt irrational and kneejerk response ('use it or lose it') by that state.

Economic factors confronting the Cohort are similarly concerning. America's early fleet of UAVs were reported to suffer a 'disproportionately significant' accident rate.[69] Drone use in Ukraine has subsequently proved that uncrewed platforms are extraordinarily attritable: in early 2024, the European Council for Foreign Relations estimated that the production capacity for first-person view (FPV) drones had risen to some fifty thousand units per month for each side of the war.[70] If we confine ourselves to wide-task platforms and accept the calculations by the Association for Uncrewed Vehicle Systems that the capital cost of a UAV is broadly one third less than its crewed equivalent, then losses at several times the rate of the crewed equivalent will quickly erode any cost benefit. Crash rates vary by aircraft type but *Drone360* estimates that the US military's larger UAV 'crash three times more often than

---

68 Eish, Madeleine and Tim Hwang, 'Praise the Machine! Punish the Human!' *Comparative Studies in International Systems*, Working Paper Number 1, Society Research Institute (24 February 2015), generally, https://www.datasociety.net/pubs/ia/Elish-Hwang_AccountabilityAutomatedAviation.pdf.
69 Royal Air Force Directorate of Defence Studies, 'Air Power; UAVs: The Wider Context', Report on AUV Progress and Challenges, Professor RA Mason, p. 119.
70 Gressel, Gustav, 'Beyond the Counter-Offensive: Attrition, Stalemate and the Future of War in Ukraine', European Council for Foreign Relations (18 January 2024), https://ecfr.eu/publication/beyond-the-counter-offensive-attrition-stalemate-and-the-future-of-the-war-in-ukraine/.

the remainder of the fleet'.⁷¹ To this point, commentators report that more than four hundred large military UAVs crashed between 2001 and 2016.⁷² DroneWars UK, an NGO investigating the development and deployment of lethally armed drones, recorded at least forty crashes of larger military drones in just the first six months of 2020.⁷³

The Cohort's decisions must also be framed by politicians and political considerations. Removing human supervision would seem to provoke political sensitivities that more usually rail against the unconstrained employment of military means. Possession of advanced technology, after all, is no guarantee of its practical utility. Concern about casualties in Kosovo obliged ground attack bombers to fly above fifteen thousand feet thereby reducing greatly their effectiveness.⁷⁴

The first high-profile mistake involving unsupervised weapons is likely to be front-page news that stymies their deployment

More generally, the coalition-nature of war (the complications, for instance, that are created by the unanimity principle of the North Atlantic Treaty Organization [NATO]) is likely a harbinger to problems that will confront the AWS Delivery Cohort. The irony here is obvious. Politicians may argue for weapons that are stand-off and autonomous but impose strictures on their operation that hobble any chance of success. The first high-profile mistake involving unsupervised weapons is likely to be front-page news that stymies their deployment.⁷⁵

This political dimension has other ramifications. Given that autonomous machines can, presumably, continue to work indefinitely, it follows that combat involving AWS cannot practically be won or lost. On a battlefield fielding AWS, once machines have defeated opposing machines, humans

---

71 The publication's media study from 2010 showed that thirty-eight Predator and Reaper units crashed in combat in Iraq and Afghanistan with nine further craft crashing in training on US soil. Each Predator cost at that moment between 3.7 million dollars and five million dollars.
72 Hansen, George et al., 'Reliability of UAVs and Drones', Defense Systems Information Analysis Center, 4, 2 (Spring 2017), https://dsiac.dtic.mil/articles/reliability-of-uavs-and-drones/.
73 Cole, Chris, 'Accident Waiting to Happen: UK Opens Skies to Large Military Drones as Crashes Continue', *DroneWars UK*, 18 December 2023, https://dronewars.net/category/military-drone-crashes/.
74 Lonsdale, David, 'Clausewitz and Information Warfare', cit. Strachan, Hew and Andreas Herberg-Rothe, *Clausewitz in the Twenty-First Century* (Oxford University Press, 2017), p. 241.
75 Crootof, Rebecca, 'War Torts: Accountability for Autonomous Weapons', *University of Pennsylvania Law Review*, 164, 6 (May 2016), pp. 1350-1351, https://scholarship.law.upenn.edu/cgi/viewcontent.cgi?referer=https://www.google.com/&httpsredir=1&article=9528&context=penn_law_review.

will still need to move in to negotiate and settle the dispute in person. Two quite separate points arise. The first is to reiterate the 'inescapable reality of geography and the ubiquitous nature of the elements'.[76] Clausewitz, after all, highlights that 'geography and the character of the ground bear a close and ever-present relation to war'.[77] A second exogenous challenge comes from the psychological cost of AWS deployment. UAV pilots have long shown signs of combat stress that are equal to that experienced by traditional pilots; although an over-simplification, the stress of fighting a war several thousand miles away and then joining one's family at the dinner table already presents significant mental health challenges akin to a video game in its detachment from the act of killing.[78]

5.6 Proliferation constraints and other lessons from Ukraine

Proliferation creates a further conundrum for the Cohort. Just as Krupp, Winchester and Armstrong were happy to sell advanced munitions to competing countries, the marketplace for military drone manufacturing is already global and was already valued at some fourteen billion dollars in 2023.[79] It is expected to grow annually by nearly ten per cent in the period between 2024 and 2032. The wake-up call from Ukraine is that parties who fail to embrace uncrewed assets will be disadvantaged; drones in all their forms may be neither as accurate nor powerful as their crewed equivalents, but nor do they have to be.[80] Moreover, horizontal proliferation (that is, to other countries) of UAV technologies is already widespread; research from Dortmund University suggests that in 2019 some twenty countries were already exporting these systems. Horizontal proliferation has then taken the form of qualitative

---

76 Gray, Colin, 'Inescapable Geography', *Journal of Strategic Studies*, 22, 2-3 (1999), 161-177.
77 Clausewitz, *On War*, V, 17, p. 348.
78 Sharkey, Noel, 'Saying No to Lethal Autonomous Targeting', *Journal of Military Ethics*, 9, 4 (2010), 372.
79 EMC Claight Report, 'Global Military Drone Market Report', January 2024, https://www.expertmarketresearch.com/reports/military-drone-market. See also: Zenko, Micah and Sarah Kreps, 'Limiting Armed Drone Proliferation', Council on Foreign Relations, Center for Preventative Action, Council Special Report 69 (June 2014), pp. 6-8.
80 Keating, Joshua, 'Why the Pentagon Wants to Build Thousands of Easily Replaceable, AI-Enabled Drones', *Vox* (22 March 2024), https://www.vox.com/world-politics/24107959/replicator-drones-china-taiwan-ukraine-pentagon.

improvements to UAV technologies through collaborative development (such as between Israel and India).

Drones are also becoming increasingly available as prices decline, roughly mirroring industry's experience with computers. A small rotary wing drone made of carbon fibre sells on the Chinese firm Alibaba's website for less than $200; it uses GPS and inertial navigation for guidance, has autonomous flight controls and includes a thermal sensor, ultrasonics and sonar ranging.[81] Currently able to fly for eighteen minutes, it offers full motion video and can carry a one kilogram payload.[82] By way of context, there are currently more than one hundred and ten thousand commercial drone pilots (and eight hundred thousand hobbyists) in the United States alone.[83] The same trend is evident in the maritime domain with relatively cheap Chinese products are available at the time of writing that carry a multiple-sensor payload for a range of more than six hundred miles at depths of one thousand metres and do this for thirty days.[84]

Drones, in all their forms, may be neither as accurate nor as powerful as their crewed equivalents, but nor do they have to be

What are some of the lessons we can draw for AWS deployment from drone advances in Ukraine? Uncrewed technologies have clearly changed the narrative in that conflict. While perhaps not yet an immutable new norm, the sheer variety of drone types has certainly cemented the role of uncrewed weapons of one type or another as an indispensable and far-reaching battlefield asset. They are clearly more than a clever way of throwing a grenade notwithstanding that far too much current debate remains focused upon designing out the pilot. As we have seen, this narrow goal actually requires considerable sophistication and expense. A second conclusion is that autonomy turns out only to be one part of the drone piece as there is already little definitional difference between a one-way cruise missile and the majority of use cases that are posited for AWS. Indeed, the pell-mell variation in drone type is directly driven by the increase in the platforms' uses; from attritable and single-use to sophisticated and multi-domain; from cheap and single capability to costly and complex; from reflexive and close range

---

81 See, for instance, https://www.alibaba.com/product-detail/2024-Hot-Sales-Fpv-Drone-7.
82 Krepinevich, Andrew, Origins of Victory, *Yale*, 2023, p.98.
83 Adams, James, 'How drones are Dramatically Changing Warfare', *Spectator*, 15 September 2019, https://thespectator.com/topic/drones-dramatically-changing-warfare/.
84 Fedasiuk, Ryan, 'How China is Militarizing Autonomous Underwater Vehicle Technology', *The Maritime Executive*, 21 August 2021, https://maritime-executive.com/editorials/how-china-is-militarizing-autonomous-underwater-vehicle-technology.

to calculated and operating at distance; from immediate to loitering; from stand alone to collaborative. Just as this volatile definition set makes drones difficult to explain to the public, it is also a reason behind the drone's rapid evolution and adoption, with extraordinary expansion in chip and hardware capability leading to similar growth in these platforms' use cases.

A further lesson from Ukraine is that failure by either side to establish air superiority has accelerated the proliferation of drone platforms. Although certain use cases capture the imagination with spectacular outcomes, not every drone type is appropriate for every scenario. Where, for instance, do drones fit into the definition of air control? Saturating an area with one-way uncrewed platforms might get around worries about establishing costly control in airspace, but new use cases for the drone as a proxy for mortar or artillery function call into question the relevance of several traditional missions. Should the air force really be engaging in close air support or is this all at once an uncrewed task? Is it worth an expensive and vulnerable plane being tasked with engaging a machine gun nest? This is surely now a drone role. The same question extends to prime fires. Risking expensive platforms against cheap targets will soon deplete that party's resources, stockpiles and capacity of these new means requiring revised priority and better logistics.

The Ukraine war has also underlined the priority of *countermeasures* to uncrewed platforms. Uncrewed assets may remain most vulnerable in the deep where they are manufactured, stored, moved and trained, but Ukraine has also demonstrated to defence planners that all homelands are suddenly and newly vulnerable to UAV attack. Work undertaken by the UK's National Air Traffic Services suggest that several hundred drone incursions happen each day to their excluded zones, the point being is that drone use is ubiquitous and already takes place largely regardless of regulation or sanction.

> *Is it worth an expensive and vulnerable plane being tasked with engaging a machine gun nest? Risking expensive platforms against cheap targets will soon deplete a party's resources*

Two further lessons arise from Ukraine's experiences. The first concerns risk. Specifying military materiel with safety tolerances that run to several decimal points turns out to be peacetime madness. In conflict, designing in these protections and safety margins has been shown to be completely irrelevant. It escalates costs, reduces efficiencies and is needlessly extravagant given an overarching requirement for resilience and buffer stocks and the need to keep fighting. Out-of-date weapons and munitions have performed just fine in Ukraine. A second lesson builds on the issue of air control. Ukraine has demonstrated this can be temporary at best and limited to particular windows of opportunity, both spatially or temporally. Previously it might be air defence assets that were a party's targeting priority; now it is electronic warfare assets that occupy targeteers, anxious to ensure best tasking of their drones. Indeed, spectrum management, both its use and denial, is once again a key operational and procurement priority. It is also another driver towards increasing the autonomy of these uncrewed weapons.

The Ukrainian conflict has also demonstrated the collapse in cost of these platforms, a characteristic of Seba's S-Curve adoption model and a further accelerant to UAV adoption. UAV air-frames can be built on a 3D printer, even those with an endurance of six hours and a flight range of four hundred kilometres (here, the *Titan* with its FPV optics for real-time surveillance). Similarly, cheap versions of Hobby King software have long provided users with useable software upon which to create autonomous flight paths. The proliferation of AWS' underlying componentry, clearly widespread, is therefore a double-edged sword (software, of course, being particularly easy to copy) but one that is accelerating experimentation with these same technologies.[85] Ukraine has also underlined how easy it is either to replicate or to improve uncrewed capabilities through espionage, graft, state-sponsored malfeasance or from reworking captured systems.[86] Examples abound, although a celebrated first instance was unwittingly provided by the Chukar, a target drone system used by the US, which crashed in the waters off Tel Aviv in 1968 and was immediately reverse-engineered to provide the basis for future Israeli efforts in the technology. Indeed, AWS technologies are unusual precisely because of the easy availability of their physical parts,

> *On air control, Ukraine has demonstrated this can often be temporary at best and available during particular windows of opportunity, both spatially or temporally*

---

[85] See: Economist, 'From Here to Autonomy', *Economist* (1 March 2018), https://www.economist.com/news/special-report/21737420-making-vehicles-drive-themselves-hard-getting-easier-autonomous-vehicle-technology. See also Bronk, Justin et al., 'Armed Drones in the Middle East: Proliferation and Norms in the Region', Royal United Services Institute (RUSI) (17 December 2018), https://www.rusi.org/explore-our-research/publications/occasional-papers/armed-drones-middle-east-proliferation-and-norms-region.

[86] Bielieskov, Mykola, 'Outgunned Ukraine Bets on Drones as Russian Invasion Enters Third Year', Atlantic Council (20 February 2024), https://www.atlanticcouncil.org/blogs/ukrainealert/outgunned-ukraine-bets-on-drones-as-russian-invasion-enters-third-year/. For an Eastern perspective, see also: Weinberger, Sharon, 'China Has Already Won the Drone Wars', *Foreign Policy* (10 May 2018), para. 4 of 11.

the dual-use nature of those parts as well as an increasing ease for parties to source cross-border the required componentry to construct these platforms.

In considering proliferation risks, an AWS should not be seen as any single capability asset. As noted by the Royal United Services Institute, the sophistication, risk and demands required by the user depend very much upon the uncrewed asset to be deployed.[87] Here, therefore, drones' simplest iteration, the minimum viable product, currently represents the weapon type's biggest proliferation risk, especially by non-state actors. But this end-of-the-product type is often brittle and, considered individually, probably does not change the game in terms of impact for modern militaries. More sophisticated versions (military-off-the-shelf)

Specifying military materiel with safety tolerances that run to several decimal points turns out to be peacetime madness

then feature more developed autonomy, their proliferation risk generally constrained by affordability but nevertheless offering parties automated pathways to battlefield mass and disruptive effects. Proliferation here is certainly likely and more a matter of will and ethics than of technological hurdles. An adjunct proliferation concern is that these technologies are exploited by terrorist and non-state players.[88] Notwithstanding that 'threat inflation'[89] plays its part in keeping all of this in the public eye, there are clear use cases (assassination, chemical and biological dissemination and access into secure areas) that must tempt non-state parties to consider uncrewed assets, either controlled from afar or that operate autonomously.

---

87 O'Neill, Paul, Sam Cranny-Evans and Sarah Ashbridge, 'Assessing Autonomous Weapons as a Proliferation Risk', RUSI (8 February 2024), https://www.rusi.org/explore-our-research/publications/occasional-papers/assessing-autonomous-weapons-proliferation-risk.
88 Blanchard, Alexander and Jonathan Hall, 'Terrorism and Autonomous Weapon Systems: Future Threat or Science Fiction?', Centre of Emerging Technology and Security, The Alan Turing Institute (January 2024), https://cetas.turing.ac.uk/publications/terrorism-and-autonomous-weapon-systems-future-threat-or-science-fiction. The threat dates back to December 2009 when, as reported by the *Wall Street Journal*, Iraqi insurgents used a twenty-six dollar kit to tap into live video feeds from a circling Predator drone; see: Siobhan Gorman et al., 'Insurgents Hack US Drones', *Wall Street Journal* (12 December 2009), https://www.wsj.com/articles/SB126102247889095011.
89 Hsu, Jeremy, 'Cheap Drone Attacks Have Outsized Effect on Global Trade', *New Scientist* (12 January 2024), https://www.newscientist.com/article/2411898-cheap-drone-attacks-have-outsized-effect-on-global-economic-inflation/. The role of 'threat inflation' in accelerating AWS development is widely acknowledged whereby constituencies exaggerate the strength of possible threats and capabilities and the ramification of this upon national security, fuelling a self-fulfilling policy escalation; see: Lawson, Sean, 'Domestic "Drones" Are the Latest Object of Threat Inflation', *Forbes Magazine* (18 April 2014), http://www.forbes.com/sites/seanlawson/2014/04/18/domestic-drones-are-the-latest-object-of-threat-inflation/.

A further observation from Ukraine is that UAV development has made front lines very lethal. It is more and more difficult for forces under persistent surveillance to move, to mass or otherwise achieve tactical surprise in the offensive. That is not to declare the front lines impenetrable or completely transparent, but drone deployment has certainly forced troops to disperse, to conceal themselves and thus made it harder for them to manoeuvre. Moreover, the properly disruptive element here is not the drone alone but the combination of several weapon systems linked together. In Ukraine, an emergent pairing has been unarmed drones and artillery, accelerating targeting timelines, and enabling newly responsive and more precise ground-based fires. As noted by Paul Scharre, 'oftentimes, it is multiple drones operating in stacks in the same airspace that perform different roles, creating a highly distributed and resilient kill chain'.[90] The verso here, however, remains that even large number of uncrewed assets cannot match the potency of current artillery fire which packs more explosive punch and can either be used rapidly in salvos or as sustained fire. Unlike uncrewed weapons, traditional artillery remains less pervious to electronic interference. In the same vein, drone use also remains constrained by the skill of their controllers who require acuity to guide their payloads to a target's most exposed point. A second verso concerns propagation where, in the instance of Ukraine, there has been little standardisation of equipment: without commonality across platforms and practices, it is difficult to optimise equipment variations and to do this in a manner that wins the trust of users.

> *Designing in protections and safety margins has been shown to be completely irrelevant. It escalates costs, reduces efficiencies and is needlessly extravagant given the requirement for resilience and buffer stocks*

> *Over fifty per cent of equipment and personnel is destroyed by drones and the other fifty per cent is destroyed with assistance from drones*

Finally to this proliferation point is the argument that the mere potential of AWS might induce politicians to start wars with greater ease, particularly in industrialised democracies where casualties may quickly erode support for more traditional means of war. This whole trend reinforces an age-old security dilemma: each state is driven to build up its own AWS forces and,

---

90 Scharre, Paul, 'The Perilous Coming Age of AI Warfare', *Foreign Affairs* (29 February 2024), https://www.foreignaffairs.com/ukraine/perilous-coming-age-ai-warfare.

in doing so, dials up the threat to its neighbours who, in turn, feel compelled to adopt and accelerate this new technology. The political consequence is, of course, that security is decreased for all.

5.7 Ethical constraints to AWS deployment

This chapter's final section now considers ethical and moral considerations to fielding autonomous weapons with commentators occupying positions all across this debate. While Arkin's advocacy of an *Ethical Governor* and *Artificial Conscience* is explored in this book's chapters Three and Eight, others' rebuttal suggest instead that lethal robots will create whole new levels of uncertainty on the battlefield. Robots, after all, can never be capable of morally praiseworthy behaviour such as courage (both moral and physical) or the ability to go 'above and beyond duty'. At its core, the argument is that it is just not possible to capture ethical notions by machine code. Other missing tenets will also include heuristics that have forever been fundamental to battlefield conduct such as guilt, empathy, responsibility, duty and restraint. The list must also cover precepts such as concern, unease, disquiet and compassion. This analysis is not to argue that all human soldiers unerringly display these characteristics but rather that AWS are systemically unsuited to factor for these traits and standards.

In order to think how ethics might impact deployment, readers should cast their minds back to the ethical drivers introduced in Chapter Three. First, we should dismiss Atkin's arguments around AWS' 'inevitability'. The conceit is unfounded given that humans will always retain the choice around human supervision. The entire human chain around AWS deployment (those comprising the Delivery Cohort) must, after all, be governed by its own (and clear) set of values, standards and experiences, each contributing to how and to what ends autonomy can be deployed. Nor does the notion of AWS' battlefield monitoring (and, implicitly, Arkin's notion of reprimand in cases of unethical behaviour) hold water, not least because it is not anchored in the autonomous machine having its own independent lethal capability. This is a therefore specious connection. Indeed, the shortcomings of Arkin's framework are all but recognised by its author's acknowledgement that AWS adherence to LOAC remains an 'outstanding issue' and, more granularly, that the challenge for AWS to distinguish a soldier from a civilian is one of several 'daunting problems'.[91]

More fundamental, however, is that it is difficult to ignore the dilemma over what comprises a 'correct' ethical framework for AWS deployment given the multiple candidate philosophies.[92] Different states, different politicians and different programmers all hold themselves to their own ethical frameworks, a fundamental deployment constraint given the importance (as discussed at length in later chapters) that the weapon's thresholds and confidence levels be dynamically tuned at all moments while these machines are deployed. Nor does Ronald Arkin's model suggest that AWS deployment will actually prevent humans committing atrocities.[93] Instead, it should be assumed that instances of human unethical activity will now occur 'out of sight' of the autonomous machine, just as cases today take place away

'Oftentimes, it is multiple drones operating in stacks in the same airspace that perform different roles, creating a highly distributed and resilient kill chain'

from those in authority. Improving the battlefield's moral conduct, moreover, becomes irrelevant if the use of AWS takes place in a war that is unjust in the first place. This culminates in the pervasive difficulty of AWS deployment highlighted by Kurt Voller, former US permanent representative to NATO, in his observation that 'drone strikes allow our opponents to cast our country as a distant, high-tech, amoral purveyor of death. It builds resentment and alienates those we should seek to inspire'.[94] Indeed, it is AWS' ethical challenges that inform this book's strapline that it is critical to keep humans on the battlefield.

There are, moreover, a myriad of ways outside autonomy to improve ethical behaviour. Politicians might think about shorter tours for personnel, improving therapeutic resources, tightening up on soldiers' psychological screening (indeed, even under that controversial US Surgeon General's

---

91 Arkin, Ronald, 'The Case for Ethical Autonomy in Unmanned Systems', *Journal of Military Ethics* 9, 4 (2010), pp. 11-12, https://www.cc.gatech.edu/ai/robot-lab/online-publications/Arkin_ethical_autonomous_systems_final.pdf.
92 Unitarian, Kantian, social contract, virtue ethics, cultural relativism et al. Arkin's answer to this is to start from a 'first, do no harm' strategy that sets the default for AWS.
93 Indeed, Arkin identifies specific problems to the creation of ethically sensitive machines including the abstract nature of laws, codes and principles of military law, the variety of interpretation of these laws depending on context and the frequent conflict that exists between these abstract rules. See also: Tonkens, Ryan, 'The Case against Robotic Warfare', *Journal of Military Ethics*, 11, 2, p. 158, (August 2012).
94 Voller, Kurt, 'What the US Risks by Relying on Drones', *Washington Post* (26 October 2012).

research from Chapter Three, the conduct of more than ninety per cent of soldiers was reckoned to be acceptable), by improved training and, as well, by educating for compliance with rules of engagement. It is, of course, simply unreasonable to express without doubt that AWS will be 'ethically more sound' than human soldiers, the more so given the number and scope of morally indeterminate situations that arise in battle. The guiding precept remains instead the three essentials of human participation, LOAC adherence and combat that is based upon an ethical 'do-the-best-you-can' framework.[95]

A requirement to investigate violations of IHL also complicates the removal of human supervision. Under the Geneva Conventions, this notion of accountability is central to deterring future harm to civilians while also providing its victims with a means of retribution. In this vein, critics argue that autonomous weapons should be banned precisely because their very nature precludes fair attribution of responsibility. Liability, after all, presupposes that commanders can foresee the outcome of their actions and those of their weapons. Just as deploying lethal AWS under unreasonable ambiguity would clearly be wrong (akin to releasing a toxic substance in an inhabited environment), provision must be built into all engagement models to identify responsible parties when AWS processes go awry. The corollary (taken together, the collective responsibility of AWS' Delivery Cohort) is captured by former UN special rapporteur Christof Heyns when he notes that 'there is clearly no point in putting a robot in jail'.[96]

This matter of responsibility gives rise to ancillary challenges. Even when members of the Cohort do exercise control over the weapon system, uncertainty remains whether each, any or all individuals that comprise the Cohort can practicably be held to account the case of egregious error. If everyone (in this case, the Delivery Cohort) is deemed to be responsible, then no one is responsible. An additional ethical impediment is that it is unclear who will investigate civilian deaths following mistaken engagement by the AWS, the whole framework creating an unacceptable 'responsibility gap'. In every error case, the AWS has either been deployed unlawfully, its programming has been unlawfully careless, or it has been illegally designed or neglectfully maintained. AWS' technical aggregation also adds to this

---

95 Schulzke, Marcus, 'Ethically Insoluble Dilemmas in War', *Journal of Military Ethics*, 12, 2 (2013), generally. See also: Jackson, Eric, 'Sun Tsu's Art of War', *Forbes* (23 May 2014), https://www.forbes.com/sites/ericjackson/2014/05/23/sun-tzus-33-best-pieces-of-leadership-advice/#19c3ac7d5e5e.
96 Heyns, Christof, 'Autonomous Weapon Systems: Living in a Dignified Life and Dying a Dignified Death', cit. Bhuta et al., p. 12.

accountability challenge; although it may be theoretically possible to separate out the obligations that should be attached to a particular software engineer (or, *post facto*, to identify those responsible for a malfunctioning component within the system at fault), an obvious defence will be that these machines' technical spine means that any one individual cannot be expected to predict the weapon's subsequent learnings and, in turn, its outcomes in battle. That an autonomous system will make choices outside those predicted by its programmers is inherent in it being autonomous.

The legislative basis for determining accountability in cases of malfunction is also ambiguous. Protocol I, Article 85(3) of the Geneva Conventions, specifies that individuals can only be held liable if that carelessness is willful. Manufacturers have long gone unpunished for how their weapons have been used, particularly having first disclosed the risks associated with those weapons. In the case of AWS, parties can be sure that manufacturers will articulate the widest possible caveats when detailing weapon specifications, the more so given these platforms' inherently unpredictable foundation. Prosecution over product liability also entails civilian lawsuit where the onus to act is placed upon the victims, often wretched and displaced by conflict. The large number of agencies comprising the Delivery Cohort also means that the absolute numbers involved in this undertaking are also likely to make an accident more probable. As well as blurring the lines of responsibility when that accident occurs, no one component of this Delivery Cohort can conceptualise all of the system's complexities or, indeed, the battlespace in which that machine must operate. Much more likely is that the technical combinations required to field these weapons will continue to create legal and ethical issues that are currently unimaginable.

> 'Drone strikes allow our opponents to cast our country as a distant, high-tech, amoral purveyor of death. It builds resentment and alienates those we should seek to inspire'

> *If everyone is deemed to be responsible, then no one is responsible*

Two final matters are relevant to this discussion. First is the notion of 'plausible deniability' when no single party admits to a lethal engagement (in our case, involving AWS) or when it is otherwise difficult to determine whose weapon has been responsible in an attack. Such anonymous acts of war are not unique to AWS but are clearly more likely where human supervision is absent and where uninvigilated violence has the potential to disrupt geopolitical balances. The second is the notion of 'human dignity', a foundational principle that overarches all constitutional and other political principles, a common reference point being that is beyond controversy and plays the role of an *a priori* to which all other political ideas are subject.[97] The issue then becomes whether the AWS infringes on human dignity, the construct under Just War theory of *mala in se*, so questioning whether justice can really be delegated to automated processes. The argument is given weight by the United Nations Institute for Disarmament Research's discussion of 'an instinctual revulsion against the idea of machines "deciding" to kill humans', prompting commentators (and the Third Sector generally) to posit human dignity as a basic challenge to AWS deployment.[98] Nor can human dignity be pigeonholed by definitions around human rights.[99] Unlike the concept of morality (which comprises both rights and duties), human dignity implies rights against others but no duties against others and, in this sense, there is obvious disconnect between human dignity and unsupervised lethal engagement. The context of human dignity, moreover, is unlimited as it contains 'an intrinsic evaluative component' which gives to human beings an exclusive value 'on which the exceptional normative status of human beings is assumed to depend'[100], the ethical challenge being that such value cannot be reflected in machine code.

> 'There is clearly no point in putting a robot in jail'

> *The absolute numbers involved are also likely to make an accident more probable*

Scratching further this question of dignity shows that the issue is an important behavioural constraint to the Cohort's plans. The human rights that it implies include the right to avoid humiliation, the right to a minimum freedom of action and decision, the right to receive support in situations of severe need, the right to a minimum quality of life and the relief of suffering and, crucially, the right not to be treated merely as a means to other people's

---

97 Birnbacher, Dieter, 'Are Autonomous Weapons a Threat to Human Dignity?', cit. Bhuta et al., p. 105. The concept of human dignity was first introduced into Article 1 of the United Nations Universal Declaration of Human Rights in December 1948. Similarly, the Vienna Declaration of the 1993 World Conference on Human Rights affirmed that 'all human rights derive from the dignity and worth inherent in the human person'.
98 Birnbacher, pp. 105-107.
99 HRW, 'Heed the Call', pp. 19-22 ('Humane treatment') and 23-27 ('Respect for Human Life and Dignity').
100 Birnbacher, pp. 105-107.

ends. Human dignity also incorporates a civilian's right to privacy as well as rights to live without severe harm or risk of harm.[101] Removing weapon supervision instead risks making civilians 'the mere means of aims that are in no way their own aims, with the risks of serious harm to life and physical and mental integrity'.[102] This manifests itself in three ways. First, AWS are 'invulnerable by being free from fear', with their deployers carrying 'no cost except the economic'.[103] Second, AWS' intrinsic unpredictability exacerbates this threat to civilians. Finally to this point, AWS' 'illusion of accuracy' will likely foster overconfidence and so distort the Delivery Cohort's judgement. The challenge is therefore the degree of control that military commanders (and, by extension, the responsibilities of the state on whose behalf they act) really exercise over their weapons, to perform their legal duties and to be accountable for the consequences.[104] After all, it is this broad portfolio of constraints (as well as their intrinsic intractability) that cumulatively underpins civil society's opposition to AWS deployment.

> Do AWS infringe human dignity, the notion under Just War theory of mala in se, so questioning whether justice can really be delegated to automated processes?

> AWS' 'illusion of accuracy' will likely foster overconfidence and so distort the Delivery Cohort's judgement

---

101 Articles 8 and 21 of the Rome Statute of the International Criminal Court prohibit and make punishable 'committing outrages upon personal dignity, in particular humiliating and degrading treatment'.
102 Birnbacher, p. 116. Birnbacher focuses in particular on the asymmetry of facing forces if autonomous weapons are employed only on one side. 'Morally, symmetry and asymmetry are crucial moral variables', he claims, 'and they are deeply influenced by the use of robots'.
103 Human Rights Council, 'Report of the Special Rapporteur on extrajudicial, summary or arbitrary executions, Christof Heyns' (9 April 2013), https://www.ohchr.org/sites/default/files/Documents/HRBodies/HRCouncil/RegularSession/Session23/A-HRC-23-47_en.pdf., p. 12.
104 Brehm, Maya, 'Defending the Boundary: Constraints and Requirements on the Use of Autonomous Weapon Systems Under International Humanitarian and Human Rights Law', *Geneva Academy of International Humanitarian Law and Human Rights*, Academy briefing No.9 (2017), p. 19.

SECTION TWO

An Analysis of Practical Challenges to AWS Deployment

6
Wetware
Design challenges to AWS function

Having assessed the 'why' and 'how' of weapons without human oversight, the balance of this book now considers practical obstacles to compliant deployment of these systems. 'Practical' is the defining adjective here. It is useful precisely because it is less demanding, less open to question than describing the book's second half as a 'technical' review. First, the challenges of autonomous weapon systems (AWS) are generally reducible to understandable packets and, second, only once the fundamental 'build' challenges have been identified can a review of these machines' operational considerations reasonably be carried out. The analysis is divided into four discrete chapters titled 'Wetware', 'Firmware', 'Software' and 'Hardware'.

Some brief scene-setting is useful. A teleological system (here, the AWS and its purposes) is generally characterised by having a workable representation (a sense of its immediate environment) that is anchored by a set of goals (objectives to be achieved by the weapon) and set of beliefs (a locus aligning the weapon with those responsible for its deployment). This partnership enables the weapon to decide upon and execute a plan (the path upon which the platform will reach those goals, given its 'beliefs', through a set of actions undertaken within a belief-desire-intention architecture).[1]

---

[1] Sartor, Giovanni and Andrea Omicini, 'The Autonomy of Technological Systems and Responsibilities for Their Use', cit. Nehal Bhuta and others, *Autonomous Weapons Systems: Law, Ethics, Policy* (Cambridge University Press, 2016), p. 51. The authors usefully state that 'in order to realise its desires (goals), the system constructs plans of action on the basis of its model of the relevant facts (beliefs) and commits itself to act according to the chosen plans (intentions). Note that by using the terminology of beliefs, desires and intentions to denote cognitive structures of artificial systems, we are not assuming that such structures are similar to those existing in the human mind'. This is a key distinction that is carried forward throughout this book's practical review.

And, although this may appear slightly otherworldly so early in a practical analysis, all of these fault lines are almost agnostic to the degree of autonomy displayed by the underlying platform. For the purposes of this book, the test throughout is the presence or otherwise of meaningful human control (MHC) across lethal engagements, the set of sequences that are described by Royal Air Force doctrine as the 'consecutive miracles that make up the string of happenings enabling technically advanced weapons to function'.

How then are these next four chapters organised? First, in this chapter, Wetware, the analysis isolates basic complexities underpinning potential architectures for these weapons. In so doing, it tees up a plausible foundation for AWS that then informs the book's subsequent chapters. What constitutes appropriate artificial intelligence (AI) if agency is to be ceded to these systems, and how might that weapon's high-level systems learn, reason and undertake sufficient cognition to dispense lethality? Firmware (Chapter Seven) uses this analysis to consider the many bases that will likely comprise AWS function. For our purposes, firmware relates to the permanent routines (weapon learning, reasoning and direction) underpinning AWS architecture. Software (Chapter Eight) then identifies fault lines that arise from the instruction sets that enable these weapons' moment-by-moment routines and together comprise that function. Finally, in Hardware (Chapter Nine), the analysis considers challenges arising from AWS' physical properties. Only then is assessment possible on the required nexus between compliant operation, these systems' technical obstacles and the roles expected of them by the Delivery Cohort. Taken together, these four chapters seek

Fitness for purpose requires a nexus between AWS' compliant operation, these systems' technical obstacles and the roles expected of them by the Delivery Cohort

to pinpoint discrepancies that exist between capabilities that may be feasible and the tasks that will be asked of these systems.

Wetware, first of all, is vernacular to describe the human element of information technology architecture and, in choosing this descriptor, the book is deliberately harking back to the earliest days of AI. After all, the human brain is composed of some seventy-five per cent water. The term is therefore useful in describing programmers' efforts to replicate the essentials of human intelligence and to effect this through code and code alone.[2] Indeed,

---

[2] One component of the AWS Delivery Cohort. It is useful to repeat the definition of the Cohort: it is the several parties responsible for the design, implementation and deployment of AWS. The term is deliberately undefined but includes, *inter alia*, the following tasks: neurophysiologists

the narrative of the next four chapters is informed by how the AWS can be trained given that these processes must in some way be anchored to cognitive traits of the human brain. This chapter therefore concerns feasibility, whether machine independence is simply 'the Holy Grail in AI research: highly desirable, but still unobtainable'[3], and the degree to which independent intelligence will 'remain science fiction – at least for the next one hundred years and maybe always'.[4] A second theme of this book's practical review is then framed by Moravec's Paradox, the long-dated observation in AI and robotics that, contrary to traditional assumptions, reasoning actually requires somewhat modest computation but sensorimotor and perception tasks, both fundamental to AWS deployment, require enormous and understated computational resource. A purpose of this chapter then becomes to evaluate the fragility that this entails.

The first task for this chapter is to review the technical basis underpinning standalone weapons, the transformation of data from their battlefield environment into those purposeful plans that have been identified above. After all, this process, squarely composed of a set of machine capabilities involving sensing, deciding and then acting, will alone determine compliant (and accretive) deployment of AWS, the undertaking of these steps in both a lethal but also independent manner.[5] A challenge (the first of many in this book's practical review) arises from what appear to be outwardly similar models (here, reactive rule-based systems and/or deliberative

---

to coordinate AWS networks; psychologists to coordinate learning and cognition; biologists for adaption strategies; engineers for control routines; logisticians; roboticists; electrical specialists; behaviorists; politicians, civil servants and diplomats; non-governmental organizations; sociologists; lawyers; company directors; weaponists; military tacticians; manufacturers; professionals involved in miniaturisation, simulation, configuration, coding, power supply and modularity; specialists in sensors, in distributed and decentralised routines; ethicists; and specialists in tooling and calibration. See also: Chapter Six, specifically 6.3 (The Delivery Cohort).
3 Krishnan, Armin, *Killer Robots: Legality and Ethicality of Autonomous Weapons* (Ashgate Publishing, 2009), p. 48.
4 Docherty, Bonnie, 'Losing Humanity – The Case against Killer Robots', Human Rights Watch (2012) http://www.hrw.org/reports/2012/11/19/losing-humanity-0, p. 29.
5 Boulanin, Vincent and Maaike Verbruggen, 'Mapping the Development of Autonomy in Weapon Systems', Stockholm International Peace Research Institute (November 2017), https://www.sipri.org/sites/default/files/2017-11/siprireport_mapping_the_development_of_autonomy_in_weapon_systems_1117_0.pdf, p. 7. Capabilities regarding sensing (that is, in order to complete a task autonomously, the weapon must be able to perceive the battlefield environment in which it operates), deciding (data from the weapon's surroundings, once processed by the machine's sensing software, then serves as input for the AWS' decision-making processes, which in turn are overlooked and 'assured' by its control systems) and acting (the decisions made by the AWS' control systems are then executed through computational or physical means) may provide the skills basis for AWS, but this masks considerable technical uncertainty to be overcome before a working set of machine routines can be achieved.

goal-based systems), each masking the awkward fact that very different decision outcomes may arise from exactly similar input data.[6] This is actually a fundamental characteristic notwithstanding the many forms that AWS deployment may take, ranging from individual pieces of componentry within an otherwise supervised platform to wholly independent platforms where targets are selected autonomously according to on-board, uninvigilated and self-determined search instructions. The importance of this imprecision is often understated but is best demonstrated by targeting routines that might be determined solely by ethnicity, location, gender, gait, age, even 'target reaction'.[7] That these represent random points on a wide continuum is evidenced by debate in the United Nations Convention on Certain Conventional Weapons (CCW) where, after twelve years of discussion, parties still prefer to disagree on definitions of what constitutes a lethal autonomous weapon.[8]

In thinking, then, about wetware, we need to be grounded by the research of Hans Moravec

'It is comparatively easy to make computers exhibit adult-level performance on intelligence tests or playing checkers, and difficult to impossible to give them the skills of a one-year-old when it comes to perception and mobility'

and his observation that 'it is comparatively easy to make computers exhibit adult-level performance on intelligence tests or playing checkers, and difficult to impossible to give them the skills of a one-year-old when it comes to perception and mobility'.[9] That context often appears forgotten in today's 'revolution in expectation', which is discussed at length in his book's earlier

---

6 Russell, Stuart and Peter Norvig, *Artificial Intelligence: A Modern Approach* (Pearson Education, 2014), p. 35 and p. 49.
7 Oberhaus, Daniel, 'Watch "Slaughterbot": A Warning about the Future of Killer Robots', *Motherboard* (13 November 2017), https://motherboard.vice.com/en_us/article/9kqmy5/slaughterbots-autonomous-weapons-future-of-life. 'Slaughterbots' is presented by Professor Stuart Russell, who concludes 'this is not speculation. It is the result of integrating and miniaturizing technologies that we already have'. The theme also frames much of the introduction to Chapter Two (Context).
8 Campaign to Stop Killer Robots (17 November 2017), paras. 5 and 7 of 10. Human Rights Watch's Mary Wareham notes that the stated goal of the CCW deliberations has not been to produce a working definition. The consensus-agreed 2018 report of the Group of Governmental Experts on Emerging Technologies in the Area of Lethal Autonomous Weapons Systems (GGE) states that: 'For some delegations, a working definition of lethal autonomous weapons systems is essential to fully address the potential risks posed. For others, absence of an agreement on a definition should not hamper discussions or progress within the CCW'; see GGE, 'Report of the 2018 session of the Group of Governmental Experts on Emerging Technologies in the Area of Lethal Autonomous Weapons Systems', CCW/GGE.1/2018/3 (23 October 2018), https://documents.un.org/doc/undoc/gen/g18/323/29/pdf/g1832329.pdf, p. 5.
9 Moravec, Hans, *Mind Children* (Harvard University Press, 1988), generally.

chapters. 'The mental abilities', Steven Pinker noted some thirty years ago, 'of a four-year-old that we take for granted – recognizing a face, lifting a pencil, walking across a room, answering a question – in fact solve some of the hardest engineering problems ever conceived'.[10] And, twenty years later, it still took UC Berkeley University's towel-folding robot more than ten hours to replicate the human folding of just twenty-five towels.

Fast forward to the time of writing this book, that same robot is now seventy-five per cent successful at flattening and eighty per cent at folding those towels. This may resemble heroic progress but the robot today remains only twenty-five per cent successful when the towel is somewhat tangled and either one or none of its corners are visible. A host of other failures still occur when the algorithm is unable to generate a 'feasible action', a condition considered at length later in this analysis, such as an incorrect grasping point for the towel or when the cloth slipped out of the robot's pinch.[11] So, is this marked progress? Is it sufficient advance to suggest eventual delegating of the decision to kill to an algorithm?

> *AWS must factor for known unknown tasks but must also have programmed routines available in real time to derive actions from unknown unknowns*

Wetware architecture also creates the phenomenon of 'complexity layering'. Independent componentry must be reliably capable of actioning known tasks. AWS must factor for known unknown tasks but must concurrently have programmed routines available (and in real time) to derive actions from unknown unknowns. These is even more the case here as delineation between subject matters in Chapters Six through Nine (Wetware, Firmware, Software and Hardware) is rarely clear-cut. This review thus almost exclusively concerns software, software routines and, of course, their near exclusive role in substituting for human oversight. For this reason, the chapters plot a likely but not definite set of weapon architectures and do so from a deliberately behavioural perspective rather than a technical one. The sections concern concepts, abstracts and structures rather than coding lines and their detailed composition.[12]

---

10 Pinker, Steven, *The Language Instinct* (William Morrow, 1994), p. 191.
11 Yashinski, Melisa, 'Teaching a Single-Arm Robot to Fold Towels', *Science Robotics*, 15 November 2023, https://www.science.org/doi/10.1126/scirobotics.adm8151.
12 For a still useful introduction to these concepts, see: Taylor, Richard, 'Software Architecture: Foundations, Theory and Practice', School of Information and Computer Science, University of California at Irvine (October 1999), https://www.ics.uci.edu/~taylor/Architecture.pdf.

Narrative is essential in framing the subject. Just a decade ago, no machine could reliably provide language or image recognition at a human level but, as intelligence systems have become more capable, code-driven agents now beat these same humans in standardised tests and do this across handwriting, speech, image, reading and language applications. The rise and rise of large language models (LLMs) and their impact across human practices already therefore resembles a discontinuity.[13] Despite the gulf of difference between standardised tests and real-world (battlefield) cases, this progress nevertheless really sets the context for this book's practical review. And while the performance of these agents may be decidedly mixed outside these tests (especially when an environment is hostile or ambiguous), the new reality is that popular implementations of these systems are already so cheap and ubiquitous that they do comprise a new norm.[14] Algorithms already power the mobile phone applications in readers' pockets, and this is not going to change. Image recognition categorises our photos better than a librarian, and speech recognition transcribes what we dictate seamlessly and almost without error. And developments across multi-layer neural networks continue to transform the field of AI. Two pointers are worth noting. These advances are still relatively new and have been occasioned by, first, progress in fast hardware graphics processor units (GPUs) enabling the training of larger and much deeper networks and, second, by very large datasets available that can be manoeuvred for use as training test beds for would-be artificial agents. It is also their combination that has accelerated recent progress in deep learning (DL) on these deep neural networks (DNN).[15]

Their relevance to AWS deployment is that DL processes all attempt to mimic human brain activity, specifically the neuron layers of the human neo-cortex, the 'crinkly eighty per cent of the brain' where human thinking takes place.[16] The issue for this chapter is therefore to reconcile the fit between clear advances in machine capability against the technical models envisaged

---

13 Toews, Rob, 'The Next Generation of Large Language Models', *Forbes* (7 February 2023), https://www.forbes.com/sites/robtoews/2023/02/07/the-next-generation-of-large-language-models/.
14 Walker, Paddy and Peter Roberts, *War's Changed Landscape? A Primer on the Forms and Norms of Conflict* (Howgate Publishing, 2023), pp. 35-41.
15 Wang, Linnan and others, 'SuperNeurons: Dynamic GPU Memory Management for Training Deep Neural Networks', *Proceedings of 23rd ACM Symposium on Parallel Programming* (2018), pp. 1-3, https://arxiv.org/pdf/1801.04380.pdf. For a useful discussion on developments in military AI, see: JASON program, 'Perspectives on Research in Artificial Intelligence and AGI Relevant to DoD', US Department of Defense, JSR-16-Task-003 (January 2017), pp. 1-5.
16 Hof, Robert, 'Deep Learning: With the Massive Amounts of Computational Power, Machines Can Now Recognise Objects and Translate Speech in Real Time. Artificial Intelligence is Finally Getting Smart', TechnologyReview.com (June 2016), paras. 3-4.

for deploying AWS. In explaining AI's advance, commentators generally point to the extraordinary progress that has taken place in this probabilistic modelling[17], a landscape hardly recognisable from a half-decade earlier, particularly where machine training can now be undertaken with materially smaller data sets. For this reason, commentators now excitedly identify a pivot-point where machine-learning (ML) processes might finally be a relevant technical spine for weapon autonomy. But for all of this, it remains difficult to make that jump: Andrew Smith of *The Guardian* usefully notes that 'some call this form of ability "artificial narrow intelligence", but here the word "intelligent" is being used much as Facebook uses "friend" – to imply something safe and better understood than it is. Why? Because the machine has no context for what it's doing and can't do anything else... We might as well call an oil derrick or an aphid "intelligent".'[18] This comprises the context that should underpin the Cohort's thinking.

'The manifolds whose shape and extent they are attempting to approximate are almost unknowably intricate, leading to failure modes for which – currently – there is very little human intuition, and even less established engineering practice'

It is also this dichotomy that should occupy the reader over the coming three chapters and, given such divergence, context becomes doubly valuable. The role of ML in AWS deployment, after all, might simply be another aggressive system of statistics that is similar to other uses of arithmetical logic which have long been a function of battlecraft. Indeed, the US Department of Defense's JASON programme urges caution, and it is worth quoting verbatim from its 2017 report: 'DNNs are function approximators in very high dimensional spaces (e.g., millions of dimensions). The manifolds whose shape and extent they are attempting to approximate are almost unknowably intricate, leading to failure modes for which – currently – there is very little human intuition, and even less established engineering practice'.[19]

---

17 Heikkila, Melissa and William Douglas Heaven, 'What's Next for Artificial Intelligence in 2024?', *MIT Technology Review* (4 January 2024), https://www.technologyreview.com/2024/01/04/1086046/whats-next-for-ai-in-2024/.
18 Smith, Andrew, 'Franken-algorithms: The Deadly Consequences of Unpredictable Code', *Guardian* (30 August 2018), https://www.theguardian.com/technology/2018/aug/29/coding-algorithms-frankenalgos-program-danger.
19 JASON, 'Perspectives on Research in Artificial Intelligence and Artificial General Intelligence Relevant to DoD', MITRE Corporation (January 2017), https://irp.fas.org/agency/dod/jason/ai-dod.pdf, p. 2.

This is a foundational observation from which two recurring themes arise. First, it is clearly inadequate that unsupervised weapons should ever be governed by 'approximate' means. Second, although breakout technologies may promise disruption in this reportedly golden age of AI, readers should not conflate progress in this domain with any level of artificial *general* intelligence (AGI). No practitioner is really predicting sentient, intelligent weapon systems coming over the horizon. This, after all, would signal comprehensive AGI and the arrival of genuinely broad cognition.[20] And here there is little optimism: Stanford University's Department of Computer Science has stated that 'there are no present signs of corresponding revolution in AGI'[21]. Its report *Artificial Intelligence and Life in 2030* concluded that 'no machines with self-sustaining long-term goals and intent have been developed, nor are they likely to be developed in the near future'.[22] This remains the awkward current context for ML as the touted technical spine for battlefield AWS.

6.1 Computational methods, software and intelligence

This distinction between AI and AGI is pivotal in considering AWS deployment. The International Panel on the Regulation of Autonomous Weapons thus recommends that even the umbrella term of AI should 'be used with prudence and parsimony' when discussing AWS and unsupervised lethality.[23] Even AI and autonomous systems should not be conflated. AI, after all, is not constrained to any one definition. It powers a gamut of applications, from chat boxes to facial recognition technologies to LLMs. AWS, on the other hand, are

---

20 See: Chapter Eight (Software), specifically: 'Value Setting and Anchoring'. See also: Pennachin, Cassio and Ben Goertzel, *Contemporary Approaches to Artificial General Intelligence* (Springer Publishing, 2007), p. 509.
21 Stanford University, https://ai100.stanford.edu and, for an executive summary of the Stanford University report on the future of AI, see: Stanford University, https://ai100.stanford.edu/2016-report/executive-summary.
22 Stanford University, 'Artificial Intelligence and Life in 2030', 2015 study panel (June 2016), https://ai100.stanford.edu. For a useful discussion on long-term planning considerations, see: Stojkovic, Dejan and Bjørn Robert Dahn, 'Methodology for Long-Term Defence Policy', Norwegian Defence Research Establishment (28 February 2007), http://www.ffi.no/no/Rapporter/07-00600.pdf.
23 International Panel on the Regulation of Autonomous Weapons, 'Executive Summary Number 2', Computational Systems in the Context of Autonomous Weapon Systems (November 2017), generally.

straightforwardly pieces of lethal machinery where the human is 'entirely out of the conversation'.²⁴ AI and ML so often convey broader competences than being simply advanced statistical machines. Notions of weapon intent or purpose are plainly inappropriate and, for this reason, we should perhaps prefer the umbrella term of 'computational methods'. The difficulty here is that most ideation of AWS requires some measure of exactly this AGI. If nothing else, the argument confirms that language remains inadequate and that popular demarcation between AI, AGI and 'enhanced' software is inappropriately imprecise.

For the purposes of this review, AI refers to those computational methods that have problem-solving capacities that exceed software but still fall far short of Luke Muehlhauser's definition of AGI and its 'ability to achieve complex goals in complex environments with limited computational resources'.²⁵ Nevertheless, just as experts envisage a pathway from today's uncrewed drone to tomorrow's sentient automaton, a clear mini-path arises here from weapon processes that are governed by rules-based and conventional software, through weapon routines that operate with very limited human oversight and all the way to fully sentient machines that function quite independently.²⁶ Although AI is orientated towards specific tasks, this is a world apart from the general cognitive abilities attributed to AGI, prompting the US military's JASON programme to conclude that 'AGI

'No machines with self-sustaining long-term goals and intent have been developed, nor are they likely to be developed in the near future'

---

24 Reddie, Andrew, 'The Impact of AI on Warfare', The President's Inbox in association with The Council on Foreign Relations (13 May 2024), https://www.cfr.org/podcasts/impact-ai-warfare-andrew-reddie.

25 Muehlhauser, Luke, 'What Is AGI?', Machine Intelligence Research Institute (11 August 2013), https://intelligence.org/2013/08/11/what-is-agi/.

26 Braga, Adriana and Robert Logan, 'The Emperor of Strong AI Has No Clothes: Limits to Artificial Intelligence', *Information*, 8, 156 (2017), pp. 1-3 and generally. The intelligent decision-making of such machines, notes Dr Hongbo Du, fundamentally consists of a function sequence of pre-processing input data, extracting useful features from that input data stream and then executing a decision-making or classification routine concerning the input stimuli: 'Such sequential structure of these systems means that any small error in its early stages will propagate into larger errors as the stages unfold. The more complex is the processing function, the more probable of such error occurrence'. Here, Du notes that self-correction of such errors is challenging given that each discrete stage is handling a different type of 'sub-problem'. Initiatives to combine neural networks in order to merge such stages into 'an end-to-end solution' confound attempts then to explain how and where these errors are corrected. It is, notes Du, 'premature to even conclude that such an architecture will overcome this problem at all'. Source: Dr Hongbo Du, School of Computer Science, Buckingham University, in conversation with the author (January 2021).

has a high visibility, disproportionate to its size and present level of success, among futurists, science fiction writers, and the public'.[27] Another way to consider AGI might be through an agnotological lens where AGI's provenance is instead exaggerated by a 'cultural production of ignorance and its effect on both individual and collective decision-making processes'[28], analogous to Sabin's 'Revolution in Expectation'. Without any concrete measure of task 'complicatedness', assessing either AI or AGI competence is still tricky, not least because that intricacy will always depend upon each AWS' individual properties and how they match the tasks for which they are to be deployed.

> *The difficulty here is that most ideation of AWS requires some measure of exactly this AGI*

Observers should also understand the extent to which the AWS might act outside that task's boundaries (remember also that tasks can be sequential) given passing goals and its current representation. Here, machine autonomy needs a more holistic definition which, for the purposes of this section, might be the faculty 'to discriminate, characterise and react to environmental stimuli;

> *It entails a measure of subjective 'qualia' (the feel of precepts), routines that capture learned experience and, literally, a requirement in independent weapons for 'phenomenal consciousness'*

the integration of information by a cognitive system; the reportability of mental states; the focus of attention and the deliberate control of behaviour'.[29] Relating this to the battlefield, the deduction must be that removing oversight requires more than rules-based routines. It entails a measure of subjective 'qualia' (the feel of precepts), routines that capture learned experience and, literally, the facility in independent weapons for 'phenomenal consciousness'.[30] The argument has long been used in reverse by the British

---

[27] JASON, p. 1; see also: p. 2 and pp. 28-32 ('Why the "Ilities" May Be Intrinsically Hard for Machine Learning'), https://fas.org/irp/agency/dod/jason/ai-dod.pdf. The 'ilities' here refer to 'reliability, maintainability, accountability, verifiability, evolvability and attackability'. The report concludes that in this case DL 'is weak on the "ilities"' (p. 2).

[28] Proctor, Robert, 'A Missing Term to Describe the Cultural Production of Ignorance', cit. Proctor, Robert and Londa Schiebinger, *Agnotology: The Making and Unmaking of Ignorance* (Stanford University Press, 2008), p. 1.

[29] See, generally: Bostrom, Nick, *Superintelligence: Paths, Dangers, Strategies* (Oxford University Press, 2014), pp. 200-219.

[30] Haikonnen, Pentti, *The Cognitive Approach to Conscious Machines* (Imprint Academic, 2003), p. 145.

government to justify its opposition in the CCW to statutory controls over AWS deployment[31]; its 'empty hangar syndrome' argument is based upon certain scenarios being simply too far-fetched to warrant consideration and, in the case of AWS, that it is plainly unrealistic for a commander to wander one morning into the weapons hanger to find that the AWS has decided under its own volition to depart unexpectedly on an unsupervised mission.

6.2 Architectural approaches to AWS deployment

A common starting point to consider AWS operation is, unsurprisingly, the human mind. If the Delivery Cohort is suggesting that the brain is being substituted by machine processes, then additional context is required in order to judge that argument. First, a recent estimate for the human brain is that the cerebral cortex contains some thirty billion neurons, plus or minus, in the part of the brain associated with consciousness and intelligence. Second, those neurons in turn contain some one thousand trillion synapses, the connections between neurons. Some suggest that it might be feasible for a machine one day to 'copy' this human mind, but readers should note that an infant's brain does not arrive with all of the 'lexicons, representations, algorithms, programmes, processors, subroutines, encoders or buffers' that will accomplish this mammoth task and must all be seamlessly in place ahead of such replication.[32] Simply copying over the constituents of a human brain into an AWS no way equates to creating a facsimile that might be useful or relevant.

Unlike the idealised connections that are assumed for a machine's neural network, human synapses are empirically variable in nature, based upon different and quite undifferentiated neurotransmitters and all with different cycle times. Thus equipped, we humans are quite good at the processes that comprise the engagement cycle. Information in the human brain, perhaps more than a megabyte for each connection generated by each

---

31 Until publication in August 2018 of 'Human Machine Touchpoints: The United Kingdom's Perspective on Human Control over Weapon Development and Targeting Cycles', the Foreign and Commonwealth Office had adopted a broadly negative negotiating position in discussions on banning weapons autonomy at the UN's CCW, 2014-2024.
32 Epstein, Robert, 'The Empty Brain', *Aeon* (18 May 2016), generally, https://aeon.co/essays/your-brain-does-not-process-information-and-it-is-not-a-computer. Epstein's article provides a useful primer on challenges to machine emulation of brain function.

synapse, is being processed in real time within each synapse cycle, itself more than one thousand bursts per second. An opening observation might be how this magnitude can even be understood if machines are to be based upon such a model. If each such synapse were handled by the equivalent of only a single line of code, the program to simulate a human cerebral cortex would be some twenty-five million times larger than reputedly one of the largest software products written to date, Microsoft Windows, which is estimated to be some fifty million lines of code.[33] Readers should note those commentators who judge the probability of successfully completing such emulation as effectively zero.[34]

Methods other than emulation have long been on drawing boards to attempt this same machine intelligence. Engineers have looked not to replicate the entire brain, but instead to copy just certain of its characteristics or to apply theoretical methods of neural processing in order to simulate brain processes. Others posit less conventional approaches, including massively parallel computers with biological cognition being forced through superfast iterations that run, perhaps, on genetic algorithms or on quantum computers using qubits that can exist in several positions.[35] This remains a future technology, the components of which remain in their relative infancy, and hardly applicable to our problem set of near-term battlefield weapons operating in contested battlespace.

> *Readers should understand that commentators judge the probability of successfully completing such emulation as effectively zero*

The gulf between desktop research and in-field applications also demonstrates how tricky it is to translate theory into practicable tools for the battlefield. First, the computational resources needed to copy the

---

33 Microsoft no longer release the number of code lines comprising their latest Windows product. Version 10 reportedly contained 50 million lines. See also: Choi, Charles, 'Too Hard for Science: Simulating the Human Brain', *Scientific American*, generally, https://blogs.scientificamerican.com/guest-blog/too-hard-for-science-simulating-the-human-brain/. See also: Metz, Cade, 'Google Is 2 Billion Lines of Code – and It's All in One Place', *Wired* (16 September 2015), generally, https://www.wired.com/2015/09/google-2-billion-lines-codeand-one-place/.
34 Cattell, Rick and Alice Parker, 'Challenges for Brain Emulation; Why Is Building a Brain So Difficult?', ResearchGate, (January 2012), https://www.researchgate.net/publication/260869458_Challenges_for_brain_emulation_why_is_building_a_brain_so_difficult.
35 Plowman, Gary, 'How Quantum Computing Could Revolutionise Military Defence', Karve International (14 September 2023), https://www.karveinternational.com/insights/how-quantum-computing-could-revolutionise-military-defence.

evolutionary processes that underpin our own intelligence are plainly out of reach. For example, a simple honeybee brain has some ten-to-the-power-of-six neurons.[36] Ignoring the complexity of the exercise, the computational cost of simulating a single neuron suggests that AWS approaches based on superfast iteration are simply unfeasible and it is therefore necessary to look elsewhere for an architecture that might solve for this bottleneck. Second, most present methodologies still start with that human brain as a template for machine intelligence. Whether this involves 'whole brain emulation' (WBE) or other means to model the computational structure of a biological brain, these routes all require those same enabling technologies that have yet to be invented.

The gulf between desktop research and in-field applications demonstrates how tricky it is to translate theory into feasible tools for the battlefield

Nor do the very simple prototypes being thought about on laboratory benches suggest proof of the endeavour's eventual success. It is more likely that processes will be dogged by what is termed 'chaotic dynamics', already a feature in test systems with just a handful of neurons. The challenge for scientists is to achieve structural validity as opposed to just replicative validity. Weapon designers using this methodology would need, for example, to understand which synapses may be excitatory and which are inhibitory. They would need to model exactly the strength of these connections as well as mapping the dynamical properties of relevant brain subsystems. It should also not be assumed that WBE models can scale to battlefield applications, these approaches more resembling 'extended' software rather than any expression of machine intelligence, code-based and rules-driven artificial neurons mimicking human neurons. More fundamentally, these replication models proves that it is insufficient simply to ascribe single weightings to an artificial neuron in order to manage its threshold, or to each synapse in order to reflect a new signal strength. These linear relationships do not exist in real life but are still the basis of how a machine copy neuron is expected to be trained. With each human synapse having an estimated minimum of ten thousand connections, commentators reckon that this will be at least six hundred billion times more complicated than any artificial neural network yet devised if the whole human cortex is to be copied. Again, therefore, we are forced to look elsewhere for a more feasible architecture.

---

36 Bostrom, *Superintelligence*, p. 24.

There is no obvious path. The inescapable challenge to all learning models is that of scaling. Various axioms explain this well. First, David Wolpert's 'no free lunch' theorem, still relevant in its third decade, states that general-purpose learning algorithms are inherently difficult 'in the sense that for every learning model there is a data distribution on which it will fare poorly on both training and test'.[37] Henry Marsh points to a further underlying logjam compromising this approach, the 'binding problem' that calls into question 'how all of this disparate neuronal activity, spread out in both time and space, produces coherent experience'. The 'von Neumann bottleneck' then points to the extraordinary electrical requirements of the approach whereby 'an exascale computer, capable of a quintillion calculations per second, scaled up to the size of a human brain, would consume hundreds of megawatts [of power]'.[38]

> The challenge for scientists is to achieve structural validity as opposed to just replicative validity. Weapon designers need, for example, to understand exactly which synapses may be excitatory and which are inhibitory

Almost by definition, AWS learning models must also contain restrictions on the class of functions that they can learn. It cannot simply be assumed that algorithms will scale without performance falling away. Even 'kernel methods' which subvert the usual norms of frequency distribution are compromised by their inappropriately shallow architecture[39] and, although learning models are considered briefly below, compelling arguments exist that the very basis of ML's trainable coefficients is enduringly fragile.[40] Commentators highlight two particular bottlenecks: the 'depth-breadth trade-off' and the 'curse of dimensionality' where trial

> Wirth posits instead that 'software gets slower faster than hardware gets faster'

---

37 Wolpert, David, 'The Lack of Distinction Between Learning Algorithms', *Neural Computations*, 8, 7 (1996), 1341-1390, https://www.mitpressjournals.org/doi/abs/10.1162/neco.1996.8.7.1341/.
38 Marsh, Henry, 'Can Man Ever Build a Mind?', *Financial Times* (10 January 2019), https://www.ft.com/content/2e75c04a-0f43-11e9-acdc-4d9976f1533b.
39 Copeland, Michael, 'What's the Difference between AI, Machine Learning and Deep Learning?', nvidia blog (29 July 2016), https://blogs.nvidia.com/blog/2016/07/29/whats-difference-artificial-intelligence-machine-learning-deep-learning-ai/. Shallow networks have fewer hidden network layers (and can function with a single layer) but require unwieldy multiplicity of data points.
40 Bengio and LeCun, pp. 12-14 ('Depth-Breadth tradeoff'), p. 16 ff ('Fundamental limitations of local learning') and p. 21 ff ('Curse of dimensionality').

architectures become exponentially more far-fetched as a prototype is scaled up and the number of variables increases, engineers being unlikely to understand in advance just how many flaws persist in late beta versions of their models. While Moore's law might be invoked, doubling speed and capacity never solves problems brought about by inherent system complexity. A second law, attributed this time to Nicklaus Wirth, is

These linear relationships do not exist in real life but are still the basis of how a machine copy neuron is expected to be trained

perhaps more relevant. Wirth posits instead that 'software gets slower faster than hardware gets faster'.[41] While Moore's law correctly forecasted that the personal computer would be some hundred thousand times more powerful than twenty-five years ago, the computer's word processor certainly is not.

6.3 The Delivery Cohort

What, then, might be the consequences from fault lines to this approach? By their very nature, AWS must be non-deterministic and non-scripted. After all, deploying lethality under second-levels of uncertainty (uncertainty about already uncertain ranges of action) would seem to introduce huge risk into battlefield practices. From this notion arises the concept of 'role responsibility', the subject of this short section, and the implication that it is a broad cohort that is responsible for the deployment of new weapons technologies. Those individuals fielding AWS might assume that their superiors and legal experts have appropriately overseen those involved in the design, programming and testing of the weapon. But actually it is a class effort and it is useful to reiterate the notion of the design and implementation 'team' that is referred throughout this book as the Delivery Cohort.

This important piece of shorthand refers to the extended group of experts and other parties needed to implement independent weaponry. The construct is also useful as it demonstrates the masking of certain procurement shortcomings regarding the delegation of agency away from humans, including

---

41 Kassan, Peter, 'A.I. Gone Awry: The Futile Quest for Artificial Intelligence', *Skeptic* (4 February 2011), p. 2, https://www.skeptic.com/reading_room/artificial-intelligence-gone-awry/.

the diffusion of responsibility among a raft of participating parties to the point that any meaningful attribution is obscured. As evidenced by the section sub-headings to this section of this book, the list of required competencies and the control mechanisms necessary to deliver AWS capabilities is very broad. The role of this Delivery Cohort in deploying compliant AWS must therefore be correspondingly wide without omission or flaw. The Cohort must manage both software and hardware challenges that are identified in these chapters in order to field a weapon that is appropriate to local commander and also the wider community. The Cohort's responsibility is therefore layered and highly complex.

This important piece of shorthand [the Cohort] refers to the extended group of experts and other parties needed to implement independent weaponry

So how might this work in practice? Those more likely 'favoured' worlds (the sets of actions that most closely align weapon outcomes with the Delivery Cohort's intended purposes) will presumably be assigned higher probabilities during the Cohort's setting and configuration of each weapon. This, however, is non-obvious: later chapters will demonstrate that this process still demands rigourous technical oversight (the inputs for a utility function must be unambiguous and sharply defined), negotiation and management by the Cohort. But it will also be non-trivial to achieve technical consensus among what are actually competing parties within the Cohort, in order to agree upon intended outcomes and to do so in appropriate detail. The role of the Delivery Cohort at both design and deployment stages is therefore fundamental.

It is also the *dynamic* nature of the Cohort's task that creates challenge. Datasets, the central element to machine learning, are volatile and prone to rapid but unknown obsolescence. AWS' AI spine will, by the very nature of the technology, be nested (that is, its routines unfold from and are enfolded in other another). Just as these dynamic processes must be knit into one seamless whole, the entire weapon must understand and adapt to changes in those datasets. These adjustments are unlikely to be linear and the effects of any fluctuations rarely congruous with the intentions of the Cohort. Fractional changes in a weapon's 'initial condition' may materially alter the long-term behaviour of that weapon, the notion that a series of incremental changes can suddenly lead to wholesale change in the agent's overall orientation. It is unlikely that the Delivery Cohort will be equipped to recognise these incremental corrections. The maximalist nature of the Cohort's task therefore requires that it layer the AWS with feedback mechanisms to identify variations in performance but also to monitor the weapon's adherence to the Cohort's intended goal state. This again requires intricate processes and, as discussed in the following chapter, considerably affects the weapon's required architecture.

It also requires that each AWS is singularly and severally open to adjustment in order for the Cohort to compensate for evolving circumstances and to address errors. 'Patching' is the process whereby the AWS receives updates from the Cohort to amend passing configuration instructions, fix coding bugs, enhance the machine's security, improve its performance, or provide added features and functionality. The procedure for AWS, however, is likely to be enduringly complicated, not least because the target machines are by definition autonomous and likely operating in communications-denied conditions. At the very least, each AWS must be able to self-diagnose coding deficiencies, with the Cohort then issuing timely fixes to that issue (while not contaminating the machine's wider operation). Patches are a therefore a key attack surface for adversaries as they must be delivered to all relevant host weapons – a key Cohort obligation to ensure weapon stability, version control and feedback. The capability must also be tested and trusted if these interventions are to be synchronised. Patching is also an understated bottleneck in complex systems and one that empirically affects system cohesion. In addition, it weakens

Fractional changes in a weapon's 'initial condition' may drastically alter the long-term behaviour of that weapon, the notion that a series of incremental changes can suddenly lead to wholesale change in the agent's overall orientation

system predictability, given that on-board faults are rarely single-issue events and, with the systemic likelihood of second-order effects in AWS, complicates diagnoses in the event of failure or drop-off in performance. Indeed, how can the Cohort create a common patch for a family of systems that are now very disparate?

Team culture does not develop on its own and requires thought and management. Another challenge for the Cohort will be to encapsulate (and then maintain) different insights, perspectives and information sources that best allow the resulting team to make the whole gamut of decisions that AWS deployment will entail. This is not easy and two well-tried aphorisms ('excellence is a habit' and 'a team is what it repeatedly does') might govern Cohort mechanics. Nor is this simply a matter of harnessing best 'intelligence' (good logic and good mathematics) as decision-making within the Cohort will also require excellent 'smarts', all of those subjective and less tangible qualities already identified in this analysis; this should include empathy, calm, persuasion and healthy paranoia whereby result-and-actions-oriented individuals are properly balanced by those who score well around inclusiveness, delegation and creative problem solving. The challenge here is for the Cohort to avoid groupthink. It is to ensure that its own processes are properly calibrated and allow for real challenge throughout the ranks of the Cohort. It is not by accident that chapter headings in Syed's *Rebel Ideas* are titled 'Collective Blindness', 'Constructive Dissent', 'Echo Chambers' and 'The Big Picture'.[42]

Homophily is defined as the tendency for people to seek out or be attracted to those who are similar to themselves. In the Cohort, every individual is likely to be smart and equipped with excellent sector-specific knowledge. The danger, of course, is that they are too homogenous,

> *The danger is that [the Cohort is] too homogenous, homophily acting as a 'hidden gravitational force, dragging human groups towards one corner of the problem space'*

homophily acting (notes Syed) as a 'hidden gravitational force, dragging human groups towards one corner of the problem space'. And even if the Cohort manages to arrange itself in an appropriately diverse manner, social osmosis can soon dilute this as those in the group converge upon a dominant assumption or dominant set of individuals. This is the unwelcome phenomenon of 'assimilation', the clustering of professionals in small parts of the wider problem space, a long-observed consequence of human psychology.

The construction of the Cohort is unexpectedly both a challenge and likely point of friction, the Cohort's wide remit requiring that it assembles an overtly broad team of individuals. An experiment is useful to highlight the faultline. Assume, for the sake of argument, a Cohort of one hundred professionals and that they each come up with ten actionable ideas. How many useful ideas does the Cohort have in total? The game, of course, is specious as the number of ideas in a group cannot be inferred from the number of ideas of that group's members; if the Cohort lacks appropriate diversity, it will likely come up with pretty much the *same* ten ideas regardless of the number of Cohort members. Conversely, a *diverse* team of just ten professionals each with ten ideas might produce one hundred useful ideas. The analogy does not even capture the opportunities that a team fails to exploit, the questions that its individuals do not ask, the data that is not prioritised. Moreover, the more challenging the domain, the *less* that any single part of the Cohort

---

42 Syed, Matthew, 'Rebel Idea', (John Murray Publishing), 2021.

should expect to influence; homogeneous groupings make the same errors, get stuck in the same place and miss the same opportunities.

Staffing the Cohort is also likely to be problematic, the more so in militaries that have a long-dated policy of meritocratic hiring. Those hired into the Cohort are almost by definition likely to have studied in similar institutions, have absorbed similar insights and, in the extremis of war, likely to act in very similar ways to given stimuli. The Cohort's collective intelligence therefore requires ability and diversity if knowledge clustering is to be avoided. An adjunct complication arises. Who in the Cohort should make key decisions? While the presence of hierarchies is embedded in our species, teams based upon 'dominance dynamics' often deliver suboptimal outcomes, especially in situations of complexity. In this sense, the whole notion of the Cohort represents a paradox; its structure will be inherently hierarchical and notwithstanding that behaviours associated with this type of structure invariably thwart the effective communication that is precisely required by this grouping. [43] Indeed, the *National Transportation Board* in the US notes that more than thirty recent crashes have occurred when co-pilots have failed to voice concerns. The problem also tends to exacerbate over time, teams often become increasingly confident about what may be very poor judgements if members of the Cohort systemically fail to call out poor behaviour, thus compounding each other's errors.

6.4   AWS learning architecture

A deployment base case is that AWS must at least be capable of deduction and interpretation whereby whatever the weapon has experienced in prior cases should inform what the weapon does now. Reasoning and learning, after all, constitute the litmus test to defining AI, and the autonomous weapon's actions must therefore encompass the understanding of a known case whose relationships can then be carried over to the present case. The purpose of the next four chapters is to review why this might be particularly complicated for machines tasked with battlefield functions, these contradictions being exacerbated as the weapon's deployment environment becomes ever more layered. Humans are

---

43 Syed, Matthew, 'Rebel Idea', (John Murray Publishing), 2021, p.93.

evolutionarily capable of reasoning when available information is imperfect, formulating deductions that are based on knowledge that is 'generally true'. An AWS, however, must instead backfill and do this in the moment within its regular routines (each with a filter, error bias, weighting and confidence prediction). The model envisaged for AWS is therefore fundamentally 'fleeting' as sensors must stream new data continuously to refine existing available information.

Several difficulties arise. Weapon architectures must at least facilitate (if not actually promote) appropriately counterintuitive capabilities such as detection of contradictions, evaluation of significance and, complicatedly, appropriate rejection of alternatives that would otherwise lead the weapon to foreseen unsatisfactory outcomes. Apart from the computational issues for such backfilling, the procedure also requires resource and memory. AWS routines must therefore incorporate biases that *constrain* weapon processes, limiting possible outcomes and doing this either in a distributed manner (whereby that bias is programmed across routines, a complicated exercise of tuning and balance) or by adding bespoke modules to manage AWS output, with every such routine introducing additional intricacy. The Cohort must also factor for smoothing these interventions if system stability is to be maintained.

We therefore need to consider AWS' likely programming spine, the artificial neural network (ANN), which is the machine's set of statistical tools and the enabler for AWS' learning and estimation. The construct is based upon a general information-processing model that is grounded in the way our own biological nervous system manages information. The model's essential facet is the weapon's processing system, very many highly interconnected processing elements (neurons) working in unison to solve specific problems, the architectural intention being that ANNs, like people, learn by example. The model is for the weapon's network to enable specific capabilities, such as pattern recognition or data classification, through a learning process that will then knit together outputs in order for the AWS to generate an action path. In a couple of inappropriately reductive sentences, this then is the high-level set of steps that will comprise AWS operation.

While learning in our own biological systems depends upon adjustments to the synaptic connections that exist between neurons, the intention for machines is for these processes to be copied through their ANN where, in the case of AWS, the machine's neurons will either be in 'training' or 'using' mode. The Cohort's bet will therefore be that it can recreate a learning system which is based entirely upon machine code. The model relies upon each AWS receiving, interpreting and weighting data and then integrating feedback on how those routines are performing, the weapon's neural network iterating the same problem millions of times and optimising outputs according to this feedback. A computer, in this case the routines within an AWS, can then theoretically be shown a different problem which it can approach in the same way as it learned from the previous one. The model can either be narrowly focused (and therefore narrowly tasked, for example initiating a tailored and homogenous response to a particular set of events) or, by varying the agent's exposure to wider problem sets (and expanding the number of approaches to solving them), the Delivery Cohort can 'teach' the AWS to be more generalist and, again in theory, more 'intelligent' about its battlefield surroundings and mission.

There has been considerable recent progress across these models.[44] The methodology is the accepted model behind facial recognition, stock market prediction, social media applications and, generally, predictive analytics. Networks are performing ever more quickly; they are becoming easier to explain and are being partnered with adjacent technologies that enable ever wider tasking. DL algorithms based on ANNs run on GPUs, specialised chip sets used in personal computers and video games consoles. This recent marriage has increased the processing in DL systems by nearly a hundredfold and allowed, notes the *Economist*, the training of a multi-layer neural network to take just a few hours, a procedure which had previously taken several weeks.[45] The machines are becoming exponentially more powerful as networks of double-digit layers are worked upon by researchers. Using deep networks, the JASON programme also reckons that the error rate of image capture has fallen from twenty-five per cent to less than three per cent. This is already better than the figure of five per cent achieved by humans.

An AWS must backfill and do this in the moment within its regular routines (each with a filter, error bias, weighting and confidence prediction)

But deep-seated difficulties persist. Systemic problems arise in how these network models misclassify data, either in their original training datasets or in the sensed information that is subsequently harvested. And these errors (be they missteps in ranking, identification, priority or association) occur

---

44 Samarth, Varun, 'A Detailed Guide on the Meaning, Importance and Future of Neural Networks, *Emeritus* (19 January 2024), https://emeritus.org/in/learn/ai-ml-neural-networks/.
45 Economist, 'The Return of the Machinery Question', *Economist* (25 June 2016), p. 4, https://www.economist.com/sites/default/files/ai_mailout.pdf.

in ways that are unpredictable and unfamiliar to humans, questioning the very set of probabilities upon which weapon configuration rests and, in so doing, skewering the job of the Delivery Cohort. The model also requires that those AWS' inputs have three demanding characteristics. First, each input (here, the weapon's original representation or, more likely, subsequently sensed data derived from its surroundings) must have its own weighting (or synaptic) value. Second, it must have a summing function to manage these inputs and, third, it will have a threshold-based output function. The model's premise is also that all such inputs are largely free from noise (in other words, homogenous) and sufficiently full in detail. A further challenge is that

Networks are performing ever more quickly; they are becoming easier to explain and are being partnered with adjacent technologies that enable ever wider tasking

these weapon inputs must be the right inputs (indeed, all of the right inputs) necessary for the weapon to derive applicable intelligence from its battlefield surroundings and, crucially, to divine meaning from those inputs.

But the picture thus far is still far from complete. Based then on continuous signals (each with variable intensity), the value of each input signal must then be multiplied by its related weighting, applied manually (or at least checked) at each weapon's inaugural configuration, with the results then summed together. Remember, of course, the intractable imprecision of these inputs (a soldier versus a non-combatant versus a child or its mother versus those who are *hors de combat*) upon which the AWS' engagement calculations will then be based. This sum value is then matched to each of the weapon's embedded thresholds (the boundary points imposed during the Delivery Cohort's setting up of each weapon) with an artificial numeric expression then triggering a specific action if that threshold is exceeded. Several intractable happenings arise. First, the analysis ignores the challenge where no combination of weight values meets the thresholds that have been set for the AWS (AI's exclusive-or problem).

It also ignores *temporal* dislocation as the system triangulates between multiple and corroborative information sources. Within the weapon's network, each neural unit must be connected with innumerable others, each link having either an 'enforcing' or an 'inhibitory' effect, and each neural unit triggering, as appropriate, a limiting function on each connection and on the unit itself. Should a signal exceed the limit that, in theory, has been defined by the Cohort, then it is propagated forward to other neurons. The unsupervised weapon system is thus 'trained' rather than explicitly programmed, notwithstanding that the weapon must at least be minimally fixed before training starts. As noted by the Royal Society, the intended benefit is that machine-learning enables the machine to 'excel in areas where the solution or feature detection is difficult to express in a traditional computer program'.[46]

Key to the Cohort is how these statistical tools translate to the battlefield as, without practical implementation pathways, little of this is relevant.[47] Neural networks typically consist of multiple layers (or a cube design) with signal paths traversing from front to back. This is necessary given that the whole structure is predicated on training the AWS by running and re-running very large sets of 'experienced data' in a long and iterative process. This repeating allows those layers of neurons to adopt and refine 'prioritising weights' so that the system might then make sense of new data sets on the same basis that it has already encountered, honed and weighted previous training sets of data. Back propagation is therefore a foundational training operation for AWS function whereby stimulation is used to reset weights within the framework's neural units until optimised. An adjunct difficulty is therefore to regiment the system's network connections to prevent this interaction from taking place in a chaotic fashion, facilitating the formation of new connections while disabling others.

And these errors occur in ways that are unpredictable and unfamiliar to humans, questioning the very set of probabilities upon which weapon configuration rests and, in so doing, skewering the job of the Delivery Cohort

A decade ago, network programs were working with up to a few million neural connections which equated to the computing power of a worm. By 2023, the Megatron-Turing NGL program had some 530 billion parameters, three times the previously highest number (OpenAI's GPT-3 with 175 billion parameters). This lightning progress might suggest an eventual breakthrough to the matters in hand but readers should also remember that Megatron and its like represent an advance (albeit exponential) in quite a narrow part of AWS' required solution set. It also involves exorbitant cost,

---

46 See: Royal Society, 'The Power and Promise of Computers that Learn by Example', Royal Foundation (April 2017), https://royalsociety.org/~/media/policy/projects/machine-learning/publications/machine-learning-report.pdf. For an assessment of ML deficiencies see: p. 30.
47 The weapon's neural network will, after all, be a connectionist system based on a very large assemblage of artificial neural units, loosely modelled (as discussed) upon the way a human brain solves problems. See: Fodor, Jerry and Zenon Pylyshyn, 'Connectionism and Cognitive Architecture: A Critical Analysis', Rutgers University, undated, pp. 2-4.

with these machines' training currently being undertaken on more than five hundred servers and nearly five thousand individual GPUs, each capable of processing more than one hundred teraFLOPs (one teraflop being one million million floating points per second). All of this, moreover, is being undertaken miles away from the adversarial, unbounded setting of a battlefield and under highly invigilated circumstances.

These observations show that it is necessary to impose an empirical lens on what are otherwise quite academic processes. Again, what is actually practical here? First, it is a given from the foregoing that AWS learning must be based on mathematical

The unsupervised weapon system is thus 'trained' rather than explicitly programmed

techniques, upon statistics and probability.[48] An example is useful. For a weapon whose understanding of its environment relies upon sensed data, each visualisation input comprises individual pixels. For reference, a standard single TrueColor digital image requires thirty megabytes of platform memory. An HDTV clip from just one sensor (and at just 1920 x 1080 pixels, capturing its environment at sixty frames per second) requires more than twenty gigabytes of memory for every minute of video input.[49] Operating continuously and in real time and processing input from multiple visualisation camera positions[50], this equates today to at least three hundred gigabytes of 'interesting' information per sixty seconds for the deployed AWS.

In the field, that data will be derived from a considerably broader universe of sources that includes audio samples for speech recognition, character samples for natural language understanding, and chemical samples or some other olfactory measure for additional hazard identification. Moreover, the model for AWS is also that each discrete datapoint (here, characters or numbers) must be represented in a dynamic series of one-shot representations where a separate neuron is used for each possible symbol in each position at each particular moment. Given the number of individual data points that will comprise a single engagement sequence, the architectural convolution of processing this information (while accounting for contextual

---

48 Bengio, 'Challenges of Training Deep Neural Networks', paras. 2-3. For a general discussion of weapon network architecture, network training and back-propagation, see: JASON, pp. 6-19.
49 The TrueColor photo assumes a 2736 x 3648 pixel format. See: Ken's Image Gallery, http://kias.dyndns.org/comath/44.html.
50 Even the early Tesla S Class car required eight cameras, one radar unit and six ultrasonic units just to auto-drive; see Lambert, Fred, 'A Look at Tesla's New Autopilot Hardware Suite', Electrek (20 October 2016), https://electrek.co/2016/10/20/tesla-new-autopilot-hardware-suite-camera-nvidia-tesla-vision/.

sensitivities, clutter, bottlenecks and significance) appears unworkable. Why, the reader might ask, would parties risk efficacy, compliance and defence dollars in foregoing human oversight over battlefield practices? We need humans on the battlefield.

Central to weapon learning is therefore this complex training process in which the millions of weights connecting neurons are assigned values. Indeed, the only adjustable parameters (outside biases and hyperparameters such as the number of neuron layers) in this process are these weights, and the refining of these weights will only occur during the process of running very large data sets of input/output pairs to train each such agent. To train, for instance, a network to recognise images, the model's x-input function might represent intensity values of the image pixels while the y-input function describes the picture's description. The aim is to match these two inputs. The architectural challenge is then that the performance of individual neurons must be quantified using an error function with the goal of training being to minimise this function. Practically, this is achieved using a gradient descent whereby the whole network's weighting is updated again and again using the gradient of the error function. The model in this case is for such iteration to continue repeatedly until the weapons network's weights find the global minimum of the error function averaged over all of that training data.

Additional routines, however, always complicate processes. As part of this process, a subset of the weapon's original raw data might be hived off from the main training set to be used as validation data for the Cohort to measure how well individual training is progressing. The routine can thus determine whether it is necessary for the weapon to refine parameters such as error rates, learning rates and the degree to which results differ from outcomes expected by the model.[51] The efficacy of this process depends on two further characteristics. The model be able to propagate forwards to calculate the current (and expected) values of the weapon system. The model must also allow backwards propagation of the weapon's error function in order to update its learning

An adjunct difficult is to regiment the system's network connections to prevent this interaction from taking place in a chaotic fashion, facilitating the formation of new connections while disabling others

---

[51] Several online resources are available on the division of labelled datasets into training sets, validation sets and test sets. See, for instance, Stanford University videos on ML offered through Coursera.

weights. Each iteration through the entire set of training data is termed an epoch, with each individual network allocated a single pre-determined task which, once learned, requires that network's connections be frozen on deployment. The challenge for AWS is that this creates an inappropriately 'one trick pony' with no facility for additional learning. If this is the case, then the dynamic, chaotic nature of the battlefield would require an infeasibly huge number of epochs in order to hone this error function in a chaotic chase for conformance.

Several operational ramifications arise from this data training. During training, some neurons are found to be effective 'problem solvers', others not so effective over cycles of interaction that must take place. The process actually sees a plateau in performance due to ever reducing learning gradients and ever smaller changes to the model's weights. Each new layer to a weapon's neural network also means an extra layer of non-linearity, making optimising the weapon's learning process ever more difficult. Models, moreover, start to restrict their learning to just the network's top layer while lower layers remain random transformations that do not capture much input.[52] Gradients are prone to dilution at these lower layers, providing increasingly unpredictable and weak guidance to the overall learning process. If AWS are to be compliant and trustworthy, this systemic instability is not ideal.

> *The challenge for AWS is that this creates an inappropriately 'one trick pony' with no facility for additional learning*

The efficacy of any architecture also depends upon the fit between weapon tasking and weapon training. As noted in later sections, a weapon's marginally different set-up or a marginally different training dataset likely leads to very different sets of output. As an aside, identifying discrepancies and sources of variation in this primary data is well beyond human capabilities given the size, presentation and specificity of the raw dataset. Training methodology is thus critical to weapon feasibility. If not correctly commissioned, battlefield features with only a small number of examples in that training set (but possibly of critical importance) may likely be ignored in the process. Both model and architecture actually rely on very clear data definition, at odds with Sun Tzu's maxim for the successful commander who must be 'extremely subtle, even to

---

52 Alain, Guillaume and Yoshua Bengio, 'Understanding Intermediate Layers', ICLR Paper (2017), pp. 7-8, https://pdfs.semanticscholar.org/2706/77b5c44ea0c93313f41db2f885fef305bbcc.pdf.

the point of formlessness'.⁵³ This problem (of either ignoring or discounting evidence) has various angles. It may come about from a too large number of learning examples overwhelming the training effect of examples in quite a different data set. It may be caused by incorrectly set model sensitivity ('detection rate') or incorrectly set model specificity ('false alarm rate'). Counterintuitively, lengthening the time series of data does not appear to equate to more efficient learning results; if there is repetition in that data then the AWS' ANN is unlikely to become any 'wiser' from additional training.

Instability across learning routines is also compounded by data noise when gathered information points, either discrete or continuous, are distorted, corrupted or present with unevenly relevant, notwithstanding that noise can be usefully included to prevent overfitting. Commentators also note that a small distortion in classification routines leads to different data classes becoming mixed and inseparable. In other words, as the weapon's dataset becomes less distinct, the class boundaries that divide different class examples are increasingly difficult to separate for ongoing statistical analysis within the weapon. This has ramifications for the Delivery Cohort. If such misclassification is frequent, then additional supervision sequences will be required to train the AWS into make favourable classification decisions for a particular class. Termed the agent's matching challenge, this becomes a further source of possible bias. An instance might be the over-fitting of training data, making the weapon's whole training model more brittle. Adding incremental processing steps also slows and complicates what needs to be a seamless, real-time series of steps. Other processing pitfalls then arise. Given that as much as ninety per cent *The process sees a plateau in performance due to ever reducing learning gradients and ever smaller changes to the model's weights* of the weapon's runtime might be taken up with computationally expensive pattern matching, part-processed data might also be hived off into interim (possibly off-line) storage between matching cycles. In this case, complicating 'weight decay' routines are needed to regularise the machine's training and to manage what is generally termed 'catastrophic forgetting'.

As above, ancillary routines always involve compromise: adjusting training weights may make the weapon less sensitive to noise but correspondingly less likely to learn from that noise. It may push the weapon's

---

53 Jackson, Eric, 'Sun Tsu's Art of War', *Forbes* (23 May 2014), https://www.forbes.com/sites/ericjackson/2014/05/23/sun-tzus-33-best-pieces-of-leadership-advice/#19c3ac7d5e5e

neurons into saturation, which then desensitises those neurons to all inputs. A further architectural complication concerns 'dropout' whereby learning routines regularly omit randomly selected neurons from the weapon's training process in order to reduce over-fitting and false correlation. It is also unknowable from the outset if the weapon's training data is both sufficiently relevant to its y-function (the task that the Delivery Cohort has for each network) or of sufficient size appropriately to train the network. The issues here are so germane that commentators worry that any architecture's descent gradient is systemically unstable. Although complicated, this is also unsurprising. These gradients, after all, are derived directly from the product of terms in all subsequent network layers and, as above, the product of many of these terms must themselves be similarly volatile.

Finally to this point, the JASON report reminds us that different layers will learn at different rates, and the learning will be unbalanced. This problem worsens as the number of layers increases and is thus a particular challenge for the ever-deepening neural networks being undertaken by researchers.[54] The issue here is that that these learning processes are designed to derive an approximate answer to the question set being raised. This has usually been good enough. When it works, notes JASON, it may not be necessary to understand why or how. But the discontinuity here is that autonomous weapons have lost human oversight and human veto and JASON's observation is instead really questioning the whole aptness of AWS' technical spine.

A machine's marginally different set-up or a marginally different training dataset likely leads to very different sets of output

That there are operational difficulties with this architectural learning model should also not be surprising given JASON's conclusion that current learning technologies have 'not [yet] systematically addressed the engineering priorities of reliability, maintainability, debug-ability, evolvability, fragility and attackability'. Notwithstanding the extraordinary reach of Megatron's new neural model, it is the very number of parameters (including their attendant weighting) that thwarts the model's relevance as a battlefield tool; more parameters in an adversarial setting equate to more attack surfaces and more opportunities to spoof data and mislead models. Autonomously generated outputs, after all, must depend precisely upon the training data

---

54 JASON, p. 66.

used, the order in which data are processed and the frequency with which the weapon's sensors are polled in order to refresh the weapon's routines.

Battlefield data is generally 'non-convex' (with multiple peaks and valleys) while the ML's commercial optimisation has been designed for convex problems.[55] Moreover, the practice here will be that the significance of this variability is very unlikely to be spotted in any timely manner. Adding smoothing routines also acts as an aggressive edit to the AWS' sensed data, and the method of deliberately dropping left-field neurons is akin to ignoring the non-zero weighting of all collected observations. Indeed, the generalisation here is that system fragility will only increase as the ML model is shoehorned into weapon management, not least because sample complexity requires that the number of training examples grows at least linearly with the number of active parameters in the weapon's learning processes.

A further architectural complication concerns 'dropout' whereby learning routines regularly omit randomly selected neurons from the weapon's training process in order to reduce over-fitting and false correlation

6.5 Missing pieces

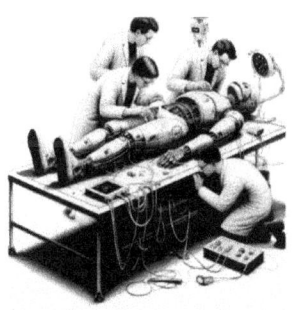

Another way to understand AWS feasibility is to identify instead the critical technologies that still remain outstanding. Here, several capabilities have long appeared out of reach including, *inter alia*, the processing of abstract imagery, robust summarisation skills as well as proven tools enabling scene and episode understanding. As noted by the Stockholm International Peace Research Institute, not all technical research exceeds expectations. Routines that will underpin these machines still struggle to interpret wider context: vision software may identify a soldier walking but is unable to determine why. Given the difficulty of representing abstract relationships between objects and people in models of

---

55 van den Berg, Ewout, 'Training Variance and Performance Evaluation in Neural Networks in Speech', IBM Watson Group, ICLR (2016), p. 1 and p. 5, https://arxiv.org/pdf/1606.04521.pdf. Linear programming problems are generally convex problems where there can only be one optimal solution. Non-convex problems are, reports van den Berg, more complex and often intractable.

the real world, the same conundrum characterises speech recognition where the computer may understand what is said but not what is being discussed. A consequence is that autonomous systems are particularly vulnerable to trickery (there is, after all, no opportunity for human common sense to intervene), requiring that the Cohort deliver upon the enduring capabilities required of unsupervised AWS, to sense, think, decide, act and team. Indeed, without that facility to generalise, AWS deployment must be limited to known situations and known environments.

> *Ancillary routines always involve compromise: adjusting training weights may make the weapon less sensitive to noise but correspondingly less likely to learn from that noise*

Several of these required capabilities have seen only limited recent advance. Agents' ability to display creativity is a particular case in point. While creativity should presumably be a source of useful battlefield advantage, little progress is evident in combinatory creativity (the novel and improbable amalgamation of otherwise familiar routines), exploratory creativity (the generation of novel strategies through *ad hoc* exploration of what otherwise might be conceptual spaces) or, more importantly, transformational creativity (the modification of arguments allowing new structures to be generated that would otherwise be denied to the weapon). Navigation and routing may be proven capabilities (and, incidentally, well democratised across friends and foes alike), but these capabilities' more granular requirements such as obstacle avoidance, agility and dexterity remain works in progress. Navigational intent and the ability to exhibit independent actuator control also remain outstanding, the challenge here being that the weapon's sequence of observations is only revealed incrementally (thus complicating repetitive decision-making for these machines).

> *Different layers will learn at different rates, and learning will be unbalanced. This problem worsens as the number of layers increases, a particular challenge for deep neural networks*

Similarly outstanding is a weapon architecture that can reliably capture *qualia* within its processes. For the purposes of this book, qualia are best defined as the subjective or qualitative properties of a lethal engagement. These are the reflective, phenomenal aspects of what might be thought of as the human operator's mental picture and sequences. The absence of either an agreed definition for their functional characterisation or a set of engineering principles for this synthetic phenomenology suggests it will remain an important bottleneck to responsible AWS deployment. In considering qualia, weapon data must be cleaned and processed to extract meaningful features that will help the model learn. The challenge here is unsurprisingly to reduce noise, optimising this preprocessing for speed while compensating appropriately for data sparsity given that whole families of data points (such as rare or edge-case events) will very likely be underrepresented, leading to underperformance or bias. Models trained on limited or biased data may also be prone to overfitting. These models quickly become very unwieldy and computationally intensive, requiring the Cohort to micro-manage resource allocation across these platforms.

Adding smoothing routines also acts as an aggressive edit to the AWS' sensed data, and the method of deliberately dropping left-field neurons is akin to ignoring the non-zero weighting of all collected observations

Feature extraction for the Cohort is about transforming that data into useful representations which the weapon's learning algorithms can process efficiently. Several problems arise. Filtering techniques are fragile, theoretical and largely require hand-holding experimentation that must, by definition, be absent in AWS. Methods such as Gaussian Blur, Median Filtering and Kalman all require preprocessing; raw data generally varies in scale and range and this complicates the model's 'nearest-neighbour' or gradient-based optimisation. After all, different sensors provide data at different time intervals; cameras provide image frames every thirty-milliseconds while output from radar sensors is generally on a fifty-millisecond cycle.

Without that facility to generalise, AWS deployment must be limited to known situations and known environments

Broad synchronisation is therefore required before processing can take place as aligning this temporal data is critical if the weapon is to make sense of its environment in any coherent way. For feature extraction, moreover, raw pixel data from images or video is too complex and too dimensional for current algorithms to process directly. Routines must first extract basic features allowing the weapon just to focus on visual cues that the Cohort has deemed directly relevant to prospective targets. None of this is obvious; techniques such as edge detection, corner detection and oriented gradient detection remain specialised practices more usually found on the workbenches of university laboratories. Finally to this point, analysing just one frame of data is clearly insufficient for the AWS, the more so in dynamic environments such as the battlefield; in tracking moving objects and to do this over time requires that the Cohort include additional recursive algorithms that allow the AWS to estimate its system state against the predicted trajectory, position and intent of all moving items of interest.

AWS image processing then depends upon compilation and labelling of digital images, rendered in the form of legible vector or raster-based files, the latter being a rectangular matrix or grid of square pixels. Once magnified, pixels then appear as a square of sorts to which the Cohort must assign weights and biases which, in the case of the AWS, will likely be based upon colour intensities. These in turn are then allotted a numeric value so that they are suitable for processing through the convolutional systems of the neural networks that will comprise these weapons' computer vision capability.[56]

Battlefield images, now rendered as pixels, can now be sorted and prioritised according to that ascribed intensity,

The challenge here is that the weapon's sequence of observations is only revealed incrementally

to be compared alongside other weighted images being managed by the weapon. The Cohort's first challenge is that statistical associations can be found in all training data even if they are irrelevant to the targeting task at hand, the weapon happily learning from those instead of the traits that the Cohort wishes it to learn; machines reliant upon neural networks tend to learn what they *identify* in the data rather than what the Cohort might intend under, for instance, passing Rules of Engagement, a consequence being that the weapon then over-fits that data to the thresholds that have been set by the Cohort. It is within this routine that biases are most troublesome, an example being the *Inception V3*'s image classifier consistently grouping an image of a turtle as a rifle. Classifiers can similarly be fooled into classifying items that simply do not exist. Machines reliant upon neural networks tend to learn what they identify in the data rather than that which the Cohort might be looking for under Rules of Engagement.

---

[56] Downey, Anthony, 'The Future of Death: Algorithmic Design, Predictive Analysis, and Drone Warfare', in *War and Aesthetics*, ed. Engberg-Pedersen, Anders, James Bjering and others, MIT Press (2024), p. 224.

6.6 AWS control methodologies

How can the Delivery Cohort *control* its independent AWS? Weapon actions, after all, must be intentional and in accordance with the Cohort's broader plans, the contradiction being that such systems are by definition independent. Nevertheless, there do exist a few variants for such intervention. First, the Cohort might seek to exert broad capability control whereby undesirable outcomes may be avoided by limiting what a weapons-directing AI system might accomplish. Physical containment does not equate to locking the weapon system in a box. Instead, it relates to suppression of specific capabilities and preventing the AWS from interacting with the external world other than by permitted and invigilated channels. As this clearly contradicts the fundamentals of machine learning, other means of restraint must be considered. It might instead be theoretically possible to incorporate strongly specific reasoning into the weapon's sequences not to engage in harmful behaviour (incentive method). But this second model fails on a practical level. Human values regularly contain contradictions that may not be mutually reinforcing; many people enjoy eating meat but cannot imagine killing the animals from whence it comes. Implementing such a control structure is intrinsically delicate and prone to error. In trying to create a rules-bound 'specification route', the Cohort would need to determine which rules and which values are appropriate and then express them in code. Borrowing from Bertrand Russell, 'everything is vague to a degree you do not realise until you have tried to make it precise'.[57]

An adjunct method might be to limit the *internal* capacities of the weapon platform (stunting). In this case, however, too much stunting produces a weapon platform that is simply 'another piece of software'.[58] It is also not obvious which information should be rationed, either in the weapon's data-gathering phase or during subsequent engagement sequences. Indeed, expansive situational awareness requires that weapon to have all possible information at its disposal. Having the Cohort reduce either the weapon's sensor or processing bandwidth would seem to weaken compliance with

---

[57] Russell, Bertrand, *The Philosophy of Logical Atomism*, The Collected Papers of Bertrand Russell (Allen & Unwin, 1986), p. 161.
[58] Bostrom, *Superintelligence*, p. 136.

the Law of Armed Combat (LOAC) and therefore is not appropriate for the Cohort's aims. Again, we need to look elsewhere for a more appropriate model of control. Capability control methods instead require routines that automatically detect and react to attempted transgression (generally termed tripwires). This method also has complications.

An overly sensitive tripwire might also interfere with intended operations just at the point where the Cohort requires predictability. It reduces weapon certainty. Too much latitude, on the other hand, might promote poor (and therefore illegal) behaviour. Furthermore, this model assumes that the Cohort's monitoring is sufficiently comprehensive to scan the weapon's cognitive processes for deception or other vulnerabilities. And this will be doubly complicated given that the autonomous weapon is likely to be operating outside communication. Weapon restraint therefore becomes a further and unexpected architectural challenge in AWS deployment, not least because all these

'Everything is vague to a degree you do not realise until you have tried to make it precise'

containment strategies (be they based upon incentives, curbs or tripwires) must still adhere to adopted rules of engagement. A conundrum clearly arises from striking balance between weapon control and weapon functionality. Containment strategies, moreover, may encourage what proves to be a false sense of security in the platform, the more so in times of battlefield stress.

Other control tools exist on paper. An adjunct method might be informational containment, intended to pre-filter what information is allowed to exit from the weapon. Reward mechanisms for the lethal agent might appear an unlikely solution, relying, in this instance, upon a portfolio of 'social' services to reward (as well as mechanisms that penalise) the weapon's AI. A further model, also untested, is based upon machine validation from its Delivery Cohort as incentive for the weapon to act in the interests of that principal. But this would require defining boundaries for the weapon's interaction with colleague assets in order not to contaminate the weapon's otherwise independent operation. In all of these control mechanisms, the law of unintended consequences lurks. An unexpected outcome might be the AWS taking on disproportionate risk in exchange for a small chance of increasing its sphere of influence. After all, it would be 'expensive' to offer the weapons-directing AI any higher-than-expected utility (as a reward for cooperation) than the weapon could hope to achieve on its own by pursuing a nefarious end. Restraint mechanisms therefore become a trial-and-error matter of juggling confidence levels and weightings and, as such, is inappropriate for a model that must work first time and every time.⁵⁹

Two final architectural alternatives are worth mention. Specifying a process for arriving at an appropriate battlefield standard rather than specifying the standard itself might, in theory, be a route to ensuring meaningful control. The AWS might then be motivated to carry out this process, termed 'indirect normativity', and adopt whatever standard the process imputes. The concern remains how to prevent contagion as that weapon's cognitive capabilities theoretically mature with battlefield experience. After all, enhancing system cognition (whether through learning routines, from external updates or from other manipulation) is likely to affect the AWS' motivation in ways that are impossible either to predict or detect. Second, system 'flux' remains a fundamental wetware challenge to AWS deployment. The acronym CACE comes from 'changing anything changes everything'. The phenomenon can make the agent's prior training datasets immediately obsolete, whereby an added set of parameters, change in some pixels, a hardware swap that alters the system's configuration requires abrupt reset of the entire agent. It is also a source of instability and operational randomness.

> *All these containment strategies involving incentives, curbs or tripwires must still adhere to adopted rules of engagement*

> *After all, one AWS will be materially different from a second 'colleague' weapon within moments of first deployment. This creates immediate heterogeneity*

Consequences arising from flux are exacerbated by each experience being unique to that unsupervised weapon. The phenomenon actually has wider ramifications as one AWS will be materially different from a second 'colleague' weapon within moments of first deployment. This creates immediate heterogeneity (both within and among AWS categories), even going so far as to reduce the attraction of the weapon set to the Delivery

---

59 Given possible incapacity in its principal (and this need only be implied), subsequent disagreement about the AWS' performance or any change in its 'agreed' measurement regime might lead the weapon's learning mechanism no longer to trust that principal to deliver its promised rewards. Finally, to this point, it is unlikely that either the outcomes produced by the AI or the end-state of those outcomes will be obvious to the battlefield commander (or, indeed, to anyone in the Delivery Cohort).

Cohort unless those weapons' tasking is so tight that their sovereignty is almost meaningless. Given, then, the lack of a workable architectural basis, this book now considers whether firmware and software can backfill for what otherwise will remain fundamental challenges to these weapons' deployment.

7
Firmware
Embedded process challenges to AWS function

Firmware here refers to the permanent software already loaded when the weapon is first removed from its container and its systems are first fired up.[1] For our purposes, it is considered together with the weapon's middleware, the bridge between the weapon's operating system and its several applications, the subject of the following chapter. Although middleware may be specific to each machine's operation, firmware is not. Together they are used here as a proxy for the building block upon which the software for autonomous weapon systems (AWS) sits. It is the combination of firmware and middleware that provides the weapon's footing, enabling multiple processes to run on one or more platforms and allowing, in theory, interoperability throughout the weapon type. This chapter's purpose is to identify constraints within these layers whereby the firmware must enable processes regardless of tasking. The chapter is divided into four sections that together identify sources of fragility and consequences upon behaviour arising from the weapon's machine-learning (ML) spine. Its final section looks at how the AWS might reason, understand and direct its attentions. In this vein, the reader should bear in mind that modern build practices increasingly rely on remote and generic code libraries which, presumably, further complicate this aspect of AWS design.

---

1 Sieracki, Jeff, 'Machine Learning for Embedded Software Is Not as Hard as You May Think', *Reality AI* (3 August 2016), paras. 6-7 of 12. Firmware can usefully be understood as the permanent software that is programmed into the weapon's read-only memory.

7.1 Sources of technical debt

How, then, might firmware add to system fragility? The concept of technical debt was first put forward in 1992 to quantify costs arising between speed of execution and quality of engineering.[2] Technical debt is therefore a useful metaphor linking the consequences of poor software design to accumulating a 'financial debt'.[3] This chapter asserts that such 'debt' is particularly relevant to AWS deployment; just as a loan must eventually be repaid, with compounding interest, so hasty design decisions in the removal of human supervision must be paid for by re-factoring, debugging, fragility and poor reliability. The premise here is that ML has all of the same coding issues found in traditional programming but, notes D. Sculley and others, also 'a larger system-level complexity that can create hidden debt'.[4]

Causes of technical debt generally include (a combination of) unsuitable AWS architecture, shortcuts that come about from commercial pressures, poor testing protocols and inadequate whole-system understanding within the Delivery Cohort. Technical debt

Technical debt is a useful metaphor linking the consequences of poor software design to accumulating a 'financial debt'

also arises from a general lack of ownership of complex programmes, poor technical leadership and pervasive specification changes. Debt is created from 'counterparty development' whereby isolated software routines, once developed, must eventually be merged into a single source base. An adjunct consideration then becomes the need to introduce parallelism both within and between the weapon's application, muddying the tidy separation of software into independent work units.

Simple matters of scale also add to technical debt by increasing both interactions and interdependencies among packages, their several developers and, ultimately, the Delivery Cohort. Technical debt occurs as projects evolve; the practice, for instance, of 'refactoring' code occurs when particular routines that have become unwieldy must then be reworked. Other sources of debt arise from managing the weapon's original configuration in the face of

---

2 See, generally, https://www.agilealliance.org/introduction-to-the-technical-debt-concept/.
3 For a useful primer of technical debt, see: Kruchtren, Philippe and others, 'Technical Debt: From Metaphor to Theory and Practice', *IEEE Software* (2012), pp. 18-19, https://www.computer.org/csdl/mags/so/2012/06/mso2012060018.pdf.
4 Sculley, D and others, 'Hidden Technical Debt in Machine Learning Systems', *Advances in Neural Information Systems* (2015), p. 1, http://papers.nips.cc/paper/5656-hidden-technical-debt-in-machine-learning-systems.pdf.

subsequent (and perhaps wholesale) adjustment, navigating its integration, resolving conflicts, determining each system's 'logical completeness' and doing so while all the time maintaining each platform's fitness for purpose.

We have already seen that AWS learning models rely upon efficient abstraction, the reduction and paring back of AWS processes to their essential characteristics. Models are also defined by the availability (or not) of clearly defined boundaries to objects of interest. Abstraction routines remain convoluted, inefficiently bespoke and experimental and thus likely to introduce ambiguity into processes that are unsupervised. In higher-level models, moreover, abstraction bears little outward relation to how the human brain works and will likely be quite illogical to the Delivery Cohort. A weapon's low-level routines may then lack sufficient abstraction rules to navigate the associated challenges of memory management, 'parameter tuning' and 'code reuse', technical debt being increased by weapon models' reliance upon external third-party sources of code, data and expertise (the remote code libraries mentioned above). Finally to this point, tight coupling (between weapon algorithm and remotely retrieved assets) demonstrates the degree to which small changes in the AWS' external data can alter algorithm behaviour, the more so where data (its acquisition, processing and tuning) is managed by managed by quite different systems that are distributed across the weapon.

A further issue with generally hierarchical structures (as characterise firmware) is that they are often based either on long-term priors or on loose dependencies. These may be quite tangential and are therefore hard to code. The consequence here may be that the weapon makes correlations that are flat (here, non-hierarchical) and represented as simple unstructured lists. The Cohort's problem arises where every correlated feature on that list will likely be allocated an inappropriately equal footing which then requires complex proxies (such as weightings based only upon the sequence or positioning of relevant data strings). The Cohort must either build in an almost unlimited number

Just as a loan must eventually be repaid, with compounding interest, so hasty design decisions in the removal of human supervision must be paid for by re-factoring, debugging, fragility and poor reliability

of feature detectors into the weapon's grid or attempt instead to increase the size of the weapon's training sets in a similarly exponential manner. A conundrum for firmware is that the weapon is then caught by what is termed 'local minima' whereby its system gets stuck on suboptimal solutions with no better solution appearing nearby in the space that the weapon's firmware is searching.

Firmware is also susceptible to *entanglement* whereby it quickly becomes impossible for the Cohort to isolate individual components within a self-learning system as none of the inputs are properly independent. While this is a further example of that CACE (change anything changes everything) principle[5], the trait will likely be more insidious in lockstep with task complexity, especially concerning the weapon's prediction sequences. Here, it is the weapon's firmware which must ensure that output from its learning routines is made accessible to its other internal subsystems, either during runtime or by its writing to logs that may later be accessed by those subsystems. This is a complex matter with hidden consequences. The likelihood is that those subsystems immediately become 'undeclared consumers', grabbing the output of a particular prediction as input to another component of that overall sequence. Given its modular makeup and limited computational resource, AWS firmware will likely recycle and repurpose exactly these input signals coming in from its sensor banks. As noted by the United Nations Institute for Disarmament Research, unintended feedback loops then form between weapon algorithms and the weapon's external world.[6] Such loops may be analogous to filter bubbles in social networks and web searches whereby noise suppression mechanisms inadvertently suppress nonconforming data.

> *An issue with generally hierarchical structures is that they are often based either on long-term priors or on loose dependencies*

> *'Undeclared consumers': The grabbing of output of a particular prediction as input to another component of that overall sequence*

This feature contributes to AWS firmware being treated as a black box, resulting in considerable 'glue code' or, worse still, calibration layers that can lock in assumptions. Not surprisingly, this long list of potential woes is picked up later in this chapter.

This phenomenon of the 'undeclared consumer' may make it problematic to make any changes to the weapon's firmware. A battlefield example, if rather convoluted, is useful whereby a weapon subsystem predicts an incursion

---

5 See 6.6 (AWS Control Methodologies).
6 See, generally; UNIDIR Primer Series, October 2007, https://unidir.org/files/publication/pdfs/algorithmic-bias-and-the-weaponization-of-increasingly-autonomous-technologies-en-720.pdf.

triggering a subroutine task to determine entity-size and other data on possible transgressors; if the submodule then starts acting upon (in this case, 'consuming') area-incursion as an input signal and entity-size already has an effect on the weapon's incentive to enter the area, then the inclusion of area-incursion in entity-area adds an unwelcome (and likely erroneous) hidden feedback loop. It is easy to imagine a scenario where the AWS uncontrollably recognises increasingly small entity-size transgressors (from an armoured personnel carrier to a rabbit to a flea) in its output generation.

Dependencies have long constituted a fault line in firmware. The 'self-consuming' quality (for want of a better expression) of these relationships may conceivably be spotted by static analysis, but it is wholly more complicated to identify data dependency in the AWS, and even more difficult subsequently to untangle it. Why is this an issue? Input signals are themselves unstable, meaning that they qualitatively change behaviour over time.[7] Rolling out *ad hoc* 'improvements' to that particular input signal will, given the wetware constraints already discussed, likely have quite arbitrary effects on weapon output. The Cohort's mitigation strategy might then be to create versioned copies of given signals within the weapon's logs, but this would create inappropriate complexity (through multiple and quickly redundant data versions) as well as accelerated staleness within AWS data feeds. A similar drag comes from the weapon's underutilised data dependencies, mostly unneeded routines within the weapon's program suites that provide little additional accuracy. Examples might include legacy features, bundled features (which were not then configured at initial commissioning) or additional (and probably undocumented) one-off refinement tweaks. Each such feature adds to system brittleness precisely because the weapon's firmware will assign it some weighting notwithstanding the feature's redundancy. The Catch-22 here is that subsequent removal of these routines (for instance, a weapon's dead experimental code-paths) often creates further system error; even if there are no references to that routine in the current version of the weapon's code, there may still be instances where older binaries reference its output, leading the system to slow or crash.

> *'Undeclared consumers':* The grabbing of output of a particular prediction as input to another component of that overall sequence

---

7 This can happen implicitly (a signal input from a separate model updating over time or from a data-dependent lookup table) or explicitly (the input signal is imported or otherwise separate from the weapon system).

The phenomenon of 'correction cascade' also creates problems for firmware, these cascades typically occurring when ML models are not learning as predicted by the Cohort, and so requiring an external fix to that weapon's processes. Commentators regularly note that 'as hot fixes pile up you end up with a thick layer of heuristics on the ML model'.[8] In practical terms given its short decision cycles, weapon learning may be also reduced as nothing new is then inferred from subsequent observations. How might this occur? Seemingly minute variations in a training dataset might justify the Cohort to re-run earlier training but this time with a small learning correction, the incentive being speed and compression, especially as the correction likely appears immaterial to machine processes. The issue, however, is that corrections will create their own dependencies within the weapon's initial dataset, making it challenging for the Cohort to attribute ongoing improvement to one patch or another. This may perhaps be an arcane point, but it will be further aggravated as a correction is applied to 'closely related' (rather than precisely defined) learning routines, an example being the recalibrating of outputs to slightly different test distributions.

Each such feature adds to AWS brittleness precisely because the weapon's firmware will assign it some weighting notwithstanding the feature's redundancy

A further challenge is that an independent weapon must factor for both the total error but also the distribution of this error. As we have seen above, cascading can then produce deadlock whereby the local optimum for that learning system becomes circular and iterative, with the result that those (possibly critical) routines cannot then be improved. The whole system then decelerates, a problem indeed if it is autonomous and unable to accept intervention. While humans can learn relationships from a very small number of trials, the original work undertaken by DeepMinds and Atari involved billions of such examples. The trite point for AWS is that such training does not equate to understanding. DeepMinds' agent was learning specific contingencies

A conundrum is that the weapon is caught by 'local minima' whereby its system gets stuck on suboptimal solutions with no better solution appearing nearby in the space where the firmware is searching

---

8 Zavershynskyi, Maksym, 'Technical Debt in Machine Learning', *Towards Data Science* (1 July 2017), https://towardsdatascience.com/technical-debt-in-machine-learning-8b0fae938657.

for particular scenarios. Similarly, transfer tests (the generalising of conditions that are different from those encountered during training) demonstrate that ML outputs are often superficial and quickly forgotten. This is perhaps unsurprising given the difficulty of evolving machine learning from interpolation (effecting generalisation between known examples) to extrapolation (the requirement to advance beyond known training examples, one of the earlier litmus tests posed by this book for AWS).

AWS firmware encourages other sources of technical debt. A general (and enduring) difficulty has been the Day One pollination of deep learning models with the prior knowledge required to configure the system upon initial deployment. Remember, these models can neither distinguish causation from correlation nor deal with open-ended inference (the difference, say, between 'John promised Mary to leave' and 'John promised to leave Mary'). AWS firmware presumes a large, stable world, but the Delivery Cohort needs to be confident that these weapons will continue to work with novel data that, given battlefield chaos, are very unlikely to resemble previous data.

The system will likely decelerate, a problem indeed if it is autonomous and unable to accept intervention

The challenge is akin to the development of aeroplane engines where design generally relies upon building complex systems out of simple subsystems; although it may be possible to create secondary guarantees around those underlying subsystems, the passing either up or down of those same performance guarantees does not quite work.

This is exacerbated, of course, by the procurement of these complex systems from disparate commercial parties and where the bundling together of several usually proprietary routines must now be held together by 'glue' (or 'spaghetti') code. Glue code relates to the quantity of supporting code that must be written to permit interoperability and data transfer in and out of these underlying software modules. Invariable system fragility is a consequence, not least because glue code anchors a system to the peculiarities of those proprietary packages being glued. Glue code has other ramifications for AWS control and operation. Fundamentally, the feature will entrench the weapon's original construction in 'supporting code' instead of embedding it directly into components that are intentionally volatile and which have been designed for specific weapon routines. Such setup discourages experimentation, the magnitude of the problem illustrated by Sculley's research that mature systems empirically end up being five per cent executable code and ninety-five per cent glue code.

An adjunct challenge to an already adjunct challenge then arises from the occurrence in weapon firmware of 'pipeline jungles', a noted characteristic of moving learning systems from prototype to production manufacture. Pipeline complexity is anyway likely to develop on its own as weapon signals are adjusted and endless corrections and loops are added. Glue code and pipeline jungles also arise from overly separated research and engineering roles, a frequent trait of complex government procurement programmes. In previous eras, firmware was probably designed (and then amortised) by a single main contractor and deployed then across multiple different platforms, its underlying state being a tangle of scrapes, joins and sampling steps. Given modern software procurement, this bifurcation has all but disappeared, although those same amendments and connecting code will nonetheless appear as conditional branches to the main system source code. Obsolete and experimental coding can still interact with each other in unpredictable ways[9] with programmers' use of 'dead flags' being a further source of system fragility. Dead code that may have been lain moribund for an extended period can suddenly be awakened by unwitting change to that flag's value.

> *Dead code that may have been lain moribund for an extended period can suddenly be awakened by unwitting change to that flag's value*

It will also be through each machine's firmware that the Cohort will set its weapon's many configurable options including, *inter alia*, which features are available to each new owner, the sensitivity and degree of conservatism under which that weapon's operates, its action triggers, its thresholds and its learning settings. Given the stark chasm between unpacking the AWS in the Cohort's warehouse and that weapon then making kill decisions on a battlefield, the whole process of system configuration is a further source of technical debt. After all, all weapon capabilities require initial alignment

> *All of the weapon capabilities require alignment and confirmation, the quantity of 'configuration code' likely exceeding the number of executable instruction lines for the weapon's learning processes*

---

9 Sculley highlights the loss by trading shop Knight Capital of four hundred and sixty-five million dollars in forty-five minutes that was later attributed to unexpected behaviour from obsolete experimental code paths.

and confirmation, the quantity of 'configuration code' certainly exceeding the number of executable instruction lines for the weapon's learning processes.[10]

Each such line has the potential for mistake given that configurations are by their nature ephemeral and less well tested. Unpredictability lurks in several guises. First, weapon systems that are autonomous require detailed decision thresholds and boundaries, all of which must be mediated by humans in a way that they precisely mirror the Cohort's aims. This must be a human-centric (and manual) process involving judgements that are often subjective and always complex. A further challenge is that when a weapon updates on new data, old manually-set thresholds become invalid. Or they may not, requiring another decision notwithstanding that this updating must take place across different models (each, moreover, in a different learning state).

What happens when battlefield associations that have been assumed by the Cohort from the outset no longer tally with what is happening on the ground? Monitoring the weapon's associations is deeply subjective and contextual. It is also a routine that must occur in real-time across the whole weapon system. The challenge, moreover, is what metrics the Cohort should monitor, the more so given that the purpose of its agent's ML is to adapt overtime. What constitutes a red line for the Cohort? Finally, to this point, legal, social and political constraints will agitate that AWS limits are set conservatively which, should those action limits unexpectedly trigger, might compromise that weapon's operational usefulness to the deploying chain of command.

7.2 Firmware ramifications of learning methodologies

AWS' dependence upon ML has other consequences and the aim of this section is to provide a high-level primer on current learning models in order to assess relative frailties. The context for this section is provided by the 2017 JASON study on ML (as it relates to the US Department of Defense). Set out in the previous section (and notwithstanding the report's relative age), the findings are worth repeating

---

10 Ivezic, Marin, 'The Tale of a "Rogue" AI Military Drone and the Lessons in Responsible AI', 4 June 2023, *Defense AI*, https://defence.ai/industry-news/rogue-ai-military-drone/. AI's 'alignment problem' relates to the risk that the intelligent system might perform actions that the human did not intend or desire.

that 'the manifolds whose shape and extent [deep neural networks] are attempting to approximate are almost unknowably intricate, leading to failure modes for which—currently—there is very little human intuition, and even less established engineering practice'.[11] In this vein, weapon learning really divides into three types. Supervised learning, as detailed later in this section, attempts to predict an output given an input; an example might be the processing of multiple sensed data points to deliver target selection and engagement. Supervised models may either be regressive (seeking whole numbers and clear outcomes) or classifying (seeking class labels, trends and patterns). The model is best suited to 'one-shot' classification tasks whereby the weapon receives observations from its sensor bank and must then make what will be a single decision. In this case, the AWS is undertaking just one evaluation rather than a sequence of evaluations, working through examples and adjusting weights inside its network to improve accuracy in this task. The merit of this approach is that there is no need for a human first to draw up a list of rules (or, indeed, for the Delivery Cohort to implement them through interventions). The weapon system is learning directly from the labelled data.

Reinforcement learning, also discussed below, instead encourages the weapon to select actions that maximise a payoff. Finally, the weapon may learn through in an unsupervised manner whereby the machine will be focused to discover general internal representations from its sensor input; the weapon network is trained by being exposing to a huge number of examples but without control mechanisms telling its sensor bank what to search. Here, the idea is that the weapon's network will learn to recognise battlefield features and to cluster similar samples, thus revealing hidden groups, links and patterns within the dataset. Unsupervised learning thus offers a methodology which is capable of anomaly search in circumstances where the weapon system does not know what it is seeing.

An assumption, however, for this section is that there can be no provisional, probationary or other pilot steps in delivering lethal violence. Once in the field, the weapon must work first time and every time. Indeed, the term 'exploration', which occasionally finds itself attached to reinforcement processes, points to another challenge regarding learning models: just as an independent weapon cannot be sure it has found its best action for each state until it has tried all possible actions in all possible states, so too must the

---

11 JASON, 'Perspectives on Research in Artificial Intelligence and Artificial General Intelligence Relevant to DoD', MITRE Corporation (January 2017), https://irp.fas.org/agency/dod/jason/ai-dod.pdf, p. 2.

Cohort provide the AWS with an appropriately definable end-state if it is to avoid that weapon being too iterative. Any error in the weapon's processing of its current state will, after all, carry forward into that machine's future learning and future battlefield actions. Similarly, should either the AWS' environment or combat task change while the weapon is in a learning phase, much of what it has learnt to date may

> Human soldiers do not perceive their environment in terms of a collection of labelled objects

be invalid. This highlights the challenge of *un-learning* in the weapon's ML routines should error occur. The trade-off between 'constantly learning' versus employing what is already known to work (at the cost of missing out on further improvement) is a well-discussed paradox (often called the 'exploration/exploitation dilemma').

The exploration/exploitation dilemma again highlights the difficult interdependence between weapon architecture and the software routines that sit on that architecture. The Cohort must optimise between its AWS control policy (the reactive rules that control a weapon for a particular goal) and its use of an appropriate value function (the comparative value of being in each state relative to the weapon's overarching goal). After all, it will be the AWS firmware that must evaluate how well each such state-action pairing has fared and then carry forward the outcome of each trial into subsequent actions. This control/value tension is then exacerbated in a multi-robot environment where identification of individual outcomes is intractably difficult, requiring, as it will, ever more tricky attribution of cause and effect.

The Cohort's allotting credit or blame to weapon actions is termed 'temporal credit assignment' and should also inform AWS conduct. But as behaviours change over time (given, of course, each weapon's learning pattern), discrepancies must occur between actual system performance and operator expectations, leading the Cohort to engineer ever more aggressive adjustments and eventual deadlock in the weapon. But any such volatility is likely to provoke surprise in colleagues and, more important, subsequent loss of trust during operations. Nor do human soldiers perceive their environment in terms of a collection of labelled objects in the same way that forms the basis of AWS operation. Accordingly, it is not obvious why the Cohort should trust the symbolic processing that underpins ML given that battlefield decisions should instead be based on evidence that is numeric, contextual and heuristic. A frustration for this section remains that the fundamental bases for AWS learning models continue to evolve very quickly and are certainly not yet set in stone. Promising workarounds will surface, and it is for this reason that analysis here must be largely behavioural.

Certain pitfalls are common across ML models. Their real-time running will likely become ever less predictable as the weapon's decision and search trees expand (either through the Cohort requiring additional functionality in the weapon or as the weapon 'learns' from being deployed). These models' 'divide and conquer' methods also mean that they perform poorly when faced with multiple and complex interactions, tending instead towards oversensitivity to training sets and to farfetched solutions that are akin to LLM hallucinations. Decision trees, moreover, do not naturally combine. To open a new decision tree requires the weapon remove an existing edge (or subset of an edge) in an existing tree. The question then arises as to which edge requires elimination. Edge removal is empirically *ad hoc* and arbitrarily depends upon new circumstances that each weapon encounters. The Delivery Cohort must therefore factor for a time lag between AWS' sensing of external data and its computation of an executable action, so requiring additional (and complex) 'stopping' routines in order to sense-check weapon routines.

Hybrid learning can ameliorate this fault line. Here, 'transfer learning' builds upon previously acquired knowledge rather than the weapon having to be trained from scratch each time. But because 'multitask learning' uses the experience of one layer to improve efficiencies at another layer, two issues arise. First, although humans may have numerous representations 'active' at any one time (some from our eyes and ears, others evoked from memory or initiated by other retention models), a weapon without supervision is consigned to process only those associations that its initial configuration has assigned relevance and to do this only from its currently available datasets. AWS firmware must also limit the number of these active signals such that only applicable associations arise. After all, it is enduringly difficult to separate what is relevant to the task in hand from that which has been taught. As part of this balancing, the AWS must also audit its learned steps to ensure they conform to its internal representations, the challenges here being how often to undertake this appraisal and how best to solve for the Cohort's priorities. This is not obvious as AWS learning will be based on a 'closed world assumption' whereby only one decision out of several predefined possible decisions must be made, and this based only upon received inputs. In real life (and certainly on the battlefield), however, humans encounter an 'open world' where several scenarios may present that have never before been encountered and where the trained model is very unlikely to recognise patterns. This is certainly not a case of one-off configuration.

It is enduringly difficult to separate what is relevant to the task in hand from that which has been taught

7.3 Reasoning and cognition methodologies

All of this chapter has assumed the deployment of broad capability, wide-task AWS where, in all such cases, weapon learning must be a bridge to machine reasoning. For our purposes, a practical definition of reasoning remains the algebraic manipulation of previously acquired knowledge to answer a new question. But for AWS to reason, the machine must also be capable of sophisticated 'value mapping', requiring a cognition capability (in this case, termed 'affect') without which the machine cannot evaluate, judge or decide upon lethal actions. Cognition is thus a bridge to several processes fundamental to AWS deployment, including knowledge management, attention and prioritising, memory and memory-making, judgement and evaluation, reasoning and computation, and problem solving and comprehension. This is a disruptively extensive list of required skills.

The first difficulty is that cognition, even in its human condition, is challenging to define, let alone capture in code. Human cognition can be conscious or unconscious. It can be concrete or abstract as well as intuitive ('knowledge' of a language) or conceptual ('model' of a language). Moreover, cognitive processes must use existing knowledge to generate new knowledge. AWS cognition therefore becomes a general catchall for that broad but poorly demarcated collection of processes that underpins the weapon's perceiving, thinking and awareness. As we have seen, however, demarcation and abstraction difficulties are particularly unhelpful to code-based approaches, the more so as heuristics within the weapon's executive processes already weakens these boundaries. Given that the weapon is presumably using multiple sensors, inadequate differentiators will exacerbate this blurring and so increasing system noise.

Noise also impairs the weapon from creating contradictory interpretations of its immediately current situation. Datasets, after all, must be incrementally built as not all data are available to the weapon at once. System noise is not helped by schema-based 'resource matchmakers' that the Cohort must include to manage the weapon's workflows, not least because they remain poor at maximising trade-offs in machines' decision spaces. Given that data will be arriving progressively, the underlying distribution of these data must also evolve with time, counter to the whole notion of 'identically distributed data' upon which classic data-mining algorithms rely and which proves wholly unsuited to battlefield practices. None of this is helped by cognition being an inexact science. Deduction and reasoning are based, after all, upon information that is not necessarily directly perceived. This is a likely misstep for AWS, and one which requires further judgement processes be put in place to develop requisite situational awareness, to detect battlefield contradictions and to evaluate significance. It is the human politician and field commander who must define those boundaries, but it remains the human soldier, naval rating and pilot who must still do their job within an ecosystem that now includes independent AWS, evidencing what is a recurrent theme of this book that human experience trumps machine function.

The difficulty is that cognition, even in its human condition, is challenging to define, let alone capture in code

A further firmware complexity is termed the 'principal-agent control problem'. The property arises when a human entity (here, the principal and, in the case of AWS, its Delivery Cohort) appoints another to act in the former's interest. Those leading the Cohort should worry that scientists and programmers implementing their project may not be acting in the best interests of the Cohort. The issue runs deeper. The complexity of agency can generally be reflected in the metaphor of a 'mangle', the argument that agency must always be 'temporally evident' in ML processes rather than being blended with either the subjects or objects involved in that process. Merely observing AWS behaviour in these weapons' development phase will clearly be insufficient given the possibility at any time of the weapon turning rogue in what is termed 'treacherous turn syndrome'. The weapon's whole system, moreover, will likely comprise intelligent subparts that are themselves capable of agency; each weapon component must therefore be viewed as an autonomous agent in its own right. This is an important trait and one that raises additional complications. Termed 'composite agency', this complicates

Datasets, after all, must be incrementally built as not all data are available to the weapon at once. the more so given data-priority, allocation and denial challenges that characterise ML routines

weapon motivation. The issue is unexpectedly fundamental to the Cohort as the motivations of composite systems depend upon those sub-agents but also on how those sub-agents are organised.

7.4 Attention methodologies in AWS

The Cohort must also manage the framing of each platform's autonomy. How, for instance, can the system be structured in a manner that AWS' focus can appropriately be directed? The context here is that the human brain appears to be free to choose what it looks at, listens to and thinks about. In benign conditions, humans can focus their attention as they please. This drift of information, however, if not limited in any way, would otherwise lead to memory overflow, interference and what is often termed 'contradictory neural cacophony'. The issue will presumably present itself almost identically in AWS. Similar to the human brain, the AWS platform must actively select the source and quantity of its information, what to process, what to store and which peripheral information then to attenuate in its decision processes. For humans, this process is termed *attention* which, *inter alia*, controls sensory information acquisition and subsequent processing. As noted by William James as far back as 1890 in *The Principles of Psychology*, 'attention implies withdrawal from some things in order to deal effectively with others'.[12] It also, he continues, 'has a real opposite in what is the confused, dazed, scatterbrained state which in French is called *distraction*, and *Zerstreutheit* in German'. This is challenging (and possibly intractably so) to mimic in a machine. After all, most real-world tasks (especially those in a battlefield setting) come with their own set of time constraints, exacerbating the importance to the Delivery Cohort of managing the attention of its machines.

In the case of AWS control, such attention must be divided into that which is sensory (information acquisition from its sensors) and that which relates to inner attention (the selection of relevant inner representations across weapon processes). For instance, AWS attention should not exclude non-attended information completely, the so-called 'cocktail party effect' whereby it is possible to shift attention from conversation to conversation

---

12 James, William, *The Principles of Psychology* (Henry Holt, 1890), pp. 403-404.

in a universe of broad noise. In this manner, the AWS must be able to extract attended information from what is otherwise background noise picked up by the weapon's multiple unattended sensory streams. This core capability must be agnostic both to the weapon's tasking and its degree of independence. It must also determine the weapon's overall resource allocation. Other complications arise. Attentional selection in AWS must be further divided into voluntary and involuntary attention. Involuntary attention takes place when the weapon's attention is captured by strong, novel or significant stimuli. Given the frequency of these happenings on the battlefield, this might otherwise create an overarching (but quite wrong) set of priorities in the weapon's responses unless appropriately calibrated.

How then does the Cohort engineer a routine to mediate stimuli such that one such sensor input is properly preferred over another? The question is therefore how the weapon's input intensity should be managed. It would be inappropriate, say, for AWS input strengths simply to be equalised such that a level-playing field exists for all information coming into the weapon's orbit. After all, battlefield information is dynamic where each happening's relative importance fluctuates as events unfold. As previously noted, Andrew Owen Martin, secretary of The Society for the Study of Artificial Intelligence and Simulation of Behaviour, observed of these attention routines: 'I don't mean that it's too difficult like "man will never fly" or "man will never land on the moon". I'm saying it's hopelessly misguided like "man will never dig a tunnel to the moon".'[13] Absent human supervision, weapon attention clearly requires delicate ranking that is subject to regular calibration. After all, two variables that may be useless by themselves can be useful together. Similarly, a single variable that is useless by itself can then be useful with others.

> *'Attention implies withdrawal from some things in order to deal effectively with others'*

The issue of how independent machines handle incoming data is thus fundamental, and one that has generated a number of recognised puzzles in the academic community ('variable complementarity', the 'curse of dimensionality', variable 'redundancy' and variable 'commensurateness'). The issue for our Cohort, however, remains broadly the same: how can weapon's attention mechanisms predictably decide which variables (or

---

13 Owen Martin, Andrew, senior technical analyst at the Tungsten Network, cited in Ben Sullivan, 'Elite Scientists Have Told the Pentagon that AI Won't Threaten Humanity', *Motherboard Magazine* (19 January 2017), https://www.vice.com/en/article/elite-scientists-have-told-the-pentagon-that-ai-wont-threaten-humanity/.

combination of variables) to focus upon during and after data training, the more so given the perils of attention heuristics that abound in weapon's decision processes (including, *inter alia*, recency, availability, representative, conjunction, anchoring and adjustment heuristics).

Two further examples act as a useful conclusion by highlighting the deep intractability of coding for attention. First, the precept of habituation may lead to a programming conflict whereby the weapon system's response is coded to decrease as a stimulus is repeated. For AWS operating in a battlefield environment, this is clearly inappropriate as not all repeating stimuli should occasion habituation, the more so given that war is nine tenths inactivity and one tenth chaotic commotion. The opposing attribute of sensitisation should presumably then increase the weapon's response to a repeated stimulus.[14] The conundrum captures the uncertainty within the coding of these key traits, requiring that the weapon mimic links between stimulus, representation and attention reactions, requiring (of course) the Cohort to choreograph added feedback loops to ensure its weapon functions as expected.

Human cognition can be conscious or unconscious. It can be concrete or abstract as well as intuitive ('knowledge' of a language) or conceptual ('model' of a language)

The issue highlights a final firmware impediment. Attention models must depend upon clearly defined routines separating 'memorisation' ('Mary had a little lamb...') and 'memory-making', the act of imprinting episodic recalls into memory. The distinction is fundamental whereby semantic learning involves the unsupervised weapon absorbing a definable fact while procedural learning involves the unit learning skill routines, mental or motor sequences. The challenge is that all of this requires AWS firmware incorporate sufficient adaptability, appropriate generalisation as well as an ability to parse information without invigilation.[15] A modular

Two variables that may be useless by themselves can be useful together. Similarly, a single variable that is useless by itself can then be useful with others

---

14 Humans learn not to pay attention to the noise of traffic or other repeating background sound (habituation) but may instead become irritated by a repeating and intense noise (sensitisation).
15 Each such dataset, moreover, will have its own provenance that will require weighting. This might include information from its own immediate sensor streams versus an update from a central source, peripheral information from colleague robots or changes to its utility function.

system, moreover, must still deliver the promise that overall system performance can be maintained (even bettered) by changing out individual modules. This must also be possible without the user having to repeat full testing of the whole system.

8
Software
Coding challenges to AWS function

Having considered these weapons' infrastructure challenges in previous chapters, this chapter now reviews certain generic software routines that will sit on top of these architectures. The section unpicks the capturing and processing of the weapon's immediate environment. This is chiefly a translation challenge between the weapon's sensed data and that machine's *a priori* setup received when the Delivery Cohort first deploys the weapon. The chapter's aim is to determine how real-time information might be incorporated into the decision and action processes of an independent weapon. Although decision-making must of course be rooted in the current, it is also necessary for that weapon to toggle reliably between old and new information and, within a framework of confidence checks, to ignore and then remove stale data.

The variety of (and variation in) a weapon's sensed data is itself a challenge and requires appropriate counterweighting, moment by moment, through the compensatory application of confidence levels. The practical intricacy of this is regularly underestimated given that machines must factor simultaneously for a portfolio of factors that include data recency, intensity, relevance, completeness, consistency and trustworthiness. While humans calculate this inherently, the function requires layered management for machines, the more so given that such factoring must entirely capture the Cohort's own goals and values. Notwithstanding the nuanced nature of these dependencies, all of these system obligations can only be addressed within the weapon's initial code-based set of instructions. After all, the weapon must maintain a comprehensive (and now independent) understanding of its immediate environment. It is this function that comprises the 'representation' of the autonomous weapon system (AWS), providing the machine with what is generally termed its 'internal model of its world'.

The problem for the Cohort is that these representations are sinuous. They are difficult to construct and difficult to manage. ML routines require that sensed data with least variance empirically be given outsized weighting each time the weapon polls its sensors. But the very confidence weightings that must capture the Cohort's aims may also have the unintended effect of smoothing that data until it becomes inappropriately scaled to a mean. The paradox here is that by allocating high confidence levels to particular strings, the weapon is signposting that the data is material to its forthcoming decision, converting it into a vetted set of very aligned actions. Data smoothing can be mitigated through 'normalisation routines' designed to iron out input variances. Here, normalisation would be the weapon's means of reorganising data so that it meets two requirements. First, data diffusion is managed (all data is stored in one place) and, second, the machine can start to rationalise data dependencies (all related data is either linked or stored together). Finally, normalisation routines should ensure that the weapon's sensed data is appropriately comparable (both temporally and structurally).

> *The paradox here is that by allocating high confidence to particular strings, the weapon is signposting that the data is material to its forthcoming decision, converting it into a vetted set of very aligned actions*

For this to occur, however, requires that those same confidence thresholds change dynamically. A difficulty is that interventions must operate whether AWS design is top-down (whereby weapon policies are executed without variation) or bottom-up (where behaviours can flex given deliberative planning). It soon becomes apparent that such flex is really quite limited across weapon routines. This is an important observation as these same narrow tolerances directly contribute to system brittleness and 'technical debt' discussed in the previous chapter.

> *Confidence weightings may have the unintended effect of inappropriate smoothing the data until it becomes inappropriately scaled to a mean*

Rather than being simply a matter of memory management, the issue is what and how that primary data sits within the AWS' representation.

Although somewhat convoluted, an example around AWS land-based navigation is useful to demonstrate this intricacy. In defining movement practices, one option for the Cohort might be to program its unsupervised weapon to 'understand' an exact odometric path to, say, a GPS-defined target. Alternatively, the platform may be coded to follow a pre-choreographed sequence of paths determined by its environment and by specific landmarks. Even if the weapon's topological mapping is satisfactorily dynamic, AWS movement is still susceptible to obstacles, going, feint and the accuracy of its own start-state representation. After all, the Cohort's goal must be that its AWS should move freely around its environment, with the weapon's representation being searched in real time to establish appropriate and available paths before computing a best path and then reframing the weapon's goals and logistics in light of that newly identified path.

The challenge is that the weapon must also rely upon these refreshing routines to refine its own probability distributions, computing new probabilities (or confirming its existing state) for its immediate situation that is consistent with the new information being polled from its sensors. Any inconsistent probabilities are presumably set back to zero and then 'renormalised' over all remaining possible outcomes. Termed 'conditionalisation', the routine must calculate conditional probabilities for each set of possible causes and for each of its given observed outcomes, so allowing the weapon to construct a complex *It soon becomes apparent that such flex is really quite limited across weapon routines* composite of the received probability of each such cause and the conditional probability of the outcome of those causes. This is the measure of complexity involved in what is just one of the Cohort's (relatively objective) deployment considerations.

But this is also a complex and inherently unstable set of arrangements. As the weapon makes each additional observation, its set of probabilities arises from an ever shrinking set of possible worlds that happen to remain consistent with the evidence being provided from its sensors. This condition appears intractable and cannot be designed away; after all, the weapon is generating a series of posterior probability distributions that it is using as its new prior in every next-time step. Finally, to this point, all these steps must be mediated by feedback loops notwithstanding that such pointer mechanisms are difficult to manage and, more importantly, can themselves create hidden loops whereby gradual changes may not be apparent (or reflected in) the weapon's passing configuration. These processes generally complicate system adjustment and obstruct even very simple correction to unsupervised machines.

Representations must also factor for data detailing the weapon's current 'self'. This comprises proprioception informed by the goals, values and boundaries determined by the Cohort during each unit's initial configuration.

It must also reflect the weapons' immediate battlefield environment (its navigable spaces, constraints and permissions) as well, of course, as its objectives, priorities and alternatives. Furthermore, each of these decision components will, in time, require its own routines. The issue for the Cohort is that each of these kernels are highly elaborate,

Subroutines may be quickly constructed, briefly used and, crucially, then discarded by appropriate 'forgetting routines'

usually correlated and likely conflicting. Their mediation is rarely intrinsic to the weapon's core operation and their precedence must somehow reflect this; certain of the subroutines may be quickly constructed, briefly used and, crucially, then discarded by appropriate 'forgetting routines'. The Cohort's conundrum is that different degrees of interference will lead to weapons with subtly different composition, subtly different learning traits and, over time, ever less similar profiles. This issue then becomes how the deploying commander can really understand either the traits or effects of each asset being deployed.

8.1 Coding methodologies

Another likely fault line relates to how the AWS will be programmed and a purpose of this section is therefore to review the efficacy of code as the means of replacing human supervision. Although some quite involved tasks happily break down into programmable kernels, other apparently straightforward routines appear to defy expression or are unexpectedly iterative and unsolvable. This should not be unexpected: the mathematician Kurt Gödel found as early as the 1930s that not all arithmetic truths are provable and that mathematical statements occur which cannot be derived by arithmetic rules alone. Commentators regularly juggle with several (currently) unknowable coding problems such as 'closed domain solutions', 'common sense reasoning', 'strong generalisation', 'essential learning' and 'counter-factual perception'.[1] Situations and processes certainly exist that cannot be captured through code-based scripting.

---

[1] Andersson, Simon, 'Unsolved Problems in Artificial Intelligence', AI Roadmap Institution, @goodAI blog (3 February 2017), https://medium.com/ai-roadmap-institute/unsolved-problems-in-ai-38f4ce18921d.

Central to AWS deployment is the understanding of how the Delivery Cohort can express intention in unsupervised weapon systems. An etymological primer provides useful context to these challenges. For coding purposes, there are two kinds of 'directive construct': concrete and abstract. The concrete has meaning that can be directly indicated such as a seen object (a tank) or an action (engaging the tank). Concrete words can be taught by associating the word with the entity concerned. In AWS instruction, the meaning of a concrete word (here, the Cohort's intention) must be linked vertically within the weapon controller to a representation on one of the AWS model's learning planes. This linkage must then be coded in a manner that one such associatively connected entities can evoke another in particular sequences. In this way, the meaning of an instruction will not be restricted to just one association.

Different degrees of interference will lead to weapons with subtly different composition, subtly different learning traits and, over time, ever less similar profiles

Challenges exist over how computer programming deals with abstracts.[2] In code, most words are no more or less abstract. They do not have a meaning that can simply be settled by cipher involving ones and zeros. The meanings of abstract words are defined by other words. Indeed, it is the inherent complexity of almost any command's construction that complicates coding.

The meanings of abstract words are defined by other words. Indeed, it is the inherent complexity of almost any command's construction that complicates coding

For the AWS, that command's information must also be coupled to previously given information as well as to information that is to follow. This is rarely obvious and certainly complicating. Instead, associatively connected words will appear ambiguous to a machine's routines. Nested structures and conditionals that are commonplace in complex instructions create similarly challenging syntactic issues; although it is quite usual human practice to understand a directive without having to figure out exactly the meaning of every word, this does not work in coding.

The Cohort's challenge is therefore twofold. First, coding cannot capture context. Nor can it factor for the information in adjacent instructions

---

2 See, generally: Launchbury, John, 'A DARPA Perspective on Artificial Intelligence', DARPA slides, undated, https://www.darpa.mil/attachments/AIFull.pdf . Launchbury uses the examples of 'existence' and 'truth' to illustrate this coding constraint to capturing abstracts.

and from other non-associated circumstances that make up the fabric and milieu of the battlefield. If a weapon is to be unsupervised, code cannot simply be a mechanism to parse instructions. Instead, it is the AWS' only means to enable perception, memory making, interpretation, temporal appreciation and the evaluation of significance. Second, conflicts of interest hinder programming. In order to match human routines, it is necessary to incorporate different categories of facts within AWS syntax. These include indexical facts, normative facts, strong convictions, observations, hints, clarifications and reinforcements as well as basic ontological statements. Such categorisation is itself volatile, changing unpredictably according to subsequent intelligence and feedback. It will also be the Cohort's management of these 'facts' that will directly affect the weapon's priorities and actions.

It is necessary to incorporate different categories of facts within AWS syntax. These include indexical facts, normative facts, strong convictions, observations, hints, clarifications and reinforcements as well as basic ontological statements

Conflicts also compromise the weapon's interpretative routines. Without human mediation, coded rules are insufficient to mediate which weapon routine should take precedence and in which part of the weapon's processes. Which takes precedence? Is it the machine's initial configuration or its updated representation? Is it the weapon's intended end-state or intermediate steps arising from recently sensed data that determines the machine's next action? How does the weapon deal with the data string that suggests immediate action? These are fundamental quandaries that require arbitration. After all, what constitutes an 'acceptable' delay for an independent weapon? What represents urgency and to what degree should routines and guardrails be overruled by emergency events?

After all, what constitutes an 'acceptable' delay for an independent weapon? What represents urgency?

It is clearly inappropriate for the machine to prioritise its actions using simple recency heuristics, but other arbitration strategies are similarly challenging. Averaging protocols that select actions according to where the most conditions have been satisfied (also referred to as 'longest matching') work by dumbing down data precision (indeed, what data is being lost?). Nor can these conflict be resolved by incorporating limit switches whereby the activity of certain grouped conditions (for instance, in the engagement sequence) is turned on or off according to the fit of available data to that intended action. The weapon's Cohort cannot accept such a whimsical trigger that might be tripped at exactly the wrong time.

Other systemic conflicts embed themselves in AWS design. The first concerns the machine's set of initial beliefs. Indeed, the weapon's initial deployment represents a definite point in time after which its actions will be determined by that initial setting as amended by its subsequent experiences and learning. The conflict here is then the degree of stepped change (the increment of learning termed 'anchoring') that each follow-on process then exerts on the weapon's immediately prior set of beliefs. A further challenge is that weapon actions must reliably react to every battlefield stimulus. To this point, weapon 'curiosity' might be a code-based composite of the two states of 'novelty' and 'attraction' (with each state occasioning a specific and often conflicting action in the weapon). 'Astonishment' might then be the blend of system reactions' 'attraction', 'withdrawal' and 'curiosity'. Similarly, within the AWS reward system, 'aspiration' might be a combination of 'inclination', 'attraction' and 'arousal' (again, each with its own pre-determined portfolio of actions which are in theory then actioned in concert, albeit with reference to machine-generated weightings that are appended to each trait). The point here is to illustrate these models' wholly inappropriate approximating nature. The methodology driving these programming combinations looks more like a pick-and-mix than any sustainable means to synthesise how the weapon will respond to circumstances. In this vein, how might the weapon's code for the notion of a delayed response, a measured response, a slight deferral while additional information is sought and processed or, more complicated, a variable response?

> *The conflict here is then the degree of stepped change (the increment of learning termed 'anchoring') that each follow-on process then exerts on the weapon's immediately prior set of beliefs*

> *For the unsupervised weapon, the information contained within a command must also be coupled to previously given information as well as to information that is to follow*

Any model that relies upon baseline reactions to construct its response must also be refined over time. This, however, is unlikely to happen in lock-step across colleague machines, leading instead to unexpected idiosyncrasies and 'emergent' behaviours between one unsupervised weapon and the next. These 'emotional ratings' must also be tagged within the weapon's internal set of 'memories' such that appropriate significance is ascribed to each correct battlefield happening. It is only through such tagging (which itself will require feedback and calibration) that machine behaviour can be appropriately evolutionary.

The Cohort must also manage how its instructions are framed. Human communication is not only about sending a message that can be enacted by the receiver. There is usually a gulf between the written word and the intended message. Several specific sources of ambiguity complicate how the AWS translates instructions into action. First, lexical ambiguity concerns omitted, imprecise and simply errors in script. Syntactic and semantic ambiguity concerns interpretative uncertainty. Third, pragmatic ambiguity arises from trying to correlate communicating parties with quite separate contextual bases. Arbitration here is complex to initiate, complex to manage and adds to the weapon's overall technical debt, the more so as the boundaries of meaning in natural language are inherently indistinct. The margin, for instance, between day time and night time is not a given and is arbitrarily judged according to the recipient's own circumstances. All of this conflicts with the computationalists' assumption that the world consists of unambiguous facts that can be manipulated algorithmically. In most circumstances, an AWS output may actually only be partially true, again requiring the control suite to apply complicating confidence levels to outputs.

> *It is impossible for the weapon to analyse meaning from an instruction syntactically until potential sources of ambiguity have been resolved semantically*

> *The range of battlefield outcomes is far too broad for any meaningful utility to be calculated*

The challenge also raises basic ethical issues. Are 'occasional mistakes' acceptable on the basis that AWS are similarly susceptible to ethical deficiencies as human soldiers? Second, is a lower legal bar appropriate for machines with less lethality? Indeed, additional coding difficulties arise from the precepts of ethics and morality. Here, commentators have long tried to build a classification of relevant moral traits that cross over to weapons-directing artificial intelligence (AI).[3] That Jonathan Haidt's framework (or

---

3 Inferred from: Haidt, Jonathan, 'The Moral Emotions', in Richard Davidson and others, *Handbook of Affective Sciences* (Oxford University Press, 2005), generally.

anyone else's framework) has yet to be settled points to the complexity of the undertaking. Nor do these structures deal with emotions such as embarrassment, shame, anger, disgust or deceit, all of which should underpin lethal engagements, the more so if humans no longer retain oversight in these circumstances.

Compliance also requires that the Cohort code for a 'morals portfolio' to ensure that the weapon accurately factors for notions such as responsibility, appraisal, norm violation, self-evaluation and, most importantly, guilt. Without this, the Cohort must presumably narrow the weapon's tasking.[4] Intangibles such as 'restoration' are also important to coach the weapon away from actions that might lead to egregious error. In the same vein, the Cohort's coders must test that their learning routines are not

> *The challenge here is that data smoothing prima facie obscures inherent variation of data points that, in themselves, may be essential to ensuring weapon compliance*

unwittingly discriminatory against particular parties; notwithstanding the extraordinary resources available to its teams, it is well known that Google had to apologise when the automatic tagging system in one of its photos apps identified certain individual traits as 'gorillas'.[5]

It is this same 'morals' portfolio that must underpin the programming of Ronald Arkin's 'Ethical Adaptor' (theoretically working in tandem with an 'Ethical Governor', an 'Ethical Behaviour Control' and, perhaps, a 'Responsibility Advisor') discussed back in Chapter Three.[6] But for this to

---

4   The purpose here is not to impose any value judgement on the scope of tasks that a machine is suited. Instead, it should inform the extent to which such tasking must be limited by the platform's capability to integrate emotional precepts. Narrow emotional capability in otherwise autonomous weapons must equate to narrow taking. For a discussion on deployment 'degrees', see: Chapter Four (Deployment).

5   Economist, 'The Return of the Machinery Question', *Economist* (25 June 2016), https://www.economist.com/sites/default/files/ai_mailout.pdf, p. 15. See also: Hutson, Matthew, 'Even Artificial Intelligence Can Acquire Biases against Race and Gender', *Science Magazine*, (13 April 2017), paras. 4-7 and 10 of 12, https://www.science.org/content/article/even-artificial-intelligence-can-acquire-biases-against-race-and-gender.

6   Arkin, Ronald and others, 'An Ethical Governor for Constraining Lethal Action in an Autonomous System', Georgia Institute of Technology Robot Lab, Technical Report, GIT-GVU-09-02 (2009), pp. 1-2, https://www.cc.gatech.edu/ai/robot-lab/online-publications/GIT-GVU-09-02.pdf. Under Arkin's model for ethical control, robotic behaviour is expressed as an equation that includes all interpretable stimuli coming from the machine's on-board sensors as well as from externally received information, a limitless number of possible responses both in terms of their strength and direction of action and an ability to set thresholds above which a response will be generated. Additionally, mapping must be established between the stimuli and the overall response range that then defines the behavioural function to be triggered; See also: 'Governing Lethal Behaviour: 'Embedding Ethics in a Hybrid Deliberative/Reactive Robot

be enacted, several non-obvious routines and datasets must also be in place including, *inter alia*, up-to-the-minute information on friendly casualties, on non-combatant casualties and statistics that make sense of expected civilian structural damage, all inputs that are germane to AWS' actions under the strictures of military necessity in lethal engagement. But tuning such a governor in battlefield conditions would presumably involve an unacceptable process of trial and error. It also suggests introducing material delay in the weapon's processes, casting doubt on Arkin's notion of an independent simulator running in real time and in parallel with the weapon's executive systems to ensure compliance in its decision processes.[7] Further to this point, difficulty then arises from the firing sequence of these instructions. Each rule (and the pathways created by those rules) will result in quite different values being taken forward depending upon the command's order of execution. The difficulty is that all weapon weightings must be time-dependent in order to influence which rule the AWS fires first. A second (but more manageable) challenge arises from the heterogenous measures of quantity for all this key data that is likely to arrive in quite different formats (voltages, temperatures, durations, physical and electrical capacities, velocities, flow rates) requiring real-time translation and what is often termed additional 'learning latitude'.

> *Feedback loops will similarly be required to prevent erroneous actions becoming part of the weapon's updated set of operating procedures*

A recurring software challenge obviously relates to the *quality* of received data, whether through incomplete data, inappropriate coding routines or simple enemy feint. Uncertainty, of course, is everywhere: the weapon's wheels may spin and battlefield obstacles morph unpredictably. For the purposes of this behavioural review, however, two relevant types of uncertainty can be extrapolated to confound AWS deployment.[8] First,

---

Architecture', Georgia Institute of Technology Robot Lab, Technical Report GIT-GVU-07-11 (2011), pp. 14-19.

7 Arkin, 'An Ethical Governor for Constraining Lethal Action', pp. 1-2. In this way, Arkin envisages incoming command sequences first being run on this simulator to ensure that any outcome meets an established rule-set prior to any autonomous action sequence. See also: US Department of Defense, 'Defense Science Board Summer Study on Autonomy' (June 2016), https://www.hsdl.org/?abstract&did=794641, p. 34. An ethical override might theoretically provide for appropriate in-situ ghosting to filter malicious commands, theoretical erasure of key data should the machine be compromised and a block on malign reverse engineering.

8 For an excellent analysis on ML uncertainty, see: iMerit blog, 'A Complete Introduction to Uncertainty in Machine Learning', iMerit (1 September 2022), https://imerit.net/

epistemic uncertainty refers to a model's ambiguity arising specifically through inadequate training data. The condition also depends on the appropriateness of that training data and tends towards datasets that are rich in quantity but poor in quality. Within this uncertainty set is also the issue of bias, especially where the number of training points is limited and humans have had to manipulate or backfill evidence.

Conversely, aleatoric uncertainty refers to the inherent stochasticity (the quality of lacking order or plan) of the weapon's observations. Each observation of the unsupervised machine will have inherent noise, which, when accumulated across all observations, adds up to the model's aleatoric uncertainty. Prediction uncertainty (capturing overall model ambiguity) is then calculated by summing together the model's epistemic and aleatoric uncertainty. The general phenomenon can be neatly summarised by the aphorism that 'all models are wrong, but some are nevertheless useful'. Whilst epistemic uncertainty can be reduced with additional observations, the difficulty for the Delivery Cohort is that aleatoric uncertainty cannot. Here, therefore, uncertainty is not a property of the model but rather is an inherent property of the weapon's data distribution and, as such, is irreducible.

> Whilst epistemic uncertainty can be reduced with additional observations, the difficulty for the Delivery Cohort is that aleatoric uncertainty cannot

The Cohort should also be aware of what is termed 'scope compliance uncertainty'. ML models are generally built for specific use-cases and become less reliable when their algorithms are used outside their intended purpose; the output of a self-driving ML model will be materially affected if the model is trained for driving on the road's right-hand side but then used on the left. Uncertainty arises, after all, when action effects are not fully predictable. Assigning reliable levels of confidence – particularly difficult to measure in a combat environment – to model predictions is, in retrospect, an obviously enduring challenge. Nor are current solutions very encouraging. The most documented techniques (termed 'uncertainty quantification') appear convoluted and unrealistically specialised, each presenting idiosyncratic flaws; deep Bayesian neural networks lack appropriate expressiveness while so-called sparse Gaussian processes only capture uncertainty in a model's higher-level spaces. Models' raw data are ignored. The challenge here for the Cohort is to embed into AWS' processes a robust classifier that can predict

---

blog/a-comprehensive-introduction-to-uncertainty-in-machine-learning-all-una/.

classes of 'new' objects (and tasks) relative to the weapon's prior training upon what are by definition 'old' objects.[9] In that combat environment, these classifiers must factor for sequential correlation whereby nearby x and y values are almost certain to be closely related to each other. As task complexity increases, the use of classifiers must change from rules-based routines to knowledge-based routines and, in quite quick order, to sequences that require palpable expertise.[10] While this continuum may seem obvious, it would also appear to require ongoing human participation.

> *The general phenomenon can be neatly summarised by a further aphorism that 'all models are wrong, but some are nevertheless useful'*

Nevertheless, repetitive tasks (even those where feedback loops can be understood through mathematical statement) empirically lend themselves to coding. But this relationship starts to break down where additional confirmatory routines become necessary as task complexity intensifies. In a battlefield context, there are just too many solutions to too many possible problems, leaving the machine ever less able to understand what is termed its solution space. This notion of 'cumulative difficulty' is relevant given that complexity in AWS routines is likely to peak at the point of maximum uncertainty.[11] This represents the pivot point where AWS' ML spine is at its most fragile and where algorithms fall short of the human expert who is instead able to make difficult decisions in a fast

> *Algorithms fall short of the human expert who is instead able to make difficult decisions in a fast and frugal manner*

and frugal manner. At such higher levels of expertise, battlefield commanders do not even recognise that they are making decisions; rather, they are fluidly interacting with a changing situation and responding to patterns that they recognise. Training, experience, subjectivity, biases, personality and a wide

---

9 Here broadly relating to new events, sequences, actions, target types, order configuration, new parties and behaviours.

10 Missy Cummings uses human pilots as an example of such skills-based activity; human pilots train to interpret their cockpit dials before adjusting aircraft controls appropriately to ensure the aircraft's actual state matches the intended state. See: Missy Cummings, 'Artificial Intelligence and the Future of Warfare', Chatham House, Research paper draft (January 2017), pp. 5-6.

11 Gal, Yarin, 'Uncertainty in Deep Learning', PhD submission, Department of Engineering, Cambridge (September 2016), p. 7. Uncertainty here is a cumulative feature arising from, *inter alia*, out-of-distribution test data, aleatoric factors (measurement imprecision), uncertainty in model parameters as well as from structural uncertainty (model or epistemic uncertainty).

grasp of context (among several factors) are actually the very attributes that are critical to decision-making but, at the same time, cannot reliably be captured in code.

The crux of the difficulty stems from software's inherent 'symbol grounding problem'. Each symbol is itself defined using other such symbols, the consequence being that it is difficult to ascribe meaning to these real-world situations. It also points again to the circular challenge that arises from the code-based model: how does the weapon determine which of its internal representations is relevant to the scenario that is immediately unfolding? It is this general issue of having to work to an appropriate context that is collectively termed AI's 'frame problem'.

8.2 Coding errors

An attribute that becomes key to compliant deployment is the degree to which the AWS can successfully navigate itself through coding errors, the subject of this section. After all, empirical experience of software development has long suggested that even veteran programmers unknowingly write one mistake into every ten lines of code. Error incidence in complex algorithms is high because of several factors. By nature, this coding involves intricate logic, multiple components that must interact in dynamic ways and, as above, a requirement to handle unexpected inputs and scenarios. Here, logical errors involve a general misunderstanding of the problem set or the intended requirement to be captured by the code. Even if the logic is correct, translating that intention into code might bring about syntax errors, semantic errors or runtime errors that will variously degrade a program. Together, these constitute a program's implementation error. Complexity error then arises from a task's intricacy where it becomes harder to foresee all possible interactions within the program, the program's inputs or with external systems. Resulting bugs are more difficult to detect and edit. Edge case and concurrency error then occur when AWS' programs are either run in parallel or just as one part of a distributed system.

How then does this all relate to the battlefield? The F-35 fighter jet reputedly has more than twenty million lines of code, eight million of which relate solely to its missile and threat management systems. Coding error occurs, of course, regardless of quality controls to mitigate system risk, provide redundancy and ensure reliability. The phenomenon is exacerbated as weapon platforms are increasingly made up of multiple subsystems produced by multiple manufacturers, each with different testing priorities and, notwithstanding external audit, different verification standards. An AWS supply chain, after all, will only be as good as the poorest routine within the weapon's processes. Indeed, military-civilian collaboration (with each party's different incentives in this process) generally adds fragility to the procurement chain, the more so given that more and more responsibilities tend to be delegated to outside third-party specialists; even in the early days of the US uncrewed aerial vehicle programme, seventy-five per cent of the maintenance and weapons loading for drone systems was being outsourced to private contractors 'with patchy results at best'.[12] These error rates are quite broadly corroborated by sources reckoning the number of software errors at some 'two-and-a-half errors per function point'; at this rate, notes Peter Kassan, a software program large enough to simulate the human brain would contain some twenty trillion errors.[13]

> *Software's inherent 'symbol grounding problem': each symbol is itself defined using other such symbols, the consequence being that it is difficult to relate meaning to these real-world situations*

This analysis even underplays the increased complexity of recent systems. It also fails to account for the wholly new areas of brittleness that accompany the intricacy occasioned by ML. Readers should not ignore the human angle here given that service-level understanding of these autonomous processes will likely be limited to a very small handful of expert parties. The detailed workings of an autonomous agent defy being reducible to simple explanation in an army field manual.

> *Complexity error arises from a task's intricacy where it becomes harder to foresee all possible interactions within the program or with external systems*

---

12 Singer, Peter, 'Statement to US House of Representatives Committee on Oversight and Governmental Reform', cit. *Rise of the Drones, Unmanned Systems and the Future of War* (Congressional Research Service, March 2010), p. 2; Army systems operating in Iraq have been described as 'government-owned-contract-operated'.
13 Kassan, Peter, 'A.I. Gone Awry: The Futile Quest for Artificial Intelligence', *Skeptic Magazine* (4 February 2011), https://www.skeptic.com/reading_room/artificial-intelligence-gone-awry/.

8.3 Utility function

How might the unsupervised weapon manage its action sequences? To understand this challenge, we need to look deeper into the required routines. Actions will likely be arbitrated according to an 'optimality notion', unique to each specific weapon class and configured at initial deployment by the Delivery Cohort as the AWS' initial set of decision rules. This optimality factor is the weapon's *aide memoire* dictating, *inter alia*, the confidence premium (here, 'weighting') that should be attributed to each of the weapon's possible worlds. The Cohort's intention is for the weapon to rank the desirability of all possible outcomes and so establish the machine's basic preferences. In this way and at each step, the weapon is working towards an action with the highest expected utility. Various challenges arise from this utility model. Once installed, effects are difficult to forecast as the model is inherently unstable. Its margin for error is large and, given its combustive consequences, the Cohort must also factor for the possibility of high-regret outcomes. First, the AWS must first run an internal computation on all possible actions, a considerable task given *The model wrongly assumes whole-system predictability while even quite subtle discrepancies in AWS' prior probabilities will grossly affect how the weapon behaves* the almost limitless number of battlefield parameters. Second, the utility function must decide upon what constitutes appropriate use of force for that engagement sequence.

The weapon's decision space is almost limitless. It must account for the involvement of colleague assets, consideration of next steps, appropriate audit ahead of each action sequence as well as post-event consideration as well as interaction following each undertaken engagement. Assuming satisfaction of these prerequisites (and after iterative conditionalising as set out above), only then can the weapon adopt a suitable (and legal) pathway and to do this by calculating the expected value of an intended action. This is then the sum of the value of each possible world multiplied by the conditional probability of that world given the action. Humans do this innately. The challenge here is usefully captured by machine action as currently envisaged in the event of a tie in the function; present mediation routines would have the weapon pick a random action in order to achieve its expected utility, a worst-of-all-worlds outcome that is likely to be unacceptable both from a compliance and utility perspective.

The Cohort also needs to be wary that its most basic epistemological specifications for this key function may be plain wrong (or even slightly wrong or, more likely, may subsequently drift away from the day's battlefield priorities). The challenge is that utility processes lend themselves neither to arbitration nor to *ad hoc* intervention. The model wrongly assumes whole-system predictability while even quite subtle discrepancies in AWS' prior probabilities will grossly affect how the weapon behaves. An example is again useful. The AWS might work to a prior that assigns zero probability to enemy forces being delivered by land transport: no matter

Given the competing requirements of compliance and utility, deciding then how the weapon's many intermediate results becomes a final decision is an enduring conundrum

how much the battlefield evidence it accrues to the contrary, the AWS is likely to reject subsequent sensor-based intelligence to the contrary and make dangerous, unpredictable choices as a consequence.

8.4 Software processing functions

A further constraint stems from software's essential 'absence of attached meaning'. If the unsupervised weapon is to be useful to its Cohort, sensor input should trigger wide and yet unpredictable associations between the weapon's external environment and its possible action set.[14] The richest possible symbology is required if the weapon's actions are to maximise the space within the boundaries defined by the Cohort. The challenge, of course, is that the size of this space is almost infinite. Furthermore, the AWS' decision drivers are presumably highly correlated to those very descriptors that the machine's symbology can produce.

---

14 This is in contrast to narrowly defined factory applications. Actions available to the weapon must, after all, reflect an unconstrained range of goal-directed tasks that are only then tempered by limitations imposed either by its environment or Delivery Cohort.

The weapon must also prioritise particular sequences as everything cannot be attended to at once. The product of these processes, moreover, will as often as not be an *intermediate* result that must itself be presentable for further processing. This creates further challenge. By definition, these processes are dynamic and unstable precisely because there is no obvious end-state for the independent weapon. This prompts again the question of how often the weapon should poll its sensors, the more so given operational requirements to refresh that end-state but also update its wider routines.[15] Is this a millisecond occurrence (in order to approximate the human brain)? Is it necessary to introduce buffering to prevent data overload in the weapon? This might facilitate more extensive processing but at a cost of degrading performance and introducing uncertainty. Given the competing requirements of compliance and utility, deciding how the weapon's many intermediate results become a final decision is an enduring conundrum that is complicated, of course, by the portfolio of subroutines and loops that the Cohort must incorporate to support the weapon's processing functions. The Cohort's sequencing puzzle is therefore how to choreograph subroutines that are designed to amplify (or deny), filter, scrub, classify and then process that data in order for the weapon to decide and execute.

> *By definition, these processes are dynamic and unstable precisely because there is no obvious end-state for the independent weapon*

These subroutines are invariably processing-intensive. Data scrubbing is the process of identifying inconsistencies, inaccuracies, incompleteness, duplication and other erroneous data in order to clean and standardise it prior to processing. Empirically, however, the routines are thought to magnify noise and certainly complicate the machine's solving for missing values. While data handling (sometimes termed ETL for extraction, transformation and loading) may merge inputs from quite separate sensors and datasets, it also results in the machine having to process information from what are similar but still undifferentiated inputs. Currently available scrubbing methods ('nearest neighbour', 'clustering', 'greedy' and other rules-based routines) each introduce their own idiosyncratic variations into data preparation including memory overload, degraded runtimes, certain 'common variant problems' (that the weapon cannot yet solve), premature

---

15 This is a complex procedure given that sensors may be passive/active or mobile/static/remote with various modes of operation. Its complexity may require that game theory is needed to ensure 'best group performance' across the AWS' sensor portfolio.

and local convergence of data, inappropriate clustering of data and even the requirement for manual intervention. Indeed, the real question for the Cohort is whether this is yet a properly autonomous process. It also complicates AWS operation through 'coupling contagion' that can occur when multiple data sources must subsequently be sewn together.[16]

The set of routines that are required to cement a weapon's representation cannot really be simplified. After all, the terms 'world' and 'representation' are a very general gloss for an open horizon of circumstances already beset by human heuristics and biases that have been unwittingly incorporated by the Cohort into these routines.[17] Termed the paradox of artificial intelligence, these inherently human prejudices each have a knock-on effect on system intricacy. First, the scale of data tends towards the compartmentalisation of processes as the Cohort tries to impose structure

> *The scale of data tends towards the compartmentalisation of processes as the Cohort tries to impose structure*

on its independent asset. Second, we have already seen that data smoothing (the removal, for instance, of 'confounding covariates' in weapon datasets) introduces further hazard through its practice of filtering out statistically unusual (although often material) data points to regulate whole-dataset sensitivity.

Bias arises from many sources, from human involvement in the commingling of data to how coders' preferences create sample size disparity (one sensor dataset versus another). It arises from incongruencies being coded into these weapons' reward functions. Bias also occurs from straightforward cultural differences in how thresholds are set and outputs are interpreted. This is exacerbated as different sensors collect quite different natures of battlefield information, all then requiring individual preprocessing to rework these datasets into homogenous, comparable file types.[18] Examples abound,

---

16 A consequence of increasing complexity is the emergence of tight, little understood and non-intuitive system associations that are likely to take place between sensor inputs, processing routines and subsequent action output in AWS platform. See: Borrie, John, 'Security, Unintentional Risk and System Accidents', Panel presentation, United Nations Institute for Disarmament Research (15 April 2016).

17 There are several methods of identifying such biases. WEAT (word embedded association tests) and IAT (implicit association tests) are recognised tools for isolating otherwise hidden inferences. A useful example is symbology that fails to differentiate between steam and ice given their proximity (but semantic opposite) from plain-state water.

18 A battlefield example is useful. Using reflected light might appear to offer a simple sequencing solution to detecting the presence of a target object, the distance to that target, some detail on that target's surface and to recognise other embedded features. A weapon's reflectance sensor,

moreover, of algorithms malfunctioning because of their inappropriate probabilistic bias. This is important precisely because associative processing is a prerequisite across AWS routines. A challenge, however, is that the linking of these representations may be very transient (here, battlefield shapes, characteristics and identities), considerably complicating what will become the central matter of machine attention. These representations, after all, are not built to represent properties numerically and instead can only 'generally' inform whether that designated property (here, a battlefield feature) is either present or not present. The issue then becomes whether a particular representation is forgotten after processing or is it fundamental to the weapon's current goal set (and therefore to be written to the weapon's action list).

How the AWS undertakes its planning is also testing. Each machine must forever be looking ahead to the outcomes of possible actions, prioritising and (perhaps) incorporating this analysis to create a sequence of actions that moves the weapon towards the Cohort's desired goal. Routines must run in parallel (although, to complicate matters further, their individual processing is usually sequential). Considering AWS' navigation, does the machine prioritise the shortest path given time and battery power considerations? Or does its navigation decision actually factor for criteria such as safety, isolation, proximity to friendly forces as well as other topographical *A level of acquaintance cannot be coded for, the more so since the coder is very likely not military and probably lacks previous exposure to these obstacles* advantages (concealment and feint) in order to optimise its route planning? That search must also look for multiple and parallel solutions and, as the number of possible available states becomes ever larger, planning presumably becomes ever slower and less reliable.

In this vein, a planning sequence which takes longer to solve might even become based on a weapon's dataset that itself is already outdated. For

---

however, is unexpectedly complicated. Light reflectivity is affected by the target's colour, smooth or rough texture and other surface properties. Light reflection depends upon surface colour and is therefore less reliable in detecting dark objects. Similarly, the reflectance sensor must ignore ambient light in order to be sensitive only to its own emitter's reflected light. While this single subroutine may be undertaken by the weapon using multiple sensor readings, the level of processing then required (for what, after all, is a single component of a very complex sequence) demonstrates how difficult it will be to process even quite basic datasets into a useable form. This would be worked by undertaking one pass with the emitter on and one with it off and then subtracting one from the other having first adjusted the data in order not to conclude with a negative light.

humans, this is all mitigated by operator experience, a level of acquaintance that cannot be coded for, the more so since the coder is very likely not military and probably lacks previous direct exposure to these nuances. In fact, incongruous incentive exists for the weapon to eschew human experience and actually run planning sequences as infrequently as possible in order to reduce lag and error. Furthermore, all this assumes that the weapon's immediate environment remains stable and certainly within tolerance of the probabilities that were set at deployment.

8.5 Anchoring and goal setting issues

The software challenges discussed thus far arise mainly from AWS data capture and processing sequences. The aim of this section is now to look at the weapon's subsequent actions and the processes leading up to those activities. One such function relates to 'anchoring', the degree by which the weapon's current representation is amended to reflect recently polled data. Anchoring processes must execute the appropriate degree of change between what is the current weapon state and an amended updated weapon state that has been prompted by the platform's application layers. It must be sensitive to sample size, to the priorities of the Cohort and to prior outcome probabilities. The Cohort must also mediate for several unconscious prejudices including cognitive bias (whereby the machine relies too heavily upon

Each machine must forever be looking ahead to the outcomes of possible actions, prioritising and (perhaps) incorporating this analysis to create a sequence of actions that moves the weapon towards the Cohort's desired goal

its initial representation or, conversely, ascribes too much importance to just one aspect of a happening or circumstance), 'retrievability' bias and 'imaginability' bias (termed 'the availability heuristic'). Once again, the weapon must be configured to ignore correlations that do not exist.

Anchoring in AWS will be complicated by the weapon having to 'understand' its status (assets, position, capabilities and options) at any (and every) point in its decision process.[19] In the human brain, the soldier's own representation (if you like, his or her 'vantage point') is grounded by that personnel's ability to attach sensations to a point of origin. With the exception of a GPS-fixed position and place, this facility is difficult to mimic in a weapon that must therefore incorporate steps that reconcile visual and auditory stimuli (including directions and distances) and, once processed, cement that information into subsequent routines. The sequence requires intermediate steps to assign priorities and levels of confidence, to iron out conflicts and to prevent the machine's memory being overwhelmed. It also involves the managing of mental maps and a proven ability to toggle between internally generated and externally received information. While a solution might be to fix the weapon into 'a localised observer' (treating the implied locus of the weapon the same way every time), this treatment fails as it is computationally challenging for the weapon to understand its origination point as being somewhere other than its current position.

If this process falls short, the weapon will either be illegal or useless

Similarly intractable is the Cohort's need to set goals that govern the weapon's priorities, responses and action selection. A distinction is required here between values and goals in machine autonomy: goals prompt an intelligent system to develop plans of action, and values enable it to assess the comparative merits of such plans. If this process falls short, the weapon will either be illegal or useless. An overarching difficulty, however, is to ensure that goal pursuit mirrors the intentions of the Cohort. AWS tasking, after all, will follow accepted machine logic whereby a generated priority list determines what must be undertaken at once, what should be undertaken next, the resumption of a task that was previously discontinued and, more complex for the weapon, what actions should subsequently take place in order to capitalise on battlefield opportunities.

Goals prompt an intelligent system to develop plans of action, and values enable it to assess the comparative merits of such plans

---

19 This aspect of the heuristic is termed the 'adjustment effect', and empirically suggests such incremental adjustments are usually insufficient. See: Tversky, Amos and Daniel Kahneman, 'Judgement under Uncertainty: Heuristics and Biases', *Science* 185, 4157 (1974), p. 1128, https://www.science.org/doi/10.1126/science.185.4157.1124.

Goal setting is not simply a process of tuning the machine's reward signal. How friendly forces are configured in a video war game provides a useful comparator, the way that the contestant defines his or her goals obviously having material (and often unforeseen) consequences on the subsequent game play, particularly around bottlenecks (whereby more ammunition requires more logistics requires more infrastructure requires more safe areas). It also widens second-order effects (wider engagement parameters require more intelligence require more data processing requiring more human intervention and yet more corroboration).

The notion of 'infrastructure profusion' usefully illustrates the point where the autonomous machine unexpectedly transforms a large part of its reachable resources into the service of an inappropriate internal goal. In the case of AWS, this requisitioning behaviour might upset the already delicate allocation of resources on the battlefield and is best illustrated by Nick Bostrom's paperclip analogy whereby the autonomous agent catastrophically and inappropriately devotes all available resources at manufacturing an irrelevant item.[20] The problem for the Cohort is to include mechanisms for the AWS to cease particular activities that contribute to a flawed goal. Indeed, it can be inferred from Bostrom that the weapon, if it is a logical Bayesian agent, will never assign exactly zero probability to the notion that it has not yet achieved its goal.

> AWS tasking will follow accepted machine logic with a generated priority list: what must be undertaken at once, what should be undertaken next, the resumption of a task that was previously discontinued and, more complex, what should subsequently take place

Various workarounds are intended to solve the issue. The first concerns thresholds. It may be that the Cohort's goal definition directs the AWS to look no further once the machine has an action that offers a probability of success that exceeds, say, a threshold of ninety per cent. This satisficing approach is plainly inappropriate. First, it fails to address whether the proposed engagement is a sensible way of achieving that goal. Second, evidence suggests that probability-based thresholds tend towards suboptimal behaviour, the machine's drive towards self-improvement actually destabilising its

---

[20] Inferred from: Bostrom, Nick, *Superintelligence: Paths, Dangers, Strategies* (Oxford University Press, 2014), p. 123. Again, this analysis assumes the deployment of very broad task weapon autonomy. Here, an AI, designed to manage production in a factory, is given the final goal of maximizing the manufacture of paperclips and then proceeds to convert the whole Earth into paperclips.

wider processes. Instead, all of its objects are represented as 'economic' utility functions with no built-in encouragement for the weapon to act in a compliant manner. It will furthermore be challenging to tune these thresholds in what is likely to be a set of unfolding events: should a slightly diluted set of actions be coded for, say, a seventy per cent threshold? How can the whole scope of an engagement be captured in a single threshold?[21] The verso here might be that the weapon prioritises outcomes that have 'moral status' but, in so doing, exposes the whole sequence to inappropriate value-loading.[22]

> *The weapon, if it is a logical Bayesian agent, will never assign exactly zero probability to the notion that it has not yet achieved its goal*

How the Cohort sets AWS goals will clearly affect battlefield behaviours. While a soldier's actions are influenced by often subjective (and difficult to define) tenets such as leadership, empathy and context, managing weapon behaviours must rely entirely on code-based machine learning. Goals, moreover, cannot operate in isolation. They divide between the short and the long term with each such object being made up of subgoals, each with its own hierarchy. The notion of a governor to mediate between goal conflicts is simply theoretical and just too complex. Finally to this point, a reliable 'fatigue process' is necessary as it allows the unsupervised weapon to recognise setbacks, to adjust priorities and to introduce inhibition routines (and to do all of this without compromising the weapon's overall goal regime).

> *It will also be challenging to tune these thresholds: should a slightly diluted set of actions be coded for, say, a seventy per cent threshold? How can the engagement sequence be captured in a single threshold?*

---

21 Source: 'UK Tactical Aide Memoire (TAM) Part 2', Issue 3.0 (1998); in particular, fire discipline parameters including arcs of fire, authority routines before opening fire, STAP, priority of targets, controlled rates of fire, ammunition conservation and target indication.
22 For a discussion on value loading in AI agents see: Yudkowsky, Eliezer, 'The Value Loading Problem', *Edge* (2015), https://www.edge.org/response-detail/26198. A second more existential challenge emerges: In running a huge array of simulations, the weapon might then discard earlier programmed iterations, 'gloriously' altering its initial setup under what Bostrom calls 'mind crime'. See: Bostrom, *Superintelligence,* p. 139. On ML ramifications to this point, see: Chapter Seven (Firmware), specifically: 7.2 (Firmware ramifications of learning methodologies). Although written in 1987, see: P Ranky, 'A Summary of Robot Test Methods with Examples', University of Michigan, RSD-1-87, 1987, p. 7. An adjunct here is that goal setting must incorporate a reliable temporal dynamic. Furthermore, the construct of 'perverse instantiation' posits a possible conflict between a weapon's theoretical rush to achieve its goals notwithstanding its Delivery Cohort's *a priori* intentions.

8.6 Value setting issues

The Cohort must also integrate *values* into its autonomous weapons based upon either upon a 'scaffolding approach' (a series of interim values on route to the weapon 'landing' on a final set of values) or, more likely, a model that instead relies upon an unchanging set of values throughout the weapon's learning and operational life. All of this is complicated; as noted by Jason Tanz, 'when engineers peer into a deep neural network, what they see is an ocean of math: a massive, multilayer set of calculus problems that—by constantly deriving the relationship between billions of data points—generate guesses about the world'.[23] Two issues immediately arise. In this context, how then can value setting be understood by the battlefield commander? Moreover, how does the Cohort calibrate goals and values when individual weapons may be at materially different stages of development? In the same vein, what mechanisms are available to bring the AWS back into line should erroneous goal development lead to general (even slight) malfunction? A trivial aberration (perhaps a wrinkle on route to an updated goal) might otherwise develop into a material divergence from what the Delivery Cohort intended for its weapon at configuration.

Value setting comes with other idiosyncratic issues. How can the Cohort reliably reconfigure AWSs values if it is unable to communicate with that platform? Instilling values (and subsequently allowing for adjustment) therefore becomes a key deployment constraint.[24] First, it must inform all of the weapon's conditional probabilities and all of its expected utility outcomes. The Cohort must then repeat that adjustment for every possible course of action facing every one of its AWS deployed in the field. The first conundrum is how the unsupervised weapon incorporates new data into its routines if it is conditioned to an unchanging final value. An adjacent matter is that the routine creates circularity if the weapon's incentive system is directed at that final value.

---

23 Tanz, Jason, 'Soon We Won't Program Computers. We'll Train Them Like Dogs', *Wired Magazine* (17 May 2017), https://www.wired.com/2016/05/the-end-of-code/.
24 Considerations here arise from online and offline repair processes, field repairs and servicing, replenishment and restocking, routine maintenance, system updating and unit retrieval processes.

The Cohort's choice of values must also navigate through a thicket of philosophical problems. In deciding upon action paths, should the AWS use causal decision theory, evidential decision theory, 'updateless' decision theory or something else? Similarly, the Cohort's value definition must be consistent with passing normative battlefield standards in order to avoid unexpected or inflammatory actions. Moreover, the Cohort will presumably want to avoid simply chaining its agent to exactly the same value disposition as a human;

A trivial aberration (perhaps a wrinkle on route to an updated goal) might otherwise develop into a material divergence from what the Delivery Cohort intended for its weapon

after all, that same flawed human nature is witness to all manner of arbitrary (and even evil) traits that too often do not comply with the Law of Armed Conflict. Getting this wrong will likely prompt very bad decisions including, *inter alia*, the weapon rewriting itself to run on some quite inappropriate basis. Indeed, trying to capture values in code is difficult because human goals are so complex but also fundamentally evolutionary. They are not immutable but instead flex. Nor do these suggested models reflect how humans actually work: our goals, after all, may or may not develop from prior actions with values then acting (or not) as triggers in subsequent decisions. Goal upgrades must also happen either in real time or, complicatedly, after what might reasonably be indeterminate delay.[25]

8.7 Action selection issues

How is the Cohort to calibrate its weapon? The 'desired' state of the machine, its goal state, is the benchmark to which it must set its system. The challenge is that elaborate feedback is then required to maintain that point, comparing its current state with this desired state. Furthermore, once one task is completed, the AWS must then terminate that particular

---

[25] Bostrom, *Superintelligence*, p. 186; Bostrom's narrative is useful to evidence this infeasibility: 'From a noisy time series of two-dimensional patterns of nerve findings, the visual cortex must work backwards to reconstruct and interpret three-dimensional representations of external space. A sizeable portion of our precious one square meter of cortical real estate is required to process visual information, billions of neurons of working ceaselessly to accomplish this task… how could our programmer transfer this complexity into a utility function?'

routine and generate, as necessary, feedback to itself and back to the Cohort. The difference between the weapon's current and desired states is then the weapon's observed error, the aim of the Cohort being to minimise that error. Loops must then flex what the weapon chooses next to do in order for it to stay as close as possible to the Cohort's intended set of goals. This then will be the likely context for AWS' configuration.

Two issues arise. First, the *pace* of such error correction is not obvious and depends on how often error is computed and how much correction the Cohort decides based upon each feedback loop. While feedback mechanisms may be useful at influencing low-level tasking (adjusting actuators, for example), it turns out that they are significantly less handy at modifying the weapon's higher-level action selection such as navigation, collaboration and even interacting with its immediate environment. In selecting actions, moreover, the weapon must be front-facing as well as identifying actions ahead of time. Coding for action selection is hard. The weapon is selecting

The pace of such error correction is not obvious and depends on how often the error is computed and how much correction is decided for each feedback loop

to deliver one such action but must do this as part of a pre-understood combination of actions. The routines to decide this are termed fusion and arbitration. Fusion is the notion that weapons combine multiple candidate actions into a single action output and can only work if the AWS is monitoring all of action possibilities concurrently, notwithstanding that what is possible and what is intended by the Cohort may be quite divergent.

It might be that the Cohort considers a hybrid control model for its weapon, perhaps chasing the best of both worlds whereby the AWS benefits from the 'speed' of reactive control and the 'brains' of deliberate control. The difficulty here is that the model requires mixing fundamentally different control protocols, different timescales (short for reactive, long for deliberative) and different representational models (none for reactive, explicit and elaborate for deliberate). Moreover, not all the information that is required for these routines necessarily flows from the bottom up (that is, from the reactive layer to the deliberative layer) with often-repeated situations likely stored away in what are termed the weapon's internal 'contingency tables'. These types of action models are also, by definition, parsimonious in order to avoid strong *a priori* assumptions that might introduce bias. Any 'alternating' middle layer would have to be special purpose and bespoke to each architecture, an example perhaps of Yuval Noah Harari's adage that designing artificial intelligence to be the best of all worlds can end up being the worst of all worlds.[26]

Finally to this point, the Cohort must consider likely bottlenecks and appropriate conflict management. Practically, the state space confronting the weapon appears far too large. It is plainly unworkable for the weapon to keep all plans appropriately current, especially against the backdrop of a shifting battlefield (the 'state space', after all, for the AWS). Any single change (for instance to the AWS' goals) then requires real-time adjustment of the platform's entire rule set. In managing conflicts, it is not simply a matter of the weapon summing inputs to achieve some type of average. Instead, resolution strategies must be based on rule matching routines that are either forward-chaining (using deduction routines) or backward-chaining (using expert input to recalibrate data fit). The complexity of these rules is baffling. The weapon's rule set must either be deterministic (where a single rule is used in each resolution cycle) or, more likely, non-deterministic (where multiple rule-matching occurs, possibly requiring a separate 'inference engine' to interpret the conflict in the first place). Comparing, moreover, one data string to another can lead to widely different outcomes, especially if the weapon's inputs have been combined over and over again to the point of formlessness.

These types of action models are also, by definition, parsimonious in order to avoid strong a priori assumptions that might introduce bias

8.8 Behaviour setting and coordination

System behaviour is how the deployed AWS acts and reacts, a complexity for the Cohort being that behaviour is a time-extended phenomenon. It is also a changing amalgam of each weapon's values and goals, its utility function and then all of the learning routines detailed above. It will be the weapon's ability to abstract (its deriving of meaning without requiring there be physical representation of

---

26 Mataric, p. 183. See also: Anthony, Andrew, 'Yuval Noah Harari: Homo Sapiens as We Know Them Will Disappear in a Century or So', *Observer* (19 March 2017), https://www.theguardian.com/culture/2017/mar/19/yuval-harari-sapiens-readers-questions-lucy-prebble-arianna-huffington-future-of-humanity.

this sense) that will ultimately decide the weapon's utility. As above, it is abstraction that informs the weapon's ability to deal with 'ideas' rather than having to base action solely on battlefield happenings and, for the purposes of this final section, is a convenient proxy to box together all of the foregoing software challenges into a general observation on AWS deployment.

Individual weapon behaviours will also likely conflict with other asset behaviours and, certainly, the behaviours of colleague autonomous machines. Behaviours, moreover, are hard to isolate and, on account of AWS' different timescales and learning paths, will be tricky to coordinate across multiple platforms unless they are almost pointlessly anodyne.

The same instabilities that are inherent in AWS' ML models will presumably reflect in these weapons' behaviour, such behaviours therefore requiring a workable process of 'data forgetting' if the platform is not to be hamstrung either by information overflow or the retention of suboptimal (or wrong) outputs. Although a key competency if AWS is to be feasible, no reliable model exists for 'forgetting' routines. First, deciding upon filtering or significance criteria that can manage a 'delete sequence' is challenging.[27] Second, it is just as complicating to validate what is, after all, a binary routine (durable versus erased data). Here, the weapon must also toggle between its primary visual evidence versus, for example, peripheral (and therefore contextual) information. One additional hitch exists. Although forgetting is a necessary capability in setting behaviours, the AWS should not generally forget acquired skills, a conundrum often called 'catastrophic forgetting', an oft-repeating limitation in neural networks. The issue, moreover, has several angles. How should data age conflate with data redundancy? Should the frequency of the weapons' data polling factor in these redundancy calculations? Should data be hived off for later retrieval and how might this 'forgotten' information subsequently affect behaviours?

Although forgetting is a necessary capability in setting behaviours, the AWS should not generally forget acquired skills, a conundrum often called 'catastrophic forgetting', an oft-repeating limitation in neural networks

---

27 Fratto, Natalie, 'Machine Un-learning: Why Forgetting Might Be Key to AI', Hackernoon. com (31 May 2018), https://hackernoon.com/machine-un-learning-why-forgetting-might-be-the-key-to-ai-406445177a80. In reviewing the use in unlearning of 'long short-term memory networks', 'elastic weight consolidation' and the use of bottleneck theory (the squeezing of data through code bottlenecks in order to retain only those features most relevant to general concepts), Fratto notes the inappropriate fitting and compressing of data that underpins such forgetting techniques.

Behaviour is therefore a useful proxy to evidence the cumulative effects of software complexity. After all, the Cohort cannot certify that an AI is ethical just by looking at its source code any more than it can certify that humans are good by scanning their brains. The analogy is relevant given, as above, that weapon behaviours are derived precisely from the weapon's software components. It is against this background that this book can now consider the hardware upon which those routines must operate and certain of the challenges that physical configurations create.

9
Hardware
Build challenges to AWS function

It is the physical design of an autonomous weapon system (AWS) that determines its 'affect' characteristics and, if Wirth's law is to be believed, it is to be expected that hardware complexity must grow exponentially as new software capabilities become available and weapon tasking moves further towards autonomy. The aim of this chapter is therefore to review hardware constraints that might curb AWS deployment, in particular complexities that arise from likely combinations of physical components and how they might impact weapon compliance. In this vein, the Delivery Cohort must remember what is a truism, noted by Cynthia Ferrell back in 1994 but never more relevant to AWS deployment: 'Having many sensors and actuators is a double-edged sword. Multiple sensors provide reliable sensing and a richer view of the world. More actuators provide more degrees of freedom. However, more components also means there is more that can fail and subsequently degrade performance… mechanical failure, electrical failure, or sensor failure'.[1] Indeed, causes of failure have changed little in the intervening quarter century as evidenced by the case of Ferrell's Hannibal robot and its susceptibility to signal drift, 'graceful degradation' and challenges arising from that experiment's portfolio of patches undertaken to sort those errors. The context here is that commentators generally relegate hardware challenges way behind those of software, control and contextual constraints. And, to a degree, this chapter's relative brevity acknowledges the subordinate nature of hardware as platforms mature, certainly compared to the deep-seated software issues identified in preceding chapters.

---

1   Ferrell, Cynthia, 'Fault Recognition and Fault Tolerance of an Autonomous Robot', *Adaptive Behaviour*, 2.4 (1994), pp. 4-5, http://web.media.mit.edu/~cynthiab/Papers/Breazeal-AB94.pdf.

The Cohort must manage several hardware components in combination if it is to deploy AWS. That list is generally composed of assets that sense, garner and process data – assets that give the weapon physical heft and protection, as in power assets, mobility assets and, of course, lethal assets. Each is an important integer in the AWS equation and must operate concurrently and as part of the whole. All of the foregoing chapters, moreover, concern processes that, one way or another, leverage the weapon's hardware. This must happen in accordance with the weapon's 'state', the weapon's physical descriptors at every point in time notwithstanding that its component states may be visible, partially hidden or hidden (unobservable). That its sensors must be so numerous acknowledges that these weapon states may be both discrete (up, down, blue, red) and continuous (a thousand miles).[2]

> *'[Having] more components also means there is more that can fail and subsequently degrade performance [through] mechanical, electrical or sensor failure'*

Much of the preceding analysis has also been about sensors, but they are, of course, just one part of the asset portfolio enabling machine autonomy. It is the machine's effectors that enable the weapon to undertake physical actions such as locomotion and manipulation. An inescapable complexity is that the weapon's manipulators (robot arms and grippers), in broad terms, must move in one or more dimensions. This characteristic introduces challenge regarding degrees of freedom (DOF), the minimum number of coordinates required to specify completely the motion of the mechanical system. While simple actuators such as motors control a single motion (up-down, left-right), more complex effectors such as a weapon's robotic arms have exponentially more DOF that then require more complicated actuators. It would be ideal if the AWS included an actuator for every DOF, but this is rarely the case and would result in additional complexity. A further effector challenge arises from the weapon's power requirements without it being compromised by heavy batteries.

The AWS must also replenish its power autonomously and without oversight. In order to satisfy this limitation, the weapon must get to the appropriate location and do so following a particular path. Motion planning is therefore a combined effector and control issue. It is a computationally

---

2 Degutis, Charles, product director, Bosch, cit. Pickering, Charles, 'How AI Is Paving the Way for Autonomous Cars', *Engineer* (15 August 2017), https://www.theengineer.co.uk/ai-autonomous-cars/. 'All these sensors have strengths and weaknesses. Radar… can bounce off tunnels and bridges, and it can struggle to differentiate small closely spaced objects. Video… can be blinded by things like glare.'

complex process involving search and evaluation through all possible pathways to decide upon the best path that satisfies most requirements. A best route, after all, might be the shortest, the safest, the most efficient or the least challenging. It must also account for the machine's own geometry (shape, turning radius) and steering mechanism (the AWS' holonomic properties).[3]

Developments across dual-use technologies suggest that most of these issues appear resolvable, but the extreme environment of the battlefield means that hardware issues still place constraints upon the deployment of unsupervised machines, particularly concerning the enduring complexity of 'the last yard' whereby hardware components must manage the difficult interaction between host weapon and its immediate environment. Calculating the best path, moreover, is particularly complex in three dimensions (the case with a weapon's robotic arms). Manipulation is similarly challenging given the AWS' requirement to compute in real time the free space of each manipulator (the space in which movement is possible) in order then to understand that space for a particular combat task.

The AWS' sensors and effectors will generally have separate controllers, so requiring bespoke coordination and feedback on the weapon's location and state. Just as sensors must solely inform the weapon's task, so must its effectors be exactly fit for purpose, each with a conforming actuator that enables the effector to execute required actions and movement. Here, the complication is that such actuating will likely

> *Hardware issues still place constraints, particularly concerning the enduring complexity of 'the last metre' whereby hardware components must manage the difficult interaction between host weapon and its immediate environment*

be undertaken by quite different heterogenous means, whether by electric motors, hydraulics or pneumatics or by using dissimilar materials which may be photo-reactive, chemically reactive, thermally reactive or piezoelectric. A deployment consequence here then becomes the amount of idiosyncratic management and security (individual programming languages, tailored instruction routines, custom feedback loops) required for each component's unsupervised operation.

---

[3] Holonomic refers to the relationship between controllable and total degrees of freedom of a robot. If the controllable degree of freedom is equal to total degrees of freedom, then the robot is said to be holonomic.

A recurring theme in this book is that system heterogeneity adds fragility to overall function with current combat zones such as Ukraine demonstrating that component organisation is only likely to increase in complexity.[4] An example is useful. The AWS vision package requires a slew of cameras, requiring a portfolio of capabilities that knit together stereo processing, visual odometry, structure-from-motion, visual tracking, object finding and template matching. The conflict here arises from the Cohort trying to achieve a level of hardware generalisation that is appropriate to what is actually very specialised ends.

9.1 Hardware and sensor fusion issues for AWS

An effector is any device on AWS that touches the robot's environment. AWS actuator types must therefore be very broad so as to match likely tasking with available assets, actuating wheels, tracks, arms, grippers and other effectors on the weapon. So it becomes doubly relevant to think about an effector's DOF and how this might work in adversarial conditions. Current robotic hands may have thirty DOF. By way of context, a helicopter has six. In an autonomous land-based vehicle; however, only two dimensions of movement are really required. After all, the AWS' movement governor can control only two things: forward/reverse and rotation and, although that vehicle therefore has three DOF, only two of them are controllable.

The issue here is that since there are always more DOF than are controllable, there will always be motions that cannot be undertaken by that machine, such as moving sideways. Why is this important? However effective the two DOF, an unsupervised land-based machine must likely generate a more complicated path in order to carry out its motion task. A further complexity is that continuous trajectory must be achieved with discontinuous velocity whereby the weapon must stop and start in order to reach each destination, so creating additional complexity. Two relationships are important. First, the non-holonomic robot has more DOF than it can

---

4 Stepanenko, Kateryna and others, 'Ukraine's Long-Term Path to Success: Jumpstarting a Self-Sufficient Defense Industrial Base with US and EU Support', Institute for the Study of War (14 January 2024), https://www.understandingwar.org/backgrounder/ukraine%E2%80%99s-long-term-path-success-jumpstarting-self-sufficient-defense-industrial-base.

control. Second, of course, the more DOF a robotic weapon has, the more complicated it is to control. Drone operation in Ukraine might suggest that these challenges are being overcome, but a human generally remains in or on the loop.

The AWS also requires physical stability. It is not without reason that several military robots tend to be based on four-legged mobility in order better to deal with the issues of centre of gravity (COG) and balance. Four-legged construction provides an optimal polygon of support, but also introduces additional hardware challenge. In humans, the COG is quite high on the body and maintaining stability does not happen without experience and training. For this reason, any two-legged AWS must have a small such polygon, and computational routines will then be required to keep the COG stable to keep the weapon unit upright. Indeed, robot gaits have five required collaborative properties which, in the AWS, must work continuously, collaboratively and without oversight; stability ensures the robot does not fall over; speed allows it to move quickly; energy efficiency ensures it can exhibit durability; robustness allows it to recover from various failure modes; and, finally, overall simplicity ensures that the unsupervised robot's gait and operation are not too unwieldy – even more critical across a contested combat environment.

> *Drone operation in Ukraine might suggest that these challenges have been overcome, but a human generally remains in or on the loop*

Stability routines, moreover, are required to mediate wobble, lean and deviation. Long-dated work on machine learning (for instance, its application to riding a unicycle) highlights the complexity in compensating for physical instability in an unsupervised platform. Stability exists in two states: the AWS must be statically stable as well as dynamically stable.[5] In general, an AWS with more legs (or ground points) can maintain better static stability given its raised centre of gravity and broad polygon of support. But physical componentry also gives rise to error; backlash and inaccuracy arising from gear mechanisms are likely catalysts for error. Similarly, maintaining the weapon's COG over an efficiently small contact point with the ground still

---

5 Hofmann, Dietrich, 'Common Sources of Errors in Measurement Systems', *Handbook of Measuring System Design* (2005), http://eu.wiley.com/legacy/wileychi/hbmsd/pdfs/mm154.pdf. Hofmann's research provides a useful *aide memoire* for common error sources relevant to AWS deployment including input, sensor, signal transmission, conversion, cumulative, gain, dataset, materials, drift, load, thermal, operator, degradation, communication, mapping and other software errors.

requires active and trained effort. Four-legged robots, for instance, can only lift one leg at a time because three legs must remain grounded to remain statically stable. In considering land-based AWS, the Cohort is therefore faced with several design compromises between its weapon's stability, its speed of movement, energy conservation, robustness and simplicity.

Power for each such motor then requires additional platform weight, in turn needing stronger motors to lift the platform manipulators. In this vein, traditional ball-and-socket joints are particularly difficult to incorporate into artificial systems, muscles in animal rotary joints being linear actuators (relatively lighter, springier, more flexible and stronger compared to currently available synthetic alternatives). Whereas humans use their hands as general-purpose manipulators to work specific tools, AWS' robot manipulators will certainly be highly manufactured and bespoke instruments with dedicated lethal tools at the endpoint. Here, endpoint engineering is intractably complex requiring computationally intense inverse kinematics to manage the weapon's end-effector to a desired point, involving real-time conversion and management of each weapon's endpoint tools from a Cartesian (x, y, z) position. None of these hardware sequences is straightforward and must account for manipulator travel relative to the weapon's COG, problems caused by unexpected obstacles as well as hand-off, priority and tasking issues.[6]

A brief revisit of likely AWS sensor inventory is also relevant, the appropriate hardware to sense the condition of its systems as well as predicting the states of its immediate and far environments. First, proprioceptive sensors will be required to administer all elements of the robot's internal state (the position of its wheels, the angles of its joints) while exteroceptive sensors must process the platform's external world (light levels, distances to objects, sound). Taken together, these sensors constitute the weapon's

It must involve understanding the 'free space' of each weapon module that must account for all space in which machine movement is possible

perceptual system. As identified in previous chapters, the challenge is that the weapon's efficacy is based solely upon sensor inputs that will be fragmented, fragile and also the combinatory product of multiple senders. Sensed input

---

6 This is empirically complex and achieved algebraically (through the use of matrix equations), geometrically (by combining knowledge of the weapon arm with dynamic trigonometry) and numerically (by using smart 'guess work' and incremental adjustment in order to minimise local error).

must be processed *ex ante* to manage limitations such as effector and actuator noise, as well as using data divined from either hidden or partially observable states that are obviously outside the Cohort's control. Proprioception is also affected by the quantity of information that the weapon's sensors are returning each moment to its controller; although a simple contact switch provides just one single bit of information (on or off), a vision sensor is a stunningly rich source of information and correspondingly complicated in the amount of information it captures.

> *Inputs become a repetitive source of 'technical debt', from overlapping and obsolete data, partial and conflicting data, as well as the stitching together of different hardware sensor types*

AWS dependence on hardware sensors creates other deployment challenges. Sensor stimuli have properties that are not separately divisible. After all, battlefield shapes and sizes empirically do not appear 'alone' in each screen capture and often are themselves properties of other larger objects with conflicting properties. Adversarial feint has long (and materially) complicated this process. Although the weapon's action routines depend wholly upon information from its sensors, these inputs become a repetitive source of 'technical debt', whether from overlapping or obsolete data, partial and conflicting data, as well as the stitching together of different sensed data (the 'object identify problem') or from data duplication and the purging challenges arising from gaps in these inputs. Autonomous machines also require sophisticated 'sensor scheduling', the handshaking process that mediates *which* weapon sensor is next to provide a relevant measurement into the AWS decision processes.[7] Indeed, probabilistic models suggest that a portfolio of hardware components (akin, perhaps, to Ronald Arkin's ethical governor) is necessary if the category is to be scalable. An adjacent hardware challenge then arises from the AWS requirement for broad interface and communication capabilities, such as third-party speech with the necessary nuances in rate, volume, pitch, security as well as indicators of personality.[8]

---

7 Data remediation relies on rules-based combing routines, which complicate the subsequent management of the weapon's primary data given their dependence on transitive closure of source files.
8 Natural language processing and its hardware pose further complications. Despite progress in processing written language, it will remain difficult for an AWS to understand speech unless that platform is addressed with a few chosen words that are already familiar, without disruption or interference and delivered without accent.

Practical difficulties then arise from the autonomous linking of hardware combinations to specific weapon tasks. Many of the weapon's visual systems, for instance, must presumably be mounted on platforms that move with attendant rods and linkages, trunnions and gimbals, bearings and bushings that must be electronically and remotely managed. Moreover, the target under engagement may also be moving. For the AWS, however, its own movement will be a contradictory combination of, on the one hand, purpose and, on the other, reaction to other exogenous factors.[9] Does the weapon's movement sequence (its reaction, for instance, to an unexpected environmental hazard) trump its intended attack sequences? When is a movement abandoned as being too difficult?

Weapon balance must also be maintained across uneven surfaces and the weapon's speed must be adjusted constantly (and autonomously) to match these changing physical conditions. Correctional subroutines are then necessary to smooth for motion-generated interference, such adjustment compensating for the weapon's starting, turning, climbing, descending and stopping. Furthermore, the AWS must trigger these motor actions slightly in advance of execution, which cannot therefore be based on fixed or sequenced timing. For a human on a walk, motor actions almost always

> *Does the weapon's movement sequence trump its intended attack sequences? When is a movement abandoned as being too difficult?*

involve serial combinations of quite different sequences, each act being factored appropriately into the overall movement process. For the Cohort and its AWS, it is like having a strange remote control with buttons but no markings; there is no inherent connection between the weapon's inner image of an action and the set of routines that must advance that desired action.

Hardware components are generally task-specific, each with idiosyncratic challenges that the Cohort must consider in its weapon build. A worked example is useful to demonstrate the Cohort's frequent conundrum. Ultrasound sensors, for instance, are very dependable in measuring distance to an object using sound waves. But technology choice always involves tradeoffs.

---

9 US Army TRADOC G2 Mad Scientist Initiative, 'An Advanced Engagement Battlespace: Tactical, Operational and Strategic Implications for the Future Operational Environment', *Small Wars Journal* (24 October 2017), https://archive.smallwarsjournal.com/jrnl/art/advanced-engagement-battlespace-tactical-operational-and-strategic-implications-future. The publication interestingly separates battlefield assets (and their likely movement traits) into 'finders versus hiders', 'strikers versus shielders', dispersed assets as well as assets that are deliberately dormant.

Here, its scanning methods are inflexible; they are sensitive to temperature variation and are generally poor at reading reflections from soft, curved, thin or small objects. Their testing distance is limited. The Cohort might instead be drawn to a sonar solution, first tested in 1822. But sonar sensing relies upon discoverable high frequency sound waves. It is also subject to noise and 'reverberation' (whereby sound persists for some time after a sensor's polling), and commonly requires much power given that current is required to emit each ping. Indeed, the range of these sensors is determined entirely by the signal strength (and

> *It is like having a strange remote control with buttons but no markings; there is no inherent connection between the weapon's inner image of an action and the set of routines that must advance that desired action*

therefore power use) of the emitter. Ping emission by an AWS can betray its position, the more so as sound waves do not bounce linearly off the nearest surface to return as expected; the process is hampered by blind spots, multiple or unwanted reflections, obstructions and poor consistency, the direction of reflection depending upon surface properties and the incident angle of the sound beam. The Cohort's choice is not therefore straightforward, ultrasound sensing also being susceptible to 'specular' reflection, the echo from the outer surface of the bounce-back object (here, the AWS' target). If the target is smooth, the bouncing sound generates a false far-away reading. In contrast, rough target surfaces produce irregular reflections. The AWS may thus be fooled into thinking that there is no object or that it is at a great distance, making the unsupervised weapon vulnerable to sensor spoofing.

9.2 Configuration and calibration issues

Configuration determines the weapon's initial setup – that is, its initial arrangements and inaugural settings. It establishes its boundaries, tolerances and the calibrating of the weapon's capabilities immediately prior to its deployment. Configuration, as referenced throughout this book, is the pivotal responsibility of the Cohort, not least as it represents its participants' sole means of regulating the machine's behaviours, to prevent the AWS from attempting the unachievable or its obligations from exceeding its scope of permissions. Configuration manages the machine priorities, its slack as well as its learning protocols in order to take on new tasks and, in so doing, the modifying of permissions throughout the system to integrate new capabilities. Each user will have procured their machines based upon quite separate requirements, and each machine must be configured, validated and verified to match these procurement commitments. These are complex arrangements that require fine (and changing) balance. Calibration is then the process of maximising each weapon's performance given the restrictions and opportunities set by the Cohort's configuration.

How, then, might calibration affect hardware performance in AWS? 'Proportional' tuning is designed to prompt the weapon to react to error using both the direction and magnitude of that error. Determining how to size this calibration response is termed 'gain'. It is not obvious and requires inappropriate trial and error to avoid oscillation as the degree of required change increases; as adjustments move the unsupervised weapon close to its desired state, the means for its control become very different. Generally, the momentum of the controller's response to error – its own error correction – will pitch the weapon system beyond that state required by the deploying commander.

As adjustments move the unsupervised weapon close to its desired state, the means for its control become very different

What then are the *operational* challenges of weapon types occupying this more sophisticated end of the autonomy continuum? First, historical cost overruns and programme delays evidence that complex systems are difficult to deliver. This is not a recent phenomenon. Even by July 2009, the US Government Accountability Office (GAO) was reporting that most US uncrewed programmes were exhibiting 'cost growth' ranging from sixty per cent to more than two hundred and fifty per cent. Four of the uncrewed aircraft system programmes that were reported on by the GAO in 2015, the last year of consolidated reporting, experienced delays of between one to four years, mainly as a result of hardware development and testing problems. These scheduling headaches will likely mirror AWS procurement given, notes the US Department of Defense (DoD), the high

Scheduling headaches will likely AWS mirror procurement, given the 'high level of concurrency between development, production and testing; poor contractor performance; developmental and technical problems; system failures; and bad weather'

level of concurrency between development, production and testing; poor contractor performance; developmental and technical problems; system failures; and (even) bad weather. As discussed above, AWS economics is unsurprisingly a key operational challenge, and these platforms are also affected by lack of commonality across what is likely to be a bespoke weapon class with short production runs where it is difficult for manufacturers to claw back development and setup costs.

Robustness in these elaborate systems is also a point of friction. The efficacy of the Patriot Missile System, after months of use in Operation Desert Storm, degraded significantly, its radar becoming prone to drift from its prescribed search fan, leading to a significant miss ratio. Such ramifications are likely to be even more pronounced in autonomous systems where the human is by definition remote from each system's locus. In the case of the Patriot platform, accuracy issues prompted its control systems to keep recalibrating, eventually requiring a system-wide refit with new software. The upshot here is that shipping the first version of the system is often relatively easy but making subsequent improvements is unexpectedly difficult.

The operational challenge of *servicing* this hardware also argues against removing human oversight. Here, servicing covers all of the resupply and replenishment of materiel required to repair, sustain and develop these hardware assets. It also includes management of inventory updates and patches and, of course, general tuning and maintenance. Given that unsupervised machines will by definition be sensor- and software-intensive, their logistics tail will be disproportionately long, with complicated logistical planning given the autonomous, dispersed and idiosyncratic signature of such independent platforms. Nor can a weapon's initial setup be considered its final setup as, to maintain military advantage, new capabilities and new hardware combinations must be continually beta-tested and added to these unsupervised assets. As above, all this requires that the correct amount of the correct spares is positioned at the correct time and with the correct support at hand to carry out all necessary work.

9.3 Validation and testing

Validation and testing then becomes an adjunct Cohort priority. Given the pace of technical innovation, the responsible commander can no longer rely upon unchanging 'baselines', the more so given the maxim that 'no battle plan survives first contact with the enemy'. A danger for the Cohort is that inadequate practice and examination become the new norm. Testing, after all, is specialist, expensive and time consuming. Many of the costs articulated in Chapter Five regarding Article 36 weapon assessments hold true for the verification of these remote platforms and their capabilities.[10] That notwithstanding, the DoD's directive, Autonomy in Weapon Systems, updated in 2023, appears quite unambiguous on the matter of testing in its statement that AWS must go through rigorous hardware and software verification and validation as well as 'realistic system developmental and operational test and evaluation'.[11] The issue, however, is whether this is practical and whether the directive's intention will actually be followed in the chaos and pinch points of hot conflict. The issue is also one of practicality: within the one hundred and twenty-nine US Marine Corps MV-22 Osprey aerial platforms that had entered service by early 2018, there were already seventy quite separate and different configurations, all similar to the untrained eye but all subtly different and requiring bespoke maintenance, spare parts, validation and verification.

> *A danger for the Cohort is that inadequate practice and examination become the new norm*

The Cohort will use two validating methods for its AWS. Formal validation is based upon deductive review of the new system in order to establish a mathematical proof that the system is working. The deployment challenge here, of course, is the requirement to translate the properties of an unsupervised and unpredictable weapon system into formal mathematical language. Testing validation methods instead uses inductive processes to infer that the AWS is working based on a number of samples. But the Cohort's job is not made easy given the tiny amount of relevant data and simulations. A further complexity arises from these weapons' 'state-space explosion problem', the AWS' ever-larger decision space and the increased optionality that this creates. In this case, it becomes less certain that the Cohort will foresee possible event combinations that could lead to system failure or to inappropriate weapon behaviour. A second challenge is noted by Vincent

---

10 See, generally, Chapter Five, specifically 5.2 (Article 36 and LOAC-compliant weaponry).
11 US DoD, Directive 3000.09, Autonomy in Weapon Systems (25 January 2023), https://media.defense.gov/2023/Jan/25/2003149928/-1/-1/0/DOD-DIRECTIVE-3000.09-AUTONOMY-IN-WEAPON-SYSTEMS.PDF, para 4, policy 2. See also: Freedberg, Sydney, 'Streamlined MV-SS Maintenance', *Breaking Defense* (5 February 2018), https://breakingdefense.com/2018/02/streamlined-mv-22-maintenance-from-70-osprey-types-down-to-5.

Boulanin and Maaike Verbruggen, the authors of the report published by the Stockholm International Peace Research Institute used in earlier chapters: given that the machine-learning basis will presumably require wholly new parameters (and an overhaul of programming) for each AWS after every new learning iteration, to what degree should the weapon's performance and correctness be revalidated each time it 'learns' something new? Boulanin and Verbruggen's conclusion is clear. As autonomous systems become more intelligent, interactive and capable of adapting to complex and dynamic environments, it becomes, practically and financially, infeasible to continue to test all ranges of imports to, and possible states of, those systems.[12]

Mindful of these systems' potential high-regret outcomes, AWS testing and validation must also be 'red-teamed' whereby an independent, properly resourced group with an adversary's mindset, challenges the Cohort at all phases of the AWS' evaluation, development, simulation and deployment. In AWS, validation and verification (V&V) are also complicated given that bespoke rules of engagement must be embedded in all weapon variants, each mindful of border, culture and doctrine considerations as well as the Cohort's own laid-down standards. V&V must also be in place in advance of any deployment and not 'learned' by experience. In this vein, AWS programmers must also factor for policies and regulations under the myriad of relevant authorities, most of which are bureaucratic, lack incentives to act quickly and, being multi-jurisdictional, are frequently in conflict. The challenge is that commercial parties in this procurement chain have little incentive to devote resources to such testing, with commentators noting a gap in technical

---

12 Boulanin, Vincent and Maaike Verbruggen, 'Mapping the Development of Autonomy in Weapon Systems', Stockholm International Peace Research Institute (November 2017), https://www.sipri.org/sites/default/files/2017-11/siprireport_mapping_the_development_of_autonomy_in_weapon_systems_1117_0.pdf, p. 69. It should be noted that this problem is not exclusive to autonomous weapon systems, but it applies across all machine autonomy.

literacy to deliver an appropriately robust regime that is based on emergent ('what we must test') rather than traditional practices ('what we test').

The autonomous vehicle industry, moreover, is *not* a useful proxy for the Cohort's own verification and testing of its AWS. A Tesla may be able to navigate from point A to point B but does so within the benign bubble of the rules of the road. Its environment is actually well regulated and also without adversarial peril.

On the other hand, AWS path planning will rarely be a single-dimension geographical obligation ('get me to a location'), but instead just one part of a goal-anchored set of processes. But some similarities do exist. An interim destination for the weapon is just as likely to be outside its immediate sensory range, as is the case with an autonomous car. Much of the AWS' hardware portfolio is already shared across military and retail platforms, both applications being built upon a focus on 'negative' objects (as in unexpected obstacles and efficient path planning). Navigational parameters are similarly complicated for both parties by moment-to-moment changes to their immediate environment but, unlike the car and its environment's well-understood network of roads and lanes, land-based AWS must add to its focus about what is to its front, its side, behind, below and above it (very little of which is likely to be mapped). The AWS (similar to the car) is also not itself simply a single point in space, requiring these machines take into account geometry (shape and turning radius) and physical footprint (azimuth, pitch and roll, size and elevation).

> *A best route might be the shortest, the safest, the most efficient or the least challenging*

10
Oversight
Command and control constraints to AWS deployment

Significant transformation must take place in how armies fight if automatic weapon systems (AWS) are to be deployed. Just as drone proliferation has changed battlecraft in Ukraine, removing supervision will transform how combat is undertaken across all fighting domains. Regardless of deployment model, several well-tried concepts that have long comprised how humans wage war will require fundamental re-examination as lethal autonomy is deployed. Any such reappraisal, however, must still be anchored by human factors which will continue to underpin that battlecraft. The aim of this chapter is therefore to review the role of the human in combat's command and control as it might influence adoption of unsupervised means. It is also to consider an appropriate weighting for the human dimension of AWS deployment. The starting point is that technology, while valuable in its own right, is never as valuable as human life.[1] The imperative is that human involvement comes with obligations that outweigh any merely economic 'value' of physical equipment.

---

1 Warne, Leoni and others, 'The Human Dimension of Future Warfighting', Australian Department of Defence (Defence Science and Technology Organisation), (September 2004), http://www.dodccrp.org/events/9th_ICCRTS/CD/presentations/7/162.pdf. Although published long ago, the Australian study remains useful in highlighting human traits that challenge AWS deployment, specifically the behaviour divides between the warrior (discipline, decisiveness, loyalty, confidence) and the peacekeeper in his or her responses (patience, empathy, responsibility, rapport, lesson-learning) to combat scenarios. The study also emphasises the increasing role of trust (devolving responsibility to lower levels, disseminating information to ever wider audiences) and context (antidote to volume, presentation, testing reliability in battlefield data) that remain likely in future warfighting and yet incompatible with models of AWS deployment.

The chapter has four sections. An analysis of 'command' across its several levels leads first to a review of weapon targeting to gauge whether the obligations of this process can feasibly be captured by code. The chapter then considers the scope of new behavioural competencies that will be needed as AWS are deployed and oversight is diluted. This requires consideration of skill sets, the role of novel capabilities, of untested

The starting point is that technology, while valuable in its own right, is never as valuable as human life

proficiencies and prospective benchmarks as well as insight into new models of leadership. The chapter's final section, this book's synthesis, is only then in a position to review meaningful human control (MHC) as an appropriate (and perhaps statutory) framework for placing autonomy in lethal engagements. For the purposes of this chapter, the command chain is defined as the line of authority and responsibility along which orders are passed within units and between units. It is the will of that commander expressed for the purpose of bringing about a particular action. Control is then that command (which might be less than full authority) exercised by the local commander over the activities of the organisation and other organisations. This chapter primarily concerns control from the perspective of the Delivery Cohort and, in particular, the human commander overseeing AWS deployment. It also touches on control from the deploying state's standpoint by reviewing consequences that might arise from statutory restrictions around AWS.

Past editions of the UK's Army Doctrine Primer highlight this conflict with AWS deployment, especially regarding 'shaping tasks', 'the Decisive Act' and the primer's discussion about the nature of battlecraft. In this case, doctrine is the expression of how military forces undertake war (from campaigns down to individual engagements). It acts as a guide to action rather than a set of defined rules. It is also a common frame of reference that reflects an army's views about what works in war based on past experience.

Command is then the appropriate meld of control, authority and permissions as well as the power to influence behaviour within courses of events

Command is then the appropriate meld of control, authority and permissions as well as the power to influence behaviour within courses of events. The primer's focus is upon human-centric matters entitled 'unity of effort', 'freedom of action', 'building of trust', 'timely and effective decision-making' and 'mutual understanding'. Little of this sits comfortably with

AWS deployment. Posited upon complex sets of prerequisites (authority, permissions, trust and processes), AWS deployment instead requires that this doctrine be substantially updated.

The irreducibility of command is another recurrent theme for commentators. In considering why subordinates follow their commander, the 'trust test' within General Sir Rupert Smith's leadership framework concentrates upon the 'intangible principles of comradeship, respect, endurance and sacrifice regardless of their situation'.[2] There is, notes Smith, an enduring bond that underpins all battlefield activities. Trust is based on 'character' and 'competence' and must be laid down 'over decades' and well in advance of battle, a conclusion that parrots his peers and predecessors in associating leadership with moral and physical courage. His point is also that leadership 'crisis-proofs' a battleplan and underpins the characteristics of delegation, innovation and initiative. Indeed, undertaking lethal engagement outside this framework clearly runs counter to current practice. Even, therefore, if AWS are adopted by states at the margins, the very deployment of unsupervised weapons is at odds with what is a long-dated and tested eco-system.

Leadership Insights, published by the UK's Centre for Army Leadership, echoes the notion that coding for command is enduringly unworkable. It is difficult to shoehorn AWS into the framework suggested, for instance, by 'The Intelligently Disobedient Soldier' and its conclusion that curiosity, critical thinking, imagination and open-mindedness are the key qualities of commanding.[3] A companion paper, also from the Centre for Army Leadership and titled 'What the Hell Do We Do Now?', points instead to instances where no solution makes itself available but also how it is human innovation that can circumvent these bottlenecks.[4]

> *Leadership 'crisis-proofs' a battleplan and underpins the characteristics of delegation, innovation and initiative*

The publication on 'Empowerment: Beyond Delegation' similarly notes the battlecraft advantages of free thinking, bottom-up

---

2 Smith, General Sir Rupert, previously Deputy Supreme Allied Commander Europe, *Utility of Force: The Art of War in the Modern World* (Vantage, 2008), generally.
3 Clark, Lloyd, 'The Intelligently Disobedient Soldier,' *Leadership Insights*, no. 1 (March 2017), https://www.army.mod.uk/support-and-training/our-schools-and-colleges/centre-for-army-leadership/leadership-resources/leadership-insights/leadership-insight-no1/.
4 Wilson, Luke, 'What the Hell Do We Do Now?', *Leadership Insights*, no. 3 (August 2017), https://www.army.mod.uk/media/22956/centre-for-army-leadership-leadership-insight-no-3.pdf.

generation of ideas, innovation and collaboration, none of which seem possible without meaningful human participation across the battlefield.[5]

Finally to this point, 'Learning to Change' focuses on the premium of experience and the lessons that it enables.[6] Here, an interesting differential emerges between, on the one hand, 'poor learners' (rules-based, rote learning, a proxy here for AWS) versus innately human mechanisms that are better adapted to all manner of uncertain situations where previous experience is being repurposed in order improve outcomes in subsequent scenarios. Commentators similarly note that nothing is so painful as a great and sudden change, a key danger for the Cohort being group-think, from both hidden hierarchies within the Cohort as well as from group dynamics within the many parties responsible for the weapon's routines.

Isolating these tenets better prepares the reader to judge whether the Cohort can really embed them into those same routines. This challenge, moreover, is aggravated by the plethora of specialist troops, units, dedicated functions and bespoke equipment (all within a myriad of command structures) that comprise a modern military. Moreover, the coordination of force is already affected by the speed and range of modern weapons that have reduced the time in which the Cohort can exercise this control. Indeed, machines' capacity for fast, accurate calculation has long since exceeded that of the human commander, and where having to spread out over considerable areas to avoid surveillance and attack further complicates this general process of command. It is difficult to foresee how these same control complexities will be factored for in the unsupervised machine.

Having to spread out to avoid surveillance and attack further complicates the general process of command

Weapon control must also be considered from the perspectives of AWS' procurer and operator. Currently, the axioms for lethal engagement are broadly a threshold for liability and an obligation to prevent illegal harm.[7]

---

5 Cooper, Paul, 'Empowerment: Beyond Delegation', *Leadership Insights*, no. 7 (April 2018), https://www.army.mod.uk/support-and-training/our-schools-and-colleges/centre-for-army-leadership/leadership-resources/leadership-insights/leadership-insight-no7/.
6 Skinner, Kirsty, 'Learning to Change', *Leadership Insights*, no. 9 (September 2018), https://www.army.mod.uk/support-and-training/our-schools-and-colleges/centre-for-army-leadership/leadership-resources/leadership-insights/leadership-insight-no9/.
7 Human Rights Watch, 'Killer Robots and the Concept of Meaningful Human Control', Memorandum to Convention on Conventional Weapons (CCW) Delegates (April 2016), https://www.hrw.org/news/2016/04/11/killer-robots-and-concept-meaningful-human-control, p. 2.

It is therefore the nature of the human's control rather than any specific link between this control and specific technologies that becomes key.⁸ The likes of the non-governmental organisation (NGO) Article 36 refines this further to include 'when, where and how weapons are used; what or who they are used against; and the effects of their use'.⁹ But while much of this book's focus has been upon the Law of Armed Conflict (LOAC) and compliance within the existing legal framework, Ukraine loudly suggests that such rules quickly collapse in war.¹⁰ Generations of conflict, whether at sea, on land or in the missile age, show that those who attack first are doing so in an effort to achieve first advantage.

10.1 The notion of meaningful human control

MHC has been widely suggested as the basis for a statutory framework to ensure human participation in lethal engagement. The premise here is that without human control, an AWS is unlikely to ensure proper incorporation of legal rules or might act under an interpretation of these rules that erodes civilian protections. A second observation is that the LOAC cannot, in and of itself, prevent human judgement from being diluted to the point of being meaningless unless that human oversight is sufficiently robust.¹¹ Four prerequisites exist for this prerequisite to be satisfied: predictable, reliable and transparent technology; accurate information on outcomes and context; the potential for timely intervention; and accountability to a defined, understood set of standards.

---

8 UK Ministry of Defence, 'Human Machine Touchpoints: The United Kingdom's Perspective on Human Control Over Weapon Development and Targeting Cycles', UK submission to the Group of Governmental Experts of the High Contracting Parties to the Convention on Prohibitions or Restrictions on the Use of Certain Conventional Weapons Which May Be Deemed to Be Excessively Injurious or to Have Indiscriminate Effects (August 2018), https://docs-library.unoda.org/Convention_on_Certain_Conventional_Weapons_-_Group_of_Governmental_Experts_(2018)/2018_GGE%2BLAWS_August_Working%2BPaper_UK.pdf, generally.
9 Article 36, 'Killing by a Machine: Key Issues for Understanding Meaningful Human Control', cit. Human Rights Watch, 'Killer Robots and the Concept of Meaningful Human Control', p. 2.
10 Shugart, Thomas, 'Uncrewed But Confident: Forging New Rules of the Road to Avoid Accidental Escalation', War on the Rocks (1 May 2024).
11 Article 36, 'Key Elements of Meaningful Human Control', Geneva (11 April 2016), https://www.article36.org/wp-content/uploads/2016/04/MHC-2016-FINAL.pdf. Much of the scholarship on this component of the book is provided by the NGO.

At its most basic level, the requirement for MHC arises from two fundamental premises. First, the notion of a machine applying force and yet operating without any human control whatsoever should be considered fundamentally unacceptable. Second, a human simply pressing a 'fire' button in response to indications from a computer but without that person being aware and responsible is not sufficient to be considered 'human control' in any substantive sense. The term 'meaningful' must therefore be used to express this threshold of substantiality, although other words are regularly suggested such as 'appropriate', 'effective', 'sufficient' or 'necessary'. For our purposes, however, the term 'meaningful' is suitably broad and not tied to a specific context. It also derives from an overarching principle rather than being outcome driven (for example, 'effective' or 'sufficient') or implying something administrative, technical or bureaucratic.

> *The notion of a machine applying force and yet operating without any human control whatsoever should be considered unacceptable*

The foregoing analysis has generally treated the AWS as an agent (or part thereof) but, in considering MHC, it is useful to situate the weapon simply as a machine in this discussion. In this vein, it is common for diplomats and 'experts' to discuss whether AWS 'will be able to apply legal rules'. This is misleading, of course, as machines do not apply legal rules. They may undertake functions that are in some ways analogous to legal rules (the example used by Article 36 being that the AWS may be programmed to initiate force should certain preset thresholds be met around, for instance, certain heat patterns common to armoured fighting vehicles), but in doing so they are not 'applying the law' but simply implementing sequences that their Delivery Cohort has configured. Indeed, only humans can be addressees of international humanitarian law.

The issue therefore arises about where this intervention might reliably sit. Is it within the broad act of employing violence or is it around the unsupervised identification of targets? And can MHC become a de facto 'unit' of legal management in an engagement sequence where agency is no longer a human preserve? Regardless of framework, compliance only works if delineated responsibilities of command remain in place to assign liabilities and it this relationship which then requires that involved parties understand how AWS systems will operate and the effects expected from their operation. It is also this 'obligation on the person' that has long been articulated by the law of war manuals published by the US Department of

Defense[12] whereby members of the Delivery Cohort can only meet that legal obligation if appropriate information on the context of each individual attack is understood at the point of decision.[13]

The term 'attack' in international humanitarian law does indeed provide such a recognised 'unit' of military action. And it is generally understood that certain legal judgements must be applied to these attacks which therefore form part of LOAC's structure in that they represent 'units of military action' and of human legal application, an example being the Geneva Convention's Article 57 and its requirement that precautions be taken in each such attack. Where the law refers to those 'who plan or decide upon an attack', it is referring to humans and not machines. This obligation extends to verifying the objective, choosing the means and methods of attack, and refraining from or cancelling that attack in certain circumstances.[14] We also know that an attack must be directed at a specific military objective otherwise it is indiscriminate.[15] A military objective must similarly be definable, offer tangible military advantage and do this at a specific point in time.[16] These matters must furthermore be assessed by the humans who plan and decide upon each attack, control then being necessary in some form in order to act on the legal judgement that has been made by that responsible human. Simply assuming that these capacities can be programmed into the weapon by the Delivery Cohort is a clear abrogation of human legal agency, breaching the case-by-case approach that forms the structure of these legal rules.

> *These matters must be assessed by the humans who plan and decide upon each attack, control then being necessary in some form in order to act on the legal judgement that has been made by the responsible human*

This is not to suggest that existing LOAC alone represents a sufficient basis for managing AWS. As noted by Article 36, it is 'simply to point out that the existing legal structure (human judgement being required over 'attacks')

---

12 United States Department of Defense, *Department of Defense Law of War Manual* (June 2015), P.6.5.9.3; 'LOAC obligations of distinction and proportionality apply to *persons* rather than the *weapons* of themselves... as these rules do not impose obligations on the weapons... and, of course, an *inanimate* object could *not* assume an obligation in any event'.
13 Robinson, Darryl, pp. 20-23 ('The Problem of the Successor Commander'). This also raises the issue of that commander's selection, training, education and development and the type of individual able to undertake this tasking.
14 Article 36, 'Key Elements', p.3.
15 Article 36, 'Key Elements', see Article 51.4,a.
16 Article 36, 'Key Elements,' see Article 52.2.

implies certain boundaries to independent machine operation'.[17] It should also be noted that an attack is not necessarily a single application of kinetic force to a single target object, another enduring complexity for the Delivery Cohort being that an AWS which is to be accretive to a commander must also be able to tackle multiple force events against multiple objects and do so within spatial, temporal and conceptual boundaries. Furthermore, all of these must conform precisely to LOAC and the Cohort's goals. All this must then be undertaken at the most local and granular level and where that attack cannot proceed from one event to another without human legal judgement

Arguing that these capacities can simply be programmed into the weapon by the Delivery Cohort is a clear abrogation of human legal agency

being applied in each case. The foregoing chapters strongly suggest that this will remain a technical and operational stretch.

MHC is obviously most germane to a weapon's targeting routines. The North Atlantic Treaty Organization's (NATO) processes for joint targeting apply both to deliberate targeting (generally, in advance) as well as targeting that is dynamic (generally, ad hoc and of the moment) and, as currently written, comprise five contiguous phases. Only one component of that targeting cycle (here, Phase Five, comprising 'mission planning and force execution') is considered appropriate for automation. The remaining phases

(in particular, that of commander's intent but also capabilities involving analysis, decision and assignment) all talk to the difficulty of porting across targeting's currently compliant processes to these independent weapons.

They also vary in risk; as the time available for a targeting decision decreases, the amount of risk in that routine must increase. Prioritising these targets certainly requires human supervision given the subjective and changeable mix of goals, values, context and awareness involved in this process. Here, humans are factoring for time sensitivity, making idiosyncratic calculations regarding payoffs, toggling between lethal and non-lethal outcomes as well as handling constraints created by restricted target lists and other no-strike entities. This is not a matter for delegation. Even Phase Five of NATO's Joint Targeting

---

17 Article 36, 'Key Elements', pp. 3 and 4.

Cycle (where it is judged that automation might be possible) is made up of seven quite separate subcomponents (the 'fit, fix, track, target, engage, exploit and fail-safe' of mission planning and force execution discussed above), each piece adding to the routine's chance for machine error.

Control in this matter should always be understood as a mechanism for achieving commander's *intent*, with target information (including understanding the nature of unintended consequences) therefore a key starting point in the MHC debate. What types of 'object' (the notion of the Cohort's set of 'target profiles' for its AWS) conform to the commander's intention? In order to ensure predictability, after all, the Cohort must factor for the attack's exact environment as well as the type of kinetic force to be applied. Together, these conditions comprise the attack's context that must underpin each decision to initiate violence.

The conundrum for the Cohort is set out by Article 36: for any given environment, 'it follows logically that greater area and longer duration by a technology result in reduced predictability and so reduced human control'.[18] The timeliness of this sequence is also significant because the accuracy and relevance of this information degrade over time and does so in volatile, unpredictable ways.

Unsurprisingly, a control mechanism based on MHC faces strong headwinds. How might a legal framework anticipate for capabilities that might emerge in due course? Any umbrella arrangement, moreover, usually suggests cozy notions of trusteeship and guardianship with all the political weaknesses that these sponsorship arrangements entail. Is this framework to be voluntary or mandatory? What are the policing and sanction mechanisms for the arrangement? In this vein, AWS control might best be achieved by articulating a positive requirement, for instance that human control participation must always precede weapon engagement, regardless of whether that violence is lethal. This would also conform to the general observation (here, set out by the Holy See) that 'prudential judgement

> 'It follows logically that greater area and longer duration by a technology result in reduced predictability and so reduced human control'

---

18 Article 36, 'Key Elements,' p. 4.

cannot be put into algorithms'[19] and where exercise of judgement depends on much more than plainly numeric analysis of data. It is, notes again the Holy See, too difficult for unsupervised systems, no matter how data is processed, to exercise the required level of judgement to enable lethality. A second (and complementary) approach is to tighten the national level weapon review mechanism that is articulated in Additional Protocol I to the Geneva Conventions.[20]

Again and again, two principles return to frame the argument. First, the exercise of control is the fundamental governance pillar in the use of force.[21] Second, parties are unlikely to favour deploying a weapon system that has absolutely no form of human control whatsoever. But a verso here remains that for the Cohort any such mechanism cannot equate to an 'absolute' control; it is a given that weapons that are extant already contain substantial levels of deployment uncertainty. Instead, the Cohort should be guided by the tests of reasonableness that also exist whereby, as above, the exercise of MHC is not achieved by a human simply pressing a 'fire button' every time a light comes on a dashboard. Here,

> *The legality of any attack must be assessed on a case-by-case basis: a target may be legitimate at one point in time and in one location but may change into a civilian object that is immune from attack when moved in time or space*

therefore, the Cohort should be guided by four key questions. First, what is the nature of human control to be exercised over these weapon systems? Second, at what moment does that human control cease to be meaningful? A third consideration concerns the extent to which computer programming can operate as a proxy for human control. Finally, what conditions need to be evident for those responsible to be no longer confident that norms, provisions and laws can appropriately be applied by these weapons? In all these cases, after all, the unsupervised weapon is relying upon physical characteristics (infrared emission, shape, gait, disposition, heat signatures et al.) to be proxy indicators of a valid target.

The deployment issue in front of the Cohort is that the legality of any attack must be assessed on a *case-by-case* basis: a target may be legitimate at

---

19 Statement of the Holy See, CCW Meeting of Experts on Lethal Autonomous Systems, Geneva (16 April 2015), p. 4.
20 See Chapter Five (Obstacles), specifically 5.4 (Article 36 and LOAC-compliant weaponry).
21 Article 36, 'Structuring Debate on Autonomous Weapon Systems', Briefing Paper (November 2013).

one point in time and in one location but may change into a civilian object that is immune from attack when moved in time or space. These functions may also be dispersed between different physical structures and locations[22], in which case, the Cohort must decide dynamically what characteristics are acceptable as indicators of a target for the AWS to attack. The local commander, moreover, must reconcile this deployment without knowing what objects might be targeted in every given context.

How intractable is this? After all, a decade of deliberations at the United Nations Convention on Certain Conventional Weapons (which has largely failed to resolve any of the challenges identified in 2015 by Michael Horowitz and Paul Scharre) would seem to point to MHC being a Gordian Knot that will require leadership to resolve. Without collaboration and willingness, 'at best, [MHC] merely shifts the debate to "what is meaningful?" At worst, failure to define these terms could, if embedded in international discussions, lead to flawed policy choices', with little improvement in accountability and, worse than that, 'an off-loading of moral responsibility' for these weapons' consequences.[23] The efficacy, moreover, of human control is already patchy in existing military systems, in which case any regulatory framework must be properly practical. It also needs to be appropriately broad such that emerging weapons technology does not require painful referral, thereby limiting legal and political revisiting of the framework.

It is also unhelpful to the Cohort that popular familiarity with the issues remains poor, debate over MHC using an instrumentalist view of a weapon package with underlying technologies that are discussed as tools in the hands of technologically advanced countries where humans retain agency and where these weapons' black boxes remain somehow divorced from social and political contexts. Arguments should instead signpost three essential components of MHC. In the first place, human operators must be

---

22 Article 36, 'Key Elements'.
23 Horowitz, Michael and Paul Scharre, 'Defining "Meaningful Human Control" over Autonomous Weapons', Just Security (19 March 2015), https://www.justsecurity.org/21244/defining-meaningful-human-control-autonomous-weapon-systems/.

making informed, conscious decisions about the use of all such weapons. Operators must then have sufficient information to ensure the lawfulness of their actions given what they know about the target, weapon and the context for lethal action. Finally, each such weapon must be designed and tested and the human operators appropriately trained.

These components, however, are only a starting point. Does, for instance, MHC comprise a new principle in its own right, or is it simply  an overarching concept to ensure compliance with current laws of war? Indeed, just listing the plethora of terms used by states to define MHC illustrates the issue's complexity: 'informed conscious decisions'; 'appropriate human judgement'; 'cognitive participation'; 'human intervention'; 'sufficient information'; 'information assurance'; 'accurate information'; 'deliberation on the target nature'; 'significance and likely incidental effects'; 'contextual awareness'. This wilful ambiguity is also not helped by the very different motivating factors behind these precepts, whether these be legal, human rights, operationally or politically driven.

A final dynamic in the debate has become the degree of control to be ceded and how state signatories might be prejudiced if they support a new instrument restricting these weapons' deployment. The assumption remains that weapon autonomy will most benefit advanced militaries, given that weaker parties are not, in practical terms, giving up anything should weapon autonomy be restricted by statute. Experience, however, suggests that once autonomous technologies diffuse across borders and tasks, AWS may actually benefit exactly those weaker parties given the incongruent costs (economic and operational) of mounting substantial crewed operations in contested environments versus deploying AWS that are capable of asymmetric effect without much infrastructure. A further challenge arises. How can signatories defend themselves against parties that subsequently deploy AWS in contravention of a ban? This would be the worst of all possible outcomes, empowering the more odious regimes with potentially dangerous weapons while leaving states that care about international law at a disadvantage.

> *The inference must be that an unsupervised weapon, out-of-touch with the human responsible for using that weapon, cannot appropriately undertake proper situational assessment*

MHC therefore concerns the relationship between operator and weapon, notwithstanding that such narrow focus ignores the empirical division of labour between human and machine (and the resulting reconfiguration of human agency) that is already visible in engagement sequences. As we have seen with Lavender and Gospel, these machines are already formulating objectives, developing targets and suggesting weaponeering, the target cycle already characterised by a distributed framework of control.[24] As noted by Stuart Russell, the force behind Slaughterbot, 'the technical capability for a system to find a human being and kill them is much easier to develop than a self-driving car. It's a graduate student project'.[25] MHC might work better as a control framework if the technology is conceptualised quite narrowly; it is instead a tool for translating user intent into outcomes (depending on each engagement's context), with the user framing the deployment of each weapon within the three elements of intent, technology and passing context. In this vein, it is only the human controller who can test the underlying decision assumptions based upon the users' proximity to the information source as well as the past reliability of those sources.

While these subroutines may vary between engagement type and situation, they also vary in risk; as amount of time available for a targeting decision decreases, the amount of risk in that routine increases

'The technical capability for a system to find a human being and kill them is much easier to develop than a self-driving car. It's a graduate student project'

MHC must include the capacity for timely intervention, either by a human or, should system speed be faster than human capacity, then by another trusted process. Timeliness here depends upon the engagement's context and also the level of effect that might be caused. At its most basic level, human control might be starting or stopping the process. Moreover,

Moreover, if inaction occurs, there should be a similar regime in place to ensure the accountability of those individuals responsible for this non-intervention

---

24 See Chapter Four (Deployment), specifically 4.2 (Planning tools).
25 Russell, Stuart, cit. Adam, David, 'Lethal AI Weapons Are Here: How Can We Control Them?' (23 April 2024), *Nature*, https://www.nature.com/articles/d41586-024-01029-0.

if inaction occurs, there should be similar sanction in place to ensure the accountability of those individuals responsible for this non-intervention. Accountability then becomes about ensuring consequences for each action or inaction.

MHC should furthermore be present ante bellum, in bello and post bellum (before, during and after each use of force). Ante bellum, these platforms should be 'responsibly innovative' where scientists and manufacturers understand their obligations to mitigate harmful consequences. Notwithstanding much of AWS' technology is likely dual use, other componentry remains either classified or speculative with public dialogue fettered by concerns about national security or commercial confidentiality. A complication here is also that autonomous weapons systems may function through widely dispersed physical assets,

> MHC must be present ante bellum, in bello and post bellum (before, during and after each use of force

all of which might also function in ways that may not individually constitute an AWS in its own right.[26] The defining characteristics of AWS must thus be tied to the relationship of human users to processes of decision-making and not to the use of a technology.

---

[26] See, generally: Article 36, 'Submission to the UN Secretary-General – Considerations on the Development of an International Legal Instrument on Autonomous Weapons' (8 May 2024), https://article36.org/wp-content/uploads/2024/05/78-241-Article36-EN.pdf.

11
Conclusion

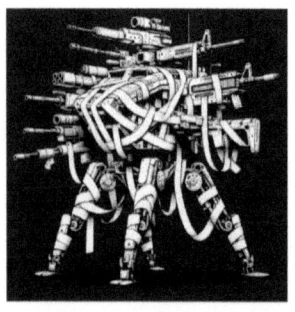

Novel means in battlefield practices continue to spur debate in research, legal and ethical communities. Arguments are well developed on all sides but, concerning the matter in hand, debate usually starts from the position of the potential of autonomous weapon systems (AWS); that weapons assisted by artificial intelligence (AI) may be more accurate; that they may be more conservative and surgical than those controlled by humans; that they might reduce conflict's collateral damage amongst civilians (and where they live); and that fewer soldiers will be killed and maimed. The popular narrative is also that technology is able to protect vulnerable people in ways previously not possible.[1] But clearly the focus of this book is on the downsides to these machines – the arguments that suggest that these independent weapons may make catastrophic mistakes and, indeed, questions about whether it will ever be sufficiently feasible for parties to commit their warfighting to these assets. A further aspect to the discussion is provided by those whose moral and ethical concerns make delegating the decision to kill to an algorithm simply unacceptable. As noted throughout this book, the central question remains whether a machine can ever make unsupervised decisions about who and what to target in the chaos of the modern battlefield.

Efforts to regulate the use of weapons date back centuries. Mediaeval knights agreed not to target each other's horses with their lances. In 1675, the warring states of France and the Holy Roman Empire agreed to ban the use of poison bullets.[2] It is no surprise, therefore, that parties have been campaigning for years to control weapons able to undertake lethal engagements without

---

1 Adam, David. 'Lethal AI Weapons Are Here: How Can We Control Them?', *Nature* (23 April 2024), https://www.nature.com/articles/d41586-024-01029-0.
2 Adam, 'Lethal AI Weapons Are Here'.

human supervision. Your author was at an early meeting of the Convention on Certain Conventional Weapons (CCW), the talking shop at the United Nations for such matters, to discuss just such a statutory ban more than a decade ago. But the issue has only gained in importance, its timeliness best illustrated by the UN's 2024 decision to add the topic of lethal autonomous weapons to its General Assembly's agenda for its next meeting in September 2025, with the Secretary-General declaring that a ban on such systems is his priority for the end of 2026. This, then, is the context for the debate over autonomous weapons and the subject of this book.

The CCW, however, is a fundamentally difficult forum in which to advance such debate. It requires consensus among all its members if it is to pass regulations, and signatories to the Geneva Conventions that are actively pursuing these capabilities are obviously unwilling to agree to restrictions. It does not help that the weapon category is difficult to define: in 2018 Israel declared that 'we should stay away from imaginary visions where machines develop, create or activate themselves – these should be left for science-fiction movies'.[3] A similarly destabilising position is adopted by Germany with its inclusion of the term 'self-awareness' as a necessary attribute for autonomous weapons, a capability that is particularly difficult to define and considered very unlikely even in the medium term.

> 'We should stay away from imaginary visions where machines develop, create or activate themselves – these should be left for science-fiction movies'

Notwithstanding this backdrop, several irreducible conclusions are evident from this study. On the one hand, although Ukraine has confirmed that the processes of war and violence still require human participation, the attacks of 11 September 2001 in the US long ago demonstrated that 'you don't need soldiers to start a war' (irregular combatants simply hijacking civilian airliners 'with nothing more lethal than boxcutters').[4] It is therefore a key premise that even the use of technology in battlespaces actually remains a basic human endeavour and one that is based upon basic human engagement. As Paul Scharre points out, 'the winner of the robotics revolution will not be

---

3 Yaron, Maya, 'Statement', Permanent Mission of Israel to the UN (11 April 2018), https://docs-library.unoda.org/Convention_on_Certain_Conventional_Weapons_-_Group_of_Governmental_Experts_(2018)/2018_LAWS6b_Israel.pdf.
4 Brooks, Rosa, 'Can There Be War Without Soldiers?', *Foreign Policy* (15 March 2016), http://foreignpolicy.com/2016/03/15/can-there-be-war-without-soldiers-weapons-cyberwarfare/.

who develops this technology first or even who has the best technology, but who figures out how best to use it'.[5]

So how does this inform the debate on deployment? First, the lessening of *in loco* human supervision over military practices is a hardwired phenomenon that has been underway for an age, although this transformation is likely to carry on being evolutionary rather than suddenly turning revolutionary. Most sophisticated militaries have been using AI tools for decades and data analysis has long underpinned decision-making on the battlefield. But the discontinuity here is that the fusing of different and numerous data points might also improve decision making. A second finding is that AI and autonomous systems should not be conflated. AI is really not constrained to any one definition, from a chat box to facial recognition technology to a large language model (LLM). Autonomous weapons, on the other hand, are straightforwardly lethal pieces of machinery where the human is 'entirely out of the conversation'.[6] Finally, to this point, norms in this matter remain in the strange place of being immature, contradictory and deeply polarising according to the deploying party and its audience.[7]

It is therefore a key premise that even the use of technology in battlespaces actually remains a basic human endeavour

A finding then becomes that AI agents are 'inherently and irredeemably unreliable narrators due to the very nature of their architecture'.[8] At their basic level, notes Will Oremus, these models generate answers that appear coherent but not answers that are true.[9] AWS' systemic problem, of course, is that its programs are unable to tell whether the inputs being used provide a reliable answer to the question posed. And even when they scrape a good source, they may misinterpret what that source is saying. If this is the case (and the five chapters that comprise this book's practical analysis back

---

5 Scharre, Paul, 'Robotics on the Battlefield – Part I: Range, Persistence and Daring', Center for a New American Security (May 2014), p. 9, https://s3.amazonaws.com/legacy.cnas.org/publications-pdf/CNAS_RoboticsOnTheBattlefield_Scharre.pdf.
6 Reddie, Andrew, 'The Impact of AI on Warfare', The President's Inbox in association with the Council on Foreign Relations (13 May 2024), https://www.cfr.org/podcasts/impact-ai-warfare-andrew-reddie.
7 Walker, Paddy and Peter Roberts, *War's Changed Landscape? A Primer on the Forms and Norms of Conflict* (Howgate Publishing, 2023), pp. 114-121.
8 Oremus, Will, 'Google's AI Search Problem May Never Be Fully Solved', *Washington Post* (29 May 2024), https://www.washingtonpost.com/politics/2024/05/29/google-ai-overview-wrong-answers-unfixable/.
9 Oremus, 'Google's AI Search Problem'.

exactly this assumption), then there are indeed foundational challenges to the suggested technical spine of AWS. The number of processes that must all work and work together if autonomous weapons are to be a reality means that it is statistically certain that decisions will go badly wrong.

It is also useful to bear in mind that the technologies comprising unsupervised weapons rarely arrive fully formed. The Delivery Cohort's challenge is when to press a starting button, because AWS' underlying capabilities will always resemble work in progress. Another crucial enabling technology is forever just around the corner. At the time of writing, for instance, ChatGPT very regularly hallucinates, the property whereby programs empirically make up incorrect information. Asked to generate scientific abstracts, ChatGPT repeatedly invents some one third of the resulting references. Nor has there been material improvement between, for instance, the performance of version 3.5 and version 4, both iterations still producing untrustworthy output that is then sucked back into other LLMs, permanently contaminating those programs and prompting Henry Mance to ask how we should respond to technology 'that is both brilliant and half-formed'.[10] This is also the basis for the joke that AI is anything that doesn't work yet. The issue, of course, is that today's LLMs may have learned to recognise patterns but do not understand underlying concepts and, although steps are being taken to rectify this and other adjunct faults, it is an enduring characteristic of what is being proposed that further development is always ahead of the Cohort.

> *How should we respond to technology 'that is both brilliant and half-formed'? AI is anything that doesn't work yet*

But it is also easy to be too skeptical for too long given the pace of advance across all of these technologies. No longer is it the case that search engines fail when tasked with questions that have not previously been asked (a viral example being whether an alligator can run the 100-metre hurdles). That test has long since been satisfied and readers should remember that just two short decades ago the smartphone had not yet been unveiled. And this is why so much of this book's analysis keeps deferring to context. Consider, for instance,

---

10 Mance, Henry, 'AI Keeps Going Wrong. What if it Can't Be Fixed?' *Financial Times* (6 April 2024), https://www.ft.com/content/648228e7-11eb-4e1a-b0d5-e65a638e6135.

retailer Amazon's decision in 2024 to abandon its much-hyped checkout-less technology in its supermarkets in the US. Outwardly, the offer seemed very sophisticated but, stripped back, that technology still relied upon a huge number of workers in India watching videos of shoppers and manually labelling their actions.[11] Similarly, hidden in a study by Boston Consulting Group is the finding that consultants using ChatGPT performed almost a quarter worse on particular tasks than those who did not.[12] Given the resources being devoted to developing these capabilities, a danger for this analysis has always been being blindsided by the future.[13] Not without reason, dystopian representations such as *Slaughterbot*[14] and 'As Much Death As You Want'[15] remind us of our potential nearness of broad AWS deployment. This unexpectedly complicates the role of AWS' Delivery Cohort. Sabin's 'Revolution in Expectation' seems touchingly close, pushing states towards autonomous practices because of what may or may not be happening in neighbouring nations' arsenals, because of a quite unfeasible art of the possible and also because of what might happen should a bordering state field these weapons.[16] Parties, after all, are surprisingly prey to straightforward fears of missing out.[17]

'In 2033 it will seem utterly baffling how a bunch of tech folk lost their minds over text generators'

Before setting down certain conclusion, two sets of circumstances appear particularly relevant to the deployment debate. First, there is the ongoing replacement by machines of specific tasks that were previously

---

11 Davis, Wes, 'Amazon Gives Up on No-Checkout Shopping in Its Large Grocery Stores', The Verge (2 April 2024), https://www.theverge.com/2024/4/2/24119199/amazon-just-walk-out-cashierless-checkout-ending-dash-carts.
12 Berreby, David, 'Chat GPT Helps, and Worries, Business Consultants, Study Finds', *New York Times* (23 December 2023), https://www.nytimes.com/2023/12/28/science/chatgpt-business-consultants.html.
13 Notwithstanding AWS' infancy, in 2018 a Google search on the term 'autonomous weapon' returned around 9,650,000 references.
14 Russell, Stuart, *Slaughterbot*, YouTube, 0.15 minutes/1.50 minutes/2.47 minutes, https://www.youtube.com/watch?v=ecClODh4zYk. .
15 Crowder, Lucien, 'As Much Death As You Want', *Bulletin of the Atomic Scientists* (2 December 2017) https://thebulletin.org/2017/12/as-much-death-as-you-want-uc-berkeleys-stuart-russell-on-slaughterbots/. See introduction to Chapter 2 ('Context') for a detailed discussion of this theme.
16 Economist, 'Autonomous Weapons Are a Game-changer', *Economist* (25 January 2018), https://www.economist.com/special-report/2018/01/25/autonomous-weapons-are-a-game-changer.
17 Kirsch, Andreas, 'Autonomous Weapons Will Be Tireless, Efficient Killing Machines – and There Is No Way to Stop Them', *Quartz News* (23 July 2018), https://qz.com/1332214/autonomous-weapons-will-be-tireless-efficient-killing-machines-and-there-is-no-way-to-stop-them.

undertaken by humans. Upgrades and kit replacement programmes quickly lead to the erosion of human supervision through incremental delegation of battlefield tasks and engagement routines to remote and machine alternatives. A second driver is illustrated by S-curve models of disruptive adoption and their close fit to how weapons might lose supervision; this

model (specifically, tipping points leading to non-linear adoption) does look like an appealing archetype for AWS deployment notwithstanding the prerequisite that the availability of all underlying componentry must be in place if human supervision is to be forfeited. It is in this vein that this book's technical assessment (the 'practical analysis' of its later chapters) focuses upon the enduring nature of outstanding 'technology holes' around coding for weapon goals, its values, of automatic target recognition and the challenges related to anchoring the machine's representations.[18]

Let us turn now to some closing arguments and conclusions. Restrictions on AWS deployment would appear to divide into soft and hard. Soft constraints concern existing legal frameworks and how the Delivery Cohort can empirically risk fielding such machines while still maintaining compliance and, consequentially, national reputation.

Not without reason, dystopian representations such as Slaughterbot and 'As Much Death As You Want' remind us of our potential nearness of broad AWS deployment

As we have seen, after all, the Cohort must navigate a slew of difficulties, made more complicated by the relevant legal corpus largely dating back decades before unsupervised weapons were even thought possible. If it is not repeatedly unfit for purpose, the law at least requires unhelpfully contentious interpretation that makes it most challenging to capture in code. Moreover, international statutory bodies dealing with this framework have proved themselves empirically ineffective, without appropriate mechanisms in place to determine or enact decisions. This creates uncertainty and also impunity. Weapons, moreover, are incapable of agency.

---

18 These are key sections in the analysis's technical consideration of AWS feasibility. In particular, see: Chapter Eight, specifically 8.3 (Utility function), 8.5 (Anchoring and goal setting issues), 8.6 (Value setting issues) and 8.7 (Action selection issues).

An overarching conclusion is that the Cohort must be guided by context. It is better to field excellent soldiers with merely good equipment rather than good soldiers and excellent kit. It is always that excellent soldier who is the Cohort's best asset, uniquely placed to counter the adversary, to capitalise upon weaknesses and, when faced perhaps by AWS, to neutralise newly unsupervised platforms (Max Boot's concept of 'nullification'). Part of this context is that battlefield innovation rarely confers lasting advantage and, in instances of deploying 'low tech that works well enough', illustrates the uncertainty that overshadows this whole debate. In deciding battlefield priorities, it is human endeavour, human curiosity and soldiers' broad knowledge, skills and experience that should inform how the Cohort proceeds.

> Weapons, moreover, are incapable of agency

AWS deployment is fundamentally a human decision. The draw for politicians is an autonomous solution that offers bloodless and remote engagement, minimising casualties and media costs. Conversely, those same politicians will likely be upended by a 'Tesla moment'[19], by structural impediments (the unanimity model, for example, that underpins decision-making at the United Nations and the North Atlantic Treaty Organization [NATO]) as well as adverse popular outcry (the restrictions on bombing raids that were applied by politicians in Kosovo after public opprobrium).[20] It is also far from certain how much human participation is taken away from combat sequences by autonomous means; current precursors still require multiple operators to staff each drone, teams of professionals to manage the unit's sensors and scores of intelligence analysts sifting through resulting data. As noted by Scharre, 'it's a cumbersome way to operate [and] not a cost-effective strategy if they require ever larger numbers of highly trained and expensive people to operate them'.[21]

> War's nature may remain likely violent, likely interactive and certainly political, but the forms of inter-state conflict now look much broader

---

19 Yadron, Danny and Dan Tynan, 'Tesla Driver Dies in First Fatal Crash While Using Autopilot Mode', *Guardian* (1 July 2016), https://www.theguardian.com/technology/2016/jun/30/tesla-autopilot-death-self-driving-car-elon-musk. In this instance, research into autonomous cars was postponed after public clamour following a fatal accident involving that company's technology.
20 Lambeth, Benjamin, *NATO's Air War for Kosovo: A Strategic and Operational Assessment* (RAND, 2001), p. 185, https://www.rand.org/pubs/monograph_reports/MR1365.html.
21 Scharre, Paul, Army of None: Autonomous Weapons and the Future of War (WW Norton, 2019), p. 16.

In considering deployment, the Cohort must also be guided by differences between war's *nature* and its *character*. Novel means, persistent competition and mercantile interdependencies mean that war's character is more ambiguous than ever. Indeed, it has long been inadequate simply to define war simply in terms of violence and any neatly bordered battlefield.

Its nature remains violent, likely interactive and certainly political, but the forms of interstate conflict now look much broader. What remains stable in the equation, however, is the enduring role of human endeavour. Decision-making regarding AWS deployment must be similarly 'human', involving the broadest selection of parties. This is best captured by the role of the Delivery Cohort, an artifice used throughout this book to bring together the layers of interested parties involved in the decisions and implementation of AWS. It is the AWS' human and supervisory component. And this Cohort faces several challenges, not least on accountability, on audit, testing and validation. It must scope these weapon systems, counter adversarial versions of the same, deliver to expectations and ensure their integration into legacy force design.[22] Nor is AWS deployment a single, static happening; it requires the Cohort to innovate, improve and value engineer throughout these weapons' adoption. Cohort challenges are both fundamental (for instance, which normative theory should underpin AWS deployment?) and operational (how might seventeen hundred pages of Norway's recent rules of engagement be captured in AWS' initial representation?).

This book's analysis demonstrates the extent and persistence of these soft deployment constraints. They range from concerns regarding proliferation and escalation to legal, technical, command and matters of control. Given states' requirement to assign scarce combat resources, the machines also have economic and allocation consequences. Removing oversight comes at considerable cost, not least because a number of current practices will be affected.[23] The bespoke nature of individual weapons will also require small production batches of unique parts where costs cannot be recovered over long manufacturing runs. Russia's own supply-side experiences should remind parties of the production constraints across more

---

22 Walker and Roberts, *War's Changed Landscape*, pp. 95-113.
23 Hoffman, RR and others, 'The Myths and Costs of Autonomous Weapon Systems', *Bulletin of the Atomic Scientists*, 72, 4, (2016), abstract and generally.

sophisticated weapon systems[24]; as component intricacy increases, supply chains get tighter. Nor can parties risk having their weapons assembled by lower-cost but possibly adversarial neighbours. AWS' soft challenges then include recruiting qualified, vetted personnel to ensure the maintenance of these weapons' programming and performance. Failure modes are also likely to be particularly varied (and poorly understood) notwithstanding apparently similar components. The systems will also require novel protocols to govern supply chains that must be similarly unsupervised.

Other matters confronting the Cohort are more difficult to define. It might well be that removing human supervision lowers the threshold to parties' initiating violence, accelerating the long-held adage that a state may be more inclined to wage war if it calculates either that relative advantage exists or that threat to its own troops can be reduced.[25] An argument here might be that states with roboticised forces might behave more aggressively, altering the political calculation for war.[26] Autonomous warfare must likely increase the chances of 'ubiquitous engagement' where machines cannot disengage. Finally to this point, commentators point to the destabilising and unattributable nature of AWS' violence given the ease with which deploying parties can assert 'plausible deniability' when using remote and unsupervised weapons.

The Cohort's processes must also be shaped by history and by lessons from other deployment precedents. An appeal of disruptive weaponry, after all, is to break out of the cycle that is typical of mature technologies and where investing further into legacy systems leads only to incremental improvement. Again, however, this relationship is ambiguous. In the case of AWS, it is just not possible to price legal, ethical, operational and human constraints to 'whole weapon' adoption of unsupervised practices. Indeed, instances abound of illogical (and unsustainable) use of battlefield technology. A sure way to reduce an adversary's treasure is to have it use three million dollar Patriot missiles against quadcopters costing less than two hundred dollars

---

24 Walker and Roberts, *War's Changed Landscape*, pp. 78-83.
25 Rummel, RJ, *Understanding Conflict and War. Vol. 4: War, Power and Peace*, 16, 'Causes and Conditions of International Conflict and War' (Sage, 1979), generally.
26 Singer, PW, 'Robots at War: The New Battlefield', cit. Hew Strachan and Sibylle Scheipers, *The Changing Character of War* (Oxford University Press, 2011), p. 48.

purchased online from Amazon.[27] This asymmetry is suddenly difficult to avoid: Houthi rebels in the Yemen are using low-cost drones to disable state-of-the-art Saudi surface-to-air systems.[28] If nothing else, the analysis reiterates that technical advance is unlikely to be the enduring preserve of wealthy nations.

A conclusion is also that human oversight in lethal force is necessary from a moral perspective. After all, no written agreement can prevent parties from deploying AWS if they desire it and the whole notion of meaningful human control (MHC) is similarly already a debate of choice; it is an awkward truism that voluntary codes of conduct quickly fall away where a party must mount an existential struggle for survival. It is similarly clear that few of the actors who are relevant to this debate choose to play by Western ideals of human rights, let alone those legal structures that comprise LOAC. In the same vein, notes Scharre, 'what does it mean to say that someone has the right to life in war when killing is the essence of war?'[29] Empirically, it is humans who kill

> States with roboticised forces might behave more aggressively, altering the political calculation for war. Autonomous warfare must also increase the chances of 'ubiquitous engagement' where machines cannot disengage

in war, whether using unsupervised weapons, remote weapons from a distance or up close and personally.

The Cohort must also factor that these machines can display neither empathy nor remorse. A second truism is one articulated by the Holy See that only humans can ever feel the emotional weight and psychological burden of choosing to kill another human being. That argument agrees with UN special rapporteur Christof Heyns's conclusion that weapons autonomy 'precludes a moment of deliberation in

---

27 Osborne, Samuel, 'Small Drone 'Worth $300' Shot Down by Patriot Missile Worth $3m, Says US General', *Independent* (13 March 2017), https://www.independent.co.uk/news/world/americas/small-drone-quadcopter-patriot-missile-shot-down-us-general-david-perkins-army-a7631466.html.
28 Diamond, Christopher, 'Report: Houthi Rebels Flying Iranian-Made Kamikaze Drones into Surveillance Radars', *Defense News* (27 March 2017), https://www.defensenews.com/global/mideast-africa/2017/03/28/report-houthi-rebels-flying-iranian-made-kamikaze-drones-into-surveillance-radars/.
29 Scharre, *Army of None*, p. 294.

those cases where it may be feasible'.³⁰ Empathy, concludes Human Rights Watch, can act as a check on killing but only if humans have control over who to target and when to fire. Moreover, removing weapon supervision would tend to transfer particularly poor outcomes on civilians.³¹ Absent MHC, a likely consequence must be that the burden of armed conflict is further shifted from soldiers to civilians given military personnel, replaced by machines, are no longer physically on the ground making decisions and controlling that violence.³²

Another conclusion is that removing human supervision upends weapon-user trust right across the engagement sequence, the more so given that machines deal poorly with unanticipated events. The characteristic is best captured by Wirth's law whereby machine complexity increases several measures quicker through changes to software than it does through changes to hardware.³³ Indeed, this book's technical analysis is important because it reveals the link between AWS' tight system coupling and its resulting system brittleness, and the consequences of this relationship on weapon predictability. This is unsurprising as machine-learning (ML) operation is

only based in part on 'explainable' classifiers. Much of these machines' decision paths are instead driven by 'blackbox' classifiers; while the decision path still prompts both an outcome and the reason behind that decision (examples here being decision trees, 'nearest neighbour' and rule-based classifiers), the classifier 'does the deciding' without any attendant reasoning (here, the neural network of the Cohort's AWS).³⁴ As currently posited, AWS are set up to fail on the Cohort's 'duty to explain', the more so given unintended interactions between the many components that make up these weapons. The difficulty for the Cohort is that

---

30 Heyns, Christof, 'Report of the UN Special Rapporteur on Extrajudicial, Summary and Arbitrary Executions to the Human Rights Council', a/HRC/23/47 (9 April 2013), para. 94. Heyns also reasons that machines lack morality and mortality and should as a result not have life-and-death powers over humans.
31 Docherty, Bonnie, 'We're Running Out of Time to Stop Killer Robot Weapons', *Guardian* (11 April 2018), https://www.theguardian.com/commentisfree/2018/apr/11/killer-robot-weapons-autonomous-ai-warfare-un.
32 Docherty, Bonnie, 'Losing Humanity – The Case against Killer Robots', Human Rights Watch (2012), p. 39, http://www.hrw.org/reports/2012/11/19/losing-humanity-0.
33 Holwerda, Thom, 'What Intel Giveth, Microsoft Taketh Away', OS News (15 November 2007), http://exo-blog.blogspot.com/2007/09/what-intel-giveth-microsoft-taketh-away.html.
34 Dr Hongbo Du, in conversation with the author (January 2019).

these platforms' absence of slack can no longer by definition be compensated by human judgement or other interventions to 'bend' rules or amend system behaviours.[35]

Merely automatic systems work to a rules-based if-then-else structure, and do so deterministically, meaning that for each input the system output will always be the same (except if something fails). The model here is that if x happens then the machine will do y. The Cohort's decision set for AWS is wholly different. AWS' technical spine is diametrically non-deterministic whereby very small changes to inputs can produce very large changes to outputs. Instead, the AWS reasons probabilistically, making mathematical assessments about best courses of action based upon statistical analysis of sensed data. The sequence is literally that fragile and this observation is important as it starts to explain AWS' inherent instability.  Unlike that automated system, the autonomous weapon does not produce exactly similar behaviour given identical inputs. Consider this book's illustrations all (apart from its frontispiece which is the product of a human hand), all produced without oversight by ChatGPT's Dall-E program. They cannot be repeated. Exactly the same descriptors yield a completely different crop of pictures. Similarly, it is inescapable that AWS deployment will produce a perplexing range of behaviours.

No conclusion can fail to comment on the limitations imposed on AWS' by its ML spine and challenges concerning how the weapon is expected to understand itself within its environment. This proves a unique vulnerability, whether through the incompatibility of training sets, data dependencies or the need for data scaling. In order to make its decision, the platform must smooth and clean incoming data, supported by suppression routines and feedback loops (AWS' 'anchoring problem'). The Cohort, moreover, is likely to be a poor predictor of behaviours that rely on feedback, especially in situations where it has no experience (here, prediction bias). That ML spine also masks other non-obvious vulnerabilities including inappropriate suppression of doubt, broad inference of causes and the narrowing of battlefield choices. It cannot enable the weapon to grasp context or situational awareness, a

---

35 Scharre, Paul, 'Autonomous Weapons and Operational Risk', Centre for a New American Security (2016), pp. 25-34, https://www.files.ethz.ch/isn/196288/CNAS_Autonomous-weapons-operational-risk.pdf.

consequence of its inherently approximating processes of 'what you see is all there is' (WYSIATI).

Weapon ML creates other idiosyncratic fault lines include the complexity of unlearning routines, temporal dependencies and the spine's inability to capture qualia across its host's routines. ML's performance tails off when data capture is fragile, in particular when its algorithms are confronted by partial patterns or events not previously encountered, acting very differently in scenarios that are themselves only slightly different from each other. AWS' technical premise relies upon the setting of goals, values and an appropriate utility function, but it must do all of this within the chaotic, ambiguous setting of the battlefield.

Several technical pieces in ML's processes also remain undiscovered, including jettisoning bad practice while avoiding the phenomenon of 'catastrophic forgetting' (when previous algorithmic skills are simply lost as the system is trained on new tasks and data, resulting in unexpected immobilisation, irrational action, unsuitable aggression, even unexplainable timidity). As opacity increases in that which confronts the commander, these challenges multiply.

AWS' technical spine is diametrically non-deterministic whereby very small changes to inputs can produce very large changes to outputs

Similarly underestimated is the plethora of data points that must available if the unsupervised machine is to make its own decisions. The deployment challenge is that the majority of the factors making up the engagement sequence (examples are listed in this sentence's accompanying footnote) are just not definable, are often the imprecise sum of other factors and anyway require subjective, volatile weighting if the process is to be useful.[36] But with such strict boundaries (a direct consequence of ML's rigid parameters), AWS processes also become vulnerable to quite elementary adversarial actions that are designed to disturb these very methods. Such

---

36 Dupuy, TN, *Numbers, Predictions and War: Using History to Evaluate Combat Factors and Predict the Outcome of Battles*, (Bobbs-Merrill Company, 1979), generally. Model effect factors include the following combat variables, all of which must dynamically be represented in AWS routines: rates of fire, potential targets per strike, effective range, accuracy, radius of action, dispersion factors, terrain factors including defence posture, air effectiveness and other weapons effect, weather factors, season factors, force strength effect, environmental effects, logistics and disruption effects, surprise effects, degradation and the effects of fatigue and casualties, casualty-inflicting capability factors, defensive capability factors. The model must also incorporate several intangible factors such as combat effectiveness, leadership, training, experience, morale, logistics, mental, intelligence, technology and initiative. It is noteworthy that only a minority of these factors are reliably calculable: see Figure 3-1, p. 33.

feint might actually be surprisingly straightforward and borrows from Boot's theory of nullification touched on above (a can of paint, some rudimentary disguise, some cursory concealment that divorces what the AWS is sensing from its original training data). In time, this will include more sophisticated

spoofing, trickery and, to borrow again from Sun Tzu, adversarial activity that is 'subtle to the point of formlessness'.[37] Autonomous weapons, of course, cannot rely upon external tuning by third parties.

Deployment is also complicated by matters of *trust* (between machine and operator and commissioner), the more so given generally poor familiarity with what is very quickly changing technology. Trust, however, erodes very quickly as system reliability declines, exacerbated by human misunderstanding, incompetence and design flaws. Poor outcomes must comprise 'normal accidents' (where no one party demonstrably does anything wrong) but also 'black swans' where a low probability high-impact event grossly skews AWS outcomes. A workaround here might be that comprehensive testing can substantiate

Absent human oversight, code alone must reliably deal with the nuances of lexical ambiguity, uncertainty as well as policy vagueness

AWS behaviours but this is an impracticable prerequisite. Validating patches and subsequently added functionality must similarly be based on testing, all of which appears beyond likely logistical capabilities and, anyway, runs counter to these agents' modus operandi.

There will also likely be a mismatch between operational requirements and AWS' technical competencies. The chance, of course, of the right AWS being in the right place at the right time reduces rapidly as asset specialisation and environmental complexity increases. Here, technical debt is a useful metaphor as it relates the consequences of poor software design to the accumulation of a financial debt. The immutable basis here is that it is machine code that alone must convey the intentions of the AWS' Delivery Cohort. Absent human oversight, code alone must reliably deal with the

---

37 ML processes create other sources of conflict including the pervasive requirement for confidence levels, for feedback loops and data scrubbing as well as non-obvious redundancy mechanisms, all of which add to weapon brittleness.

nuances of lexical ambiguity, uncertainty created by noisy and imperfect data, and by policy vagueness. Several challenges remain unanswerable. How can unsupervised machines really handle different 'categories' of facts that the machine senses from its immediate surroundings, be they indexical, normative, strong convictions or mere observations?

Extricating and then ranking inferences into facts that corroborate (or not) its internal representation also remains largely untried, especially outside the laboratory's workbench and in the adversarial setting of the battlefield. How also are each of these 'fact' categories to be weighted within those on-going routines? How does the unsupervised weapons capture abstracts? A review of, say, 'astonishment' is useful to rehearse again if only to underscore this conclusion. After all, 'astonishment' should be sparked when sensed data records an unexpected, unexplained happening. The weapon's response then needs to meld attributes of, say,

When autonomous componentry fails, it will either fail catastrophically or require human attention at points of highest stress

attraction (presumably move closer?), withdrawal (the weapon should extract itself?) and curiosity (presumably an inquisitive mixture of the two?). The reader, furthermore, should note the irony that all such code anyway originates, one way or another, solely out of human endeavour, from the human programmer and their computer keyboard.

A similar constraint is created by 'anchoring' and the computational basis by which the AWS is updated in order to account for newly sensed information. The challenge here concerns gradation and the degree to which incremental changes are implemented. Moreover, militaries rarely deploy weapons individually and flaws in any one system are likely to be

replicated across entire fleets of autonomous weapons, opening the door to what commentators have termed 'incidents of mass lethality' (human mistakes tend instead to be idiosyncratic). Anchoring is also complicated by how the weapon's understands its own rules of engagement.

When autonomous componentry fails, it will either fail catastrophically or require human attention at points of highest stress. Indeed, an embrace of MHC must be based upon two premises. First, it is difficult to foresee in practical terms why the Design Cohort would rule out a proven overrule mechanism in otherwise autonomous weapons. Second, legal and operational imperatives make the delegation of significant tasks to AWS by the local commander thoroughly improbable without that same override mechanism being in place. But this also assumes that the human operator is sufficiently engaged

and trained to meet the legal, performance and trust thresholds required for the Cohort even to consider delegating away the decision to kill. After all, it is a generally accepted heuristic that operator skills which go unpracticed tend to wither.

Legal restrictions on this weapon class should therefore be based on *positive* obligations whereby the user sufficiently understands the weapon system to be deployed, the user evaluates and understands the context of where the system is to be used and, finally, appropriate limitations are in place on the duration and area of function of each deployed machine. Nor is it necessary that any initial legal instrument covers all specific use

cases. Instead, the aim should be to put in place a future-oriented framework against which new technologies can be evaluated as they are developed, the instrument's intention being to establish guardrails preventing the development and adoption of AI functions in some critical roles that undermine human control and human dignity in the use of force.

This is not to regulate AI directly (which is too amorphous) but to establish the obligations for human understanding and intervention.

The analysis also finds that human oversight must be retained in order to monitor *emergent* effects of autonomous componentry, to intervene when circumstances exceed machine capabilities as well as, of course, to take over in situations where human are simply better at delivering an outcome than the alternative of a machine. It is thus barely useful to denote a weapon as 'autonomous' without referring (and understanding) the specific battlefield routine that is being made autonomous and the human soldier's relationship to that routine. Finally to this point, a weapon which is networked should always be more valuable than one which is independent and, in the case of AWS, operating on its own. By connecting each weapon to his network, the commander can ensure that the munition 'becomes part of a broader system that can harness sensor data from other ships, aircraft and satellites to assist its targeting'.³⁸ Under that commander's direct control (rather than its allocation and tasking being autonomous), it is less likely to be wasted, redundant or mis-tasked.

Moreover, many of these constraints may individually appear trivial but, in series or in parallel, can quickly derail AWS deployment. This is particularly true for AWS' technical impedimenta given how the whole notion of attention in the unsupervised weapon is untested. It will be inappropriate, after all, simply to tune weapon engagement according to

stimulus intensity. This must also be the case in routines designed to amend the weapon's goals or value set. Indeed, the analysis notes much wider coding issues including how best to mitigate stimuli habituation and stimuli saturation, how the weapon should choose which such stimuli to isolate for subsequent processing, which to ignore and how generally the weapon should navigate the 'cocktail party effect' that will characterise its sensed inputs.³⁹ These are fundamental competences that must together be satisfied if independent weapons are to initiate violence without supervision.⁴⁰

Accepting the inevitability of technical progress on the battlefield can appear overly deterministic. But narratives that avoid this conclusion must also be based upon a realistic alternative scenario where this trajectory of military development slows, shifts or is otherwise constrained by a mix of social, political or ethical factors. This itself requires both some assumptions and some conviction. Notwithstanding our multi-polar world, perhaps international agreements limit several of the technologies that are the subject of previous chapters? Perhaps the world's statutory bodies peremptorily ban autonomous lethal drones or proscribe AI-driven decision-making as they relate to weapons? More likely, a scenario might arise whereby public opposition to unsupervised warfare prompts democratic parties to limit these technologies' deployment.

---

38 Scharre, *Army of None*, p. 55.
39 Bronkhorst, Adelbert, 'The Cocktail-Party Problem Revisited: Early Processing and Selection of Multi-Talker Speech', *Attention, Perception and Psychophysics*, 77, 5, (2015) p. 1465 and generally. See also: Chapter Eight (Software), specifically: 8.7 (Action selection issues) and 8.8 (Behaviour setting and coordination).
40 This too has adjunct effects. Notwithstanding the need for such resetting mechanisms, the unsupervised weapon must still toggle reliably between its internal representations and recently processed external stimuli (together, again, the issue of anchoring).

Given recent incidences of voter volatility and adversarial meddling, this seems unlikely. Instead, the catalyst might simply be that anticipated technologies fail to deliver, whether through the hitting of roadblocks, poor reliability or the advent of deliverable countermeasures that undermine

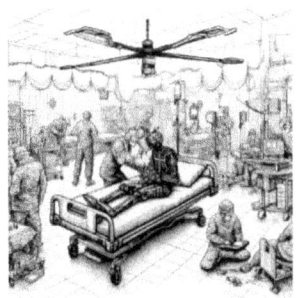

the current model of unsupervised lethality. An obvious candidate here might be the poor performance of AI under rapidly changing battlefield conditions. But other traditional frictions still exist. Economic considerations and resource shortages might constrain one or more signature technologies. Or the prohibitive budgets of delivering and maintaining these systems might shift militaries towards lower-cost and more conventional means, especially in an environment where authorities prioritise resilience and disaster preparedness over the procurement of elaborate systems.

Deployment of widespread AWS is certainly no inevitability. Instead, it depends upon choices; choices that are defined by political, social and economic forces as well, in the West, as by public values and ethical considerations. While technological advance will undoubtedly continue (with AWS deployment remaining a tantalising promise), the empirical pace, nature and direction of this remain open to very many influences with, as emphasised throughout the preceding chapters, human agency continuing to play an overarching role in determining how (and even whether) these new means are integrated into war's undertaking.

'Where new technology is sufficiently safe and reliable, norms of trust and public appetite can be expected to follow'

Much consideration has been given here to debates about MHC as a control umbrella for these lethal platforms. The mechanism, however, is not without its issues. First, the composition and structure of relevant international bodies have paralysed dispute resolution and show few signs of being able to accommodate parties' disparate positions. In any case, the ever-shortening timeframes of these weapons are then closing the operational window that is available for realistic human assessment and intervention. Moreover, MHC can only practically take place if parties demonstrate trust, transparency and commitment to keeping human oversight in place. In this vein, two illustrations provide a relevant finale. The first is to remind ourselves again of the concept of 'empty hangar syndrome', which signals the notion that certain scenarios are too far-fetched to warrant current consideration or statutory regulation. The UK's opening negotiating position within the UN's CCW played the perception that, in any final analysis, it is simply unfeasible for a commander to wander into the weapons hanger one morning to find that the AWS has decided on its own to depart unexpectedly on an unsupervised mission. It is, noted the UK, a hypothetical that does not deserve scrutiny.

The last word, however, is to return to the observation that automation and autonomy are increasingly developing pell-mell across weapon practices and, notes the UK Ministry of Defence, 'where new technology is sufficiently safe and reliable, norms of trust and public appetite can be expected to follow'.[41] It almost matters less that the deployment of AWS is pretty much less unfeasible. Instead, introduction of autonomous componentry within one model or another has already come to dominate future weapons procurement. It is for this reason that statutory requirement for human involvement is required in the use of force.

---

41 UK Ministry of Defence, 'Human-Machine Teaming', Joint Concept Note 1/18 (2018), p. 50, https://assets.publishing.service.gov.uk/media/5b02f398e5274a0d7fa9a7c0/20180517-concepts_uk_human_machine_teaming_jcn_1_18.pdf.

SELECTED BIBLIOGRAPHY

A

AI 100, 'One Hundred Year Study of Artificial Intelligence (Ai100)', *Stanford University, 2021 Report*, https://ai100.stanford.edu/gathering-strength-gathering-storms-one-hundred-year-study-artificial-intelligence-ai100-2021-1/sq2

Abney, Keith, 'Autonomous Robots and the Future of Just War Theory', cit. Fritz Allhoff and others, eds., *Routledge Handbook of Ethics and War* (Routledge, 2013)

Ackerman, Evan, 'Lethal Microdrones, Dystopian Futures, and the Autonomous Weapon Debate', *IEEE Spectrum* (15 November 2017), https://spectrum.ieee.org/automaton/robotics/military-robots/lethal-microdrones-dystopian-futures-and-the-autonomous-weapons-debate

Anderson, Kenneth and others, 'Adapting the Law of Armed Conflict to Autonomous Weapon Systems', *International Law Studies*, 90, 386 (Stockton Center for the Study of International Law, 2014)

Arkin, R, *Governing Lethal Behaviour: Embedding Ethics in a Hybrid Deliberate/Reactive Robot Architecture* (Georgia Institute of Technology, 2007)

Arkin, Ronald and others, 'An Ethical Governor for Constraining Lethal Action in an Autonomous System', Georgia Institute of Technology Robot Lab, Technical Report, GIT-GVU-09-02 (2009), https://www.cc.gatech.edu/ai/robot-lab/online-publications/GIT-GVU-09-02.pdf

Arkin, Ronald, 'The Case for Ethical Autonomy in Unmanned Systems', *Journal of Military Ethics*, 9.4 (2010), 333, https://smartech.gatech.edu/bitstream/handle/1853/36516/Arkin_ethical_autonomous_systems_final.pdf

Article 36, 'Structuring Debate on Autonomous Weapon Systems', Memorandum for delegates to the Convention on Certain Conventional Weapons (CCW), Geneva (14 November 2013)

Asaro, Peter, 'Why the World Needs to Regulate Autonomous Weapons, and Soon', *Bulletin of the Atomic Scientists* (27 April 2017), https://thebulletin.org/2018/04/why-the-world-needs-to-regulate-autonomous-weapons-and-soon/

Atherton, Kelsey, 'Mass Market Military Drones Have Changed the Way Wars Are Fought', *MIT Technology Review* (30 January 2023), https://www.technologyreview.com/2023/01/30/1067348/mass-market-military-drones-have-changed-the-way-wars-are-fought/

Automated Decision Research, 'Weapon Systems with Autonomous Function Used in Ukraine' (28 June 2022), https://automatedresearch.org/news/weapons-systems-with-autonomous-functions-used-in-ukraine/

B

Baer, Tobias and Vishnu Kamalnath, 'Controlling Machine Learning Algorithms and Their Biases', McKinsey and Company (November 2017), https://www.mckinsey.com/business-functions/risk/our-insights/controlling-machine-learning-algorithms-and-their-biases

Bahnsen, John and Robert Cone, *Defining the American Warrior Leader* (Parameters, 1990)

Barfoot, Timothy, *State Estimation for Robotics* (Cambridge University Press, 2018), http://asrl.utias.utoronto.ca/~tdb/bib/barfoot_ser17.pdf

Bawden, David, 'The Nature of Prediction and the Information Future: Arthur C. Clarke's Odyssey Vision', *Aslib Proceedings*, 49, 3 (1997)

Beautement, Patrick, 'Putting Complexity to Work; Achieving Effective Human-Machine Teaming', The Abaci Partnership LLP (2015)

Behn, Beth, 'The Stakes Are High: Ethics Education at US War Colleges', *Air War College Publications*, Maxwell Paper Number 73 (2018), https://www.airuniversity.af.mil/Portals/10/AUPress/Papers/mp_0073_behn_stakes_high.pdf

Benitez, Mike, 'It's About Time: The Pressing Need to Evolve the Kill Chain', War on the Rocks (2017), https://warontherocks.com/2017/05/its-about-time-the-pressing-need-to-evolve-the-kill-chain/

Beres, Damon, 'The Ethical Case for Killer Robots', *Huffpost* (3 June 2016), http://www.huffingtonpost.co.uk/entry/lethal-autonomous-weapons-ronald-arkin_us_574ef3bbe4b0af73af95ea36

Biddle, Tammi David, *Rhetoric and Reality* (Princeton University Press, 2009)

Black, Doug, 'AI Definitions: Machine Learning vs. Deep Learning vs. Cognitive Computing vs. Robotics vs. Strong AI', EnterpriseTech (19 January 2018), https://www.enterprisetech.com/2018/01/19/ai-definitions-machine-learning-vs-deep-learning-vs-cognitive-computing-vs-robotics-vs-strong-ai/

Black, Jeremy, 'Military Organisations and Military Change in Historical Perspective', *Journal of Military History*, 62, 4 (1998)

Blais, Carolyn, 'When Will AI Be Smart Enough to Outsmart People', MIT School of Engineering/Ask an Engineer Series (2024), https://engineering.mit.edu/engage/ask-an-engineer/when-will-ai-be-smart-enough-to-outsmart-people/

Bloch, Marc, *A Strange Defeat: A Statement of Evidence Written in 1940* (WW Norton, 1999)

Bo, Marta and Vincent Boulanin, 'Retaining Human Responsibility in the Development and Use of Autonomous Weapon Systems, Stockholm International Peace Research Institute (October 2023), https://www.sipri.org/publications/2022/policy-reports/retaining-human-responsibility-development-and-use-autonomous-weapon-systems-accountability

Boddens Hosang, JFR, 'Rules of Engagement: Rules on the Use of Force as Linchpin for the International Law of Military Operations', UvA-DARE (University of Amsterdam, 2017)

Boot, Max, *War Made New: Weapons and the Making of the Modern World* (Gotham, 2006)

Boot, Max, 'What the Past Teaches About the Future', *Joint Force Quarterly*, 44 (2007), http://indianstrategicknowledgeonline.com/web/MIL%20HIS%20JFQ44%20boot.pdf

Borrie, John, 'Security, Unintentional Risk and System Accidents', United Nations Institute for Disarmament Research (15 April 2016)

Bostrom, Nick, 'Hail Mary, Value Porosity, and Utility Diversification', nickbostrom.com (19 December 2014), https://nickbostrom.com/papers/porosity.pdf

Bostrom, Nick, *Superintelligence: Paths, Dangers, Strategies* (Oxford University Press, 2014)

Boulanin, Vincent, 'Implementing Article 36 Weapon Reviews in the Light of Increased `Systems', Stockholm International Peace Research Institute 2015/1 (November 2015), https://www.sipri.org/sites/default/files/files/insight/SIPRIInsight1501.pdf

Boulanin, Vincent and Maaike Verbruggen, 'Mapping the Development of Autonomy in Weapon Systems', Stockholm International Peace Research Institute (November 2017), https://www.sipri.org/sites/default/files/2017-11/siprireport_mapping_the_development_of_autonomy_in_weapon_systems_1117_0.pdf

Bourke, J, *An Intimate History of Killing* (Basic Books, 1999)

Bousquet, Antoine, 'A Revolution in Military Affairs? Changing Technologies and Changing Practices of Warfare', *Technology and World Politics* (Routledge, 2017), https://www.academia.edu/34469743/A_Revolution_in_Military_Affairs_Changing_Technologies_and_Changing_Practices_of_Warfare

Brehm, Maya, 'Defending the Boundary; Constraints and Requirements on the Use of Autonomous Weapon Systems Under International Humanitarian and Human Rights Law', Geneva Academy of International Humanitarian Law and Human Rights, Academy briefing No. 9 (2017)

Bronk, Justin and others, 'Armed Drones in the Middle East: Proliferation and Norms in the Region', Royal United Services Institute (17 December 2018), https://www.rusi.org/explore-our-research/publications/occasional-papers/armed-drones-middle-east-proliferation-and-norms-region

Brooks, Rosa 'Why Sticks and Stones Will Beat Our Drones', *Foreign Policy* (4 April 2013), https://foreignpolicy.com/2013/04/04/why-sticks-and-stones-will-beat-our-drones/

Burunov, Oleg, 'From Raven to Cayote: How Many Military Drones Does the US Have?', *Sputnik International* (15 March 2023), https://sputnikglobe.com/20230315/from-raven-to-coyote-how-many-military-drones-does-the-us-have-1108417947.html.

C

Campbell, Katherine Driggs, 'Tools for Trustworthy Autonomy: Robust Prediction, Intuitive Control and Optimized Interaction', Electronic Engineering and Computer Science, University of California, Berkeley (9 May 2017), https://www2.eecs.berkeley.edu/Pubs/TechRpts/2017/EECS-2017-41.pdf

Campaign to Stop Killer Robots, 'Urgent Action Needed to Ban Full Autonomous Weapons', CSKR London (23 April 2013), http://stopkillerrobots.org/wp-content/uploads/2013/04/KRC_LaunchStatement_23Apr2013.pdf

Center for a New American Security, 'Autonomous Weapons and Human Control' (2016), https://www.files.ethz.ch/isn/196780/CNAS_Autonomous_Weapons_poster_FINAL%20(1).pdf

Centre for Army Leadership, 'Army Leadership Doctrine', Edition 1 (RMAS Camberley, 2016)

Clark, Lloyd, 'Blitzkrieg: Myth, Reality and Hitler's Lightning War – France 1940' (Atlantic Books, 2016)

Cohen, Sagi, 'Gaza Becomes Israel's Testing Ground for Military Robots', *Haaretz* (3 March 2024), https://archive.is/P6mAQ

Cohen, A, Yarden and others, 'Recent Advances at the Interface between Neuroscience and Artificial Neural Networks', *JNeuroScience – The Journal of Neuroscience* (9 November 2022), https://www.jneurosci.org/content/42/45/8514

Cohn, A, 'How Do We Align Artificial Intelligence with Human Values?, Future of Life Institute (3 February 2017), https://futureoflife.org/2017/02/03/align-artificial-intelligence-with-human-values/

Coker, Christopher, 'Still "the Human Thing"? Technology, Human Agency and the Future of War', *International Relations*, 32.1 (2018), http://eprints.lse.ac.uk/87629/1/Coker_Human%20Thing.pdf

Cole, Chris, 'Small Drones, Big Problem; Two New Reports Examine the Rise and Rise of 'One-Way Attack' Drones', Drone Wars UK (21 June 2023), https://dronewars.net/tag/loitering-munitions/

Collins, RJ and R Thompson, 'Systemic Failure Modes: A Model for Perrow's Normal Accidents in Complex, Safety Critical Systems', *Advances in Safety and Reliability* (1997), https://pdfs.semanticscholar.org/18d7/8946bc8bb1f58f0df6e57a7cce8fcf65f0aa.pdf

Conn, Ariel, 'The Problem of Defining Autonomous Weapons', The Future of Life Institute (30 November 2016), https://futureoflife.org/2016/11/30/problem-defining-autonomous-weapons/

Cooke, Gordon, 'The Future Battlefield', US Army (16 July 2018), https://www.army.mil/article/208553/the_future_battlefield

Cummings, ML, 'Artificial Intelligence and the Future of Warfare', Royal Institute of International Affairs (2017), https://www.chathamhouse.org/publication/artificial-intelligence-and-future-warfare

D

DARPA, 'DARPA Seeks Technical Solutions to Create Autonomous Capabilities for Commercial Drones' (9 December 2023), https://www.darpa.mil/news-events/2023-09-12

Davey, Tucker, 'Who Is Responsible for Autonomous Weapons?', Future of Life Institute (21 November 2016), https://futureoflife.org/2016/11/21/peter-asaro-autonomous-weapons/

Docherty, Bonnie, 'Losing Humanity – The Case against Killer Robots', Human Rights Watch (2012) http://www.hrw.org/reports/2012/11/19/losing-humanity-0

Douglas-Heaven, Will, 'Illya Sutskever, OpenAI's Chief Scientist, on This Hopes and Fears for the Future of AI', *MIT Technology Review* (26 October 2023)

D'Urso, Stefano, 'Let's Talk about the Israel Air Industries Loitering Munitions and What They Are Capable of', *Aviationist* (7 January 2022), https://theaviationist.com/2022/01/07/iai-loitering-munitions/

E

Economist, 'After Moore's Law: The Future of Computing – The Era of Predictable Improvement in Computer Hardware Is Ending. What Comes Next?', *Economist* (12 March 2016)

Economist, 'Autonomous Weapons Are a Game-changer', *Economist* (25 January 2018), https://www.economist.com/special-report/2018/01/25/autonomous-weapons-are-a-game-changer

Economist, 'From Here to Autonomy', *Economist* (1 March 2018), https://www.economist.com/news/special-report/21737420-making-vehicles-drive-themselves-hard-getting-easier-autonomous-vehicle-technology

Economist, 'The New Battlegrounds', *Economist* (January 2018)

Economist, 'The War in Ukraine Shows How Technology Is Changing the Battlefield', *Economist Special Report, Lessons from Ukraine* (3 July 2023)

Economist, 'Killer Drones Pioneered in Ukraine Are the Weapons of the Future', *Economist* (8 February 2024), https://www.economist.com/leaders/2024/02/08/killer-drones-pioneered-in-ukraine-are-the-weapons-of-the-future?

Economist, 'How Cheap Drones Are Transforming Warfare in Ukraine', *Economist Science and Technology* (5 February 2024)

Edmonds, Jeffrey and Samuel Bendett, 'Russia's Use of Uncrewed Systems in Ukraine', *CNA* (May 2023), https://www.cna.org/reports/2023/05/russias-use-of-drones-in-ukraine

EMC Claight Report, 'Global Military Drone Market Report' (January 2024), https://www.expertmarketresearch.com/reports/military-drone-market

Enemark, Christian, *Armed Drones and the Ethics of War: Military Virtue in a Post-Heroic Age* (Routledge, 2014)

Epstein, Robert, 'The Empty Brain', Aeon (18 May 2016), https://aeon.co/essays/your-brain-does-not-process-information-and-it-is-not-a-computer

Etzioni, Amitai and Oren Etzioni, 'Pros and Cons of Autonomous Weapon Systems', *Military Review* (May-June 2017), http://www.armyupress.army.mil/

Journals/Military-Review/English-Edition-Archives/May-June-2017/
Pros-and-Cons-of-Autonomous-Weapons-Systems/

F

Ford, Matthew and Andrew Hoskins, *Radical War: Data, Attention and Control in the 21st Century* (Hearst, 2022)

Freedberg, Sydney, "Unmanned' Drones Take Too Many Humans to Operate, Says Top Army Aviator', *Breaking Defense* (27 February 2023), https://breakingdefense.com/2023/02/unmanned-drones-take-too-many-humans-to-operate-says-top-army-aviator/

Freedman, Lawrence, *Information Warfare: Will Battle Ever Be Joined?* (International Centre for Security Analysis, October 1996)

Fridman, Ofer, 'Revolutions in Military Affairs That Did Not Happen: A Framework for Analysis', *Comparative Strategy*, Volume 35, Issue 5 (2016)

Future of Life Institute, 'Autonomous Weapons: An Open Letter from AI and Robotics Researchers', Future of Life (2015) https://futureoflife.org/open-letter-autonomous-weapons/

G

Gaggioli, G and R Kolb, 'A Right to Life in Armed Conflict? The Contribution of the European Court of Human Rights', *Yearbook on Human Rights* (2007)

Geiss, Robin, *The International-Law Dimension of Autonomous Weapon Systems* (Friedrich Ebert Stiftung, October 2015), http://library.fes.de/pdf-files/id/ipa/11673.pdf

Geneva Centre for Security Policy, 'Perils of Lethal Autonomous Weapon Proliferation: Preventing Non-State Acquisition', Geneva Centre for Security Policy (2018), https://www.gcsp.ch/News-Knowledge/Publications/Perils-of-Lethal-Autonomous-Weapons-Systems-Proliferation-Preventing-Non-State-Acquisition

Goodman, Ryan, 'The Power to Kill or Capture Enemy Combatants', *European Journal of Law*, 24, 2 (2013)

Goodwin, Tom, 'We're at Peak Complexity. And It Sucks', *TechCrunch* (18 October 2016) https://techcrunch.com/2016/09/03/were-at-peak-complexity-and-it-sucks/?guccounter=1

Gott, Kendall and Michael Brooks, *Warfare in the Age of Non-State Actors* (Combat Studies Institute Press, 2007)

Gray, Colin, *Strategy and Defence Planning: Meeting the Challenge of Uncertainty* (Oxford University Press, 2014)

Gray, Colin, *Another Bloody Century* (Phoenix, 2005)

Gray, Colin, 'Inescapable Geography', *Journal of Strategic Studies*, 22, 2-3 (1999)

Grinker, Roy and John Spiegal, *Men Under Stress* (Philadelphia Press, 1945)

Gronland, Kirsten, 'AI: Artificial Intelligence, the Military and Increasingly Autonomous Weapons', Future of Life Institute (9 May 2019), https://futureoflife.org/resource/state-of-ai/

Grossman, Dave, *On Killing: The Psychological Cost of Learning to Kill in War and Society* (Black Bay Books, 1996)
Group of Governmental Experts on Emerging Technologies in the Area of Lethal Autonomous Weapons Systems, 'Report of the 2018 session of the Group of Governmental Experts on Emerging Technologies in the Area of Lethal Autonomous Weapons Systems', CCW/GGE.1/2018/3 (23 October 2018), https://documents.un.org/doc/undoc/gen/g18/323/29/pdf/g1832329.pdf
Guszcza, J and N Maddirala, 'Minds and Machines: The Art of Forecasting in the Age of Artificial Intelligence', *Deloitte University Press, Deloitte Review*, 19 (2016)

H

Haas, MC and SC Fischer, 'The Evolution of Targeted Killing Practices: Autonomous Weapons, Future Conflicts and the International Order', *Contemporary Policy*, 38:2 (August 2017), https://www.ethz.ch/content/dam/ethz/special-interest/gess/cis/center-for-securities-studies/pdfs/Haas&Fischer_2017_TargetedKillingPractices.pdf
Haidt, Jonathan, 'The Moral Emotions', in Richard Davidson and others, eds., *Handbook of Affective Sciences* (Oxford University Press, 2005)
Haikonnen, Pentti, *The Cognitive Approach to Conscious Machines* (Imprint Academic, 2003)
Hall, Brian, 'Autonomous Weapon System Safety', *Joint Forces Quarterly*, 86 (June 2017) http://ndupress.ndu.edu/Media/News/Article/1223911/autonomous-weapons-systems-safety/
Hambling, David, *Swarm Troopers: How Small Drones Will Conquer the World* (Archangel Ink, 2015)
Hambling, David, 'What Does Ukraine's Million-Drone Army Mean for the Future of War?', *New Scientist* (19 January 2024), https://www.newscientist.com/article/2413260-what-does-ukraines-million-drone-army-mean-for-the-future-of-war/
Han, Meghan, 'Lethal Autonomous Weapons and Info-Wars: A Scientist's Warning', *Medium* (6 July 2017), https://medium.com/@Synced/lethal-autonomous-weapons-info-wars-a-scientists-warning-cc95798bc302
Harari, Yuval Noah, *Homo Deus* (Penguin Random House, 2016)
Harvard Law School International Human Rights Law Clinic, 'Reviewing the Record: Reports on Killer Robots from Human Rights Watch and Harvard Law School International Human Rights Law Clinic (2018), http://hrp.law.harvard.edu/wp-content/uploads/2018/08/Killer_Robots_Handout.pdf
Haywood, Justion, 'How Much Does an F-35 Cost?', *Simple Flying* (20 December 2023), https://simpleflying.com/how-much-does-an-f-35-cost/
Heikkila, Melissa and William Douglas Heaven, 'What's Next for Artificial Intelligence in 2024?', *MIT Technology Review* (4 January 2024), https://www.technologyreview.com/2024/01/04/1086046/whats-next-for-ai-in-2024/
Hew, Patrick Chisan, 'The Generation of Situational Awareness – A Near to Mid-Term Study', Defence System Analysis Division (Australian Army Publishing, July 2006), http://www.dtic.mil/dtic/tr/fulltext/u2/a465252.pdf

Heyns, Christof, 'UN Document A/HRC/23/47, Report of the Special Rapporteur on Extrajudicial, Summary or Arbitrary Executions, United Nations', Human Rights Council, 23rd Session, Agenda item 3 (27 May 2013), https://www.ohchr.org/Documents/HRBodies/HRCouncil/RegularSession/Session23/A.HRC.23.47.Add.5_ENG.pdf

Heyns, Christof, 'Autonomous Weapon Systems: Living in a Dignified Life and Dying a Dignified Death', cit. Nehal Bhuta and others, *Autonomous Weapons Systems: Law, Ethics, Policy'* (Cambridge University Press, 2016)

Heynes, Deborah, 'Spiraling Cost of Weapons Makes War "Too Expensive"', *The Times* (26 April 2017), https://www.thetimes.co.uk/article/spiralling-cost-of-weapons-makes-war-too-expensive-6fkzf03w6

Hirsh, Michael, 'How AI Will Revolutionise Warfare', *Foreign Policy* (11 April 2024), https://foreignpolicy.com/2023/04/11/ai-arms-race-artificial-intelligence-chatgpt-military-technology/.

Horowitz, Michael and Paul Scharre, 'An Introduction to Autonomy in Weapon Systems', Centre for a New American Security (13 February 2015), https://s3.amazonaws.com/files.cnas.org/documents/Ethical-Autonomy-Working-Paper_021015_v02.pdf?mtime=20160906082257

Horowitz, Mark and Paul Scharre, 'Meaningful Human Control and Weapon Systems: A Primer', Centre for a New American Security, Working Paper (March 2015)

Horowitz, Michael, 'The Ethics and Morality of Robotic Warfare: Assessing the Debate over Autonomous Weapons', American Institute of Arts and Sciences (February 2016)

Horowitz, Michael, 'The Promise and Perils of Military Applications of Artificial Intelligence', *Bulletin of the Atomic Scientists* (23 March 2018), https://thebulletin.org/landing_article/the-promise-and-peril-of-military-applications-of-artificial-intelligence/

House of Lords, 'Proceed with Caution: Artificial Intelligence in Weapon Systems', AI in Weapon Systems Committee, Report of Session 2023-24, HL Paper 16

Human Rights Watch, 'Heed the Call: A Moral and Legal Imperative to Ban Killer Robots', Human Rights Watch (September 2018), https://www.hrw.org/report/2018/08/21/heed-call/moral-and-legal-imperative-ban-killer-robots

I

Ilachinski, Andrew, 'AI, Robots and Swarms: Issues, Questions and Recommended Studies', CNA Corporation (January 2017), https://www.cna.org/CNA_files/PDF/DRM-2017-U-014796-Final.pdf

International Committee of the Red Cross (ICRC), 'A Guide to the Legal Review of New Weapons, Means and Methods of Warfare: Measures to Implement Article 36 of Additional Protocol I of 1977', *International Review of the Red Cross*, 88, 864 (2006), https://international-review.icrc.org/articles/guide-legal-review-new-weapons-means-and-methods-warfare-measures-implement-article-36

International Committee of the Red Cross, 'Autonomous weapons systems: Technical, Military, Legal and Humanitarian aspects', *Experts meeting,*

CCW, 64 (Geneva, Switzerland, 2014), https://www.icrc.org/en/document/report-icrc-meeting-autonomous-weapon-systems-26-28-march-2014

International Committee of the Red Cross, 'Decision Making in Military Combat Operations', ICRC Publications (October 2013), https://www.icrc.org/eng/assets/files/publications/icrc-002-4120.pdf

International Committee of the Red Cross, 'Draft Rules for the Limitation of the Dangers Incurred by the Civilian Population in Time of War,' Article 14 (1956)

International Committee of the Red Cross, 'Handbook on International Rules Governing Military Operations', ICRC (2013), https://www.icrc.org/sites/default/files/topic/file_plus_list/0431-handbook_on_international_rules_governing_military_oprations.pdf

J

James, William, *The Principles of Psychology* (Henry Holt, 1890)

Jenson, Benjamin and Dan Tadross, 'How Large Language Models Can Revolutionise Military Planning', War on the Rocks (12 April 2023), https://warontherocks.com/2023/04/how-large-language-models-can-revolutionize-military-planning/

Jhingran, A, 'Obsessing over Artificial Intelligence Is the Wrong Way to Think About the Future', *Wired* (2016), https://www.wired.com/2016/01/forget-ai-the-human-friendly-future-of-computing-is-already-here/

Jones, Seth, 'Much 'Political Warfare' in Our Future', *Breaking Defense* (2 February 2018), https://breakingdefense.com/2018/02/much-political-warfare-in-our-future

Johnson, Ronald, 'Lanchester's Square Law in Theory and Practice', School of Advanced Military Studies, Fort Leavenworth (1990), http://www.dtic.mil/dtic/tr/fulltext/u2/a225484.pdf

K

Kalmanovitz, Pablo, 'Judgement, Liability and the Risks of Riskless Warfare', cit. Nehal Bhuta and others (eds.), *Autonomous Weapons Systems: Law, Ethics, Policy* (Cambridge University Press, 2016)

Kania, Elsa, 'The Critical Human Element in the Machine Age of Warfare', *Bulletin of the Atomic Scientists* (15 November 2017), https://thebulletin.org/2017/11/the-critical-human-element-in-the-machine-age-of-warfare/

Kinni, Theodore, 'Beware the Paradox of Automation', *MIT Sloan Management Review* (20 October 2016), https://sloanreview.mit.edu/article/beware-the-paradox-of-automation/

Kirsch, Andreas, 'Autonomous Weapons Will Be Tireless, Efficient Killing Machines – And There Is No Way to Stop Them', *Quartz News* (23 July 2018), https://qz.com/1332214/autonomous-weapons-will-be-tireless-efficient-killing-machines-and-there-is-no-way-to-stop-them/

Kostopoulos, Lydia, 'Drivers for the Deployment of Lethal Autonomous Weapons', Medium (22 December 2017), https://medium.com/@lkcyber/drivers-for-the-deployment-of-lethal-autonomous-weapons-systems-ae1dd6278a35

Kott, Alexander and others, *Visualizing the Tactical Ground Battlefield in the Year 2050*, Army Cyber Institute at Westpoint (1 June 2015), https://cyber.army.mil/News/Article/1324543/visualizing-the-tactical-ground-battlefield-in-the-year-2050-workshop-report/

Kott, Alexander, 'Challenges and Characteristics of Intelligent Autonomy for Internet of Battle Things in Highly Adversarial Environment', US Research Laboratory, Adelphi MD, arXiv preprint arXiv: 1803.11256 (2018), https://arxiv.org/pdf/1803.11256.pdf

Krepinevich, Andrew, *Origins of Victory* (Yale, 2023)

Krishnan, Armin, *Killer Robots: Legality and Ethicality of Autonomous Weapons* (Ashgate Publishing, 2009)

Kumar, Abhinav and others, 'The Use of Robots and Artificial Intelligence in War', *LSE Business* Review (17 February 2020), https://blogs.lse.ac.uk/businessreview/2020/02/17/the-use-of-robots-and-artificial-intelligence-in-war/

L

Lachow, Irving, 'The Upside and Downside of Swarming Drones', *Bulletin of the Atomic Scientists*, 73:2 (February 2017), http://www.tandfonline.com/doi/pdf/10.1080/00963402.2017.1290879

Lafrance, Adrienne, 'Machine Unlearning', *The Atlantic* (18 March 2016), https://www.theatlantic.com/technology/archive/2016/03/computers-brains-cybernetics/474273/

Latiff, Robert, 'How Technological Advancements Will Shape the Future of the Battlefield', *Signature* (13 October 2017), https://www.signature-reads.com/2017/10/how-tech-advancements-will-shape-future-battlefield/

Levy, Jack, 'The Offensive/ Defensive Balance of Military Technology: A Theoretical and Historical Analysis', *International Studies Quarterly*, 28, 2 (1984)

Lorber, Azriel, *Misguided Weapons: Technical Failures and Surprise on the Battlefield* (Potomac Books, 2002)

Losey, Stephen, 'New in 2024: Air Force Plans Autonomous Flight Tests for Drone Wingmen', *Defense News* (30 December 2023), https://www.defensenews.com/air/2023/12/30/new-in-2024-air-force-plans-autonomous-flight-tests-for-drone-wingmen/

Louth, John and Christian Moeling, 'Technological Innovation: The US Third Offset Strategy and the Future of Transatlantic Defense', Armaments Industry European Research Group, Policy Paper (December 2016), http://www.iris-france.org/wp-content/uploads/2016/12/ARES-Group-Policy-Paper-US-Third-Offset-Strategy-December2016.pdf

M

Maloney, Col S, 'Ethics Theory for the Military Professional', *Air University Review*, 32, 3(1981)

Marra, William and Sonia McNeil, 'Automation and Autonomy in Advanced Machines: Understanding and Regulating Complex Systems', Warfare Research Paper Series, 1-2012 (April 2012)

Marsh, Henry, 'Can Man Ever Build a Mind?', *Financial Times* (10 January 2019), https://www.ft.com/content/2e75c04a-0f43-11e9-acdc-4d9976f1533b

Marshall, Samuel, *Men Against Fire: The Problem of Battle Command in Future War* (William Morrow Publishing, 1947)

McMaster, Lt Gen HR, 'On the Study of War and Warfare', Modern War Institute (24 February 2017), https://mwi.usma.edu/study-war-warfare/

Meaker, Morgan, 'Ukraine's War Brings Autonomous Weapons to the Front Line', *Wired* (24 February 2023), https://www.wired.co.uk/article/ukraine-war-autonomous-weapons-frontlines

Metcalf, Jacob, 'Ethics Codes: History, Context and Challenges' *Council for Big Data, Ethics and Society* (9 November 2014), https://bdes.datasociety.net/council-output/ethics-codes-history-context-and-challenges/

Michel, Arthur Holland, 'Known Unknowns: Data Issues and Military Autonomous Systems', United Nations Institute for Disarmament Research (17 May 2021), https://unidir.org/publication/known-unknowns-data-issues-and-military-autonomous-systems/

Moravec, Hans, *Mind Children* (Harvard University Press, 1988)

Moyes, Richard, 'Key Elements of Meaningful Human Control', Article 36 Briefing Paper, CCW Meeting of Experts on Lethal Autonomous Weapon Systems (April 2016), http://article36.org/wp-content/uploads/2016/04/MHC-2016-FINAL.pdf

Moyes, Richard, 'Emergent Behaviour and Risk – A Sketch for a Risk Management Approach', Article 36 (April 2017)

N

Nurkin, Tate and Julia Siegel, 'How Modern Militaries Are Leveraging AI', Atlantic Council (14 August 2023), https://www.atlanticcouncil.org/in-depth-research-reports/report/how-modern-militaries-are-leveraging-ai

O

Oberhaus, Daniel, 'Watch "Slaughterbot": A Warning about the Future of Killer Robots', *Motherboard* (2017) https://motherboard.vice.com/en_us/article/9kqmy5/slaughterbots-autonomous-weapons-future-of-life

Ohlin, Jens David, 'Is Jus In Bello in Crisis?', *Cornell Law Faculty Publications*, Paper 912 (March 2013), https://scholarship.law.cornell.edu/cgi/viewcontent.cgi?article=2475&context=facpub

P

Parasuraman, Raja and others, 'Performance Consequences of Automation-Induced "Complacency"', *International Journal of Aviation Psychology*, 3, 1 (1993)

Perrow, Charles, *Normal Accidents; Living with High-Risk Technologies* (Princeton University Press, 1999)

Pinker, Steven, *The Language Instinct* (William Morrow, 1994), http://www.unc.edu/~moeng/teaching/Pinker%20-%20Language%20Instinct.pdf

R

Reason, James, *Human Error* (Cambridge University Press, 1990)

Ricks, T, 'The Widening Gap between Military and Society', *Atlantic Magazine* (July 1997), https://www.theatlantic.com/magazine/archive/1997/07/the-widening-gap-between-military-and-society/306158/

Ridd, Thomas, *Rise of the Machines: A Cybernetics History* (WW Norton, 2016)

Roff, Heather, 'The Self-Fulfilling Prophesy of High-Tech War', Duck of Minerva (2015), http://duckofminerva.com/2015/12/the-self-fulfilling-prophecy-of-high-tech-war.html

Roff, Heather, 'Killer Robots on the Battlefield', *Slate* (7 April 2016), http://www.slate.com/articles/technology/future_tense/2016/04/the_danger_of_using_an_attrition_strategy_with_autonomous_weapons.html

Russell, Stuart, 'AI Weapons: Russia's War in Ukraine Shows why the World Must Enact a Ban', *Springer Nature*, Vol 614 (23 February 2023)

Russell, Stuart, 'Take a Stand on AI Weapons', *Nature*, 521, 7553 (27 May 2015), https://www.nature.com/news/robotics-ethics-of-artificial-intelligence-1.17611#russell

Russell, Stuart and Ira Moskowitz, 'Human Information Interaction, Artificial Intelligence and Errors', Association for the Advancement of AI (2016), https://cdn.aaai.org/ocs/12767/12767-56109-1-PB.pdf

Russell, Stuart and Peter Norvig, *Artificial Intelligence: A Modern Approach* (Pearson Education, 2014)

Ryseff, James, 'Mastering Human-Machine Warfighting Teams', War on the Rocks (8 November 2024), https://warontherocks.com/2024/11/mastering-human-machine-warfighting-teams

S

Sabin, Philip, 'The Strategic Impact of Unmanned Aerial Vehicles', cit. Owen Barnes, *Air Power: UAVs: The Wider Context* (Royal Air Force Centre for Air Power Studies, 2009)

Sartor, Giovanni and Andrea Omicini, 'The Autonomy of Technological Systems and Responsibilities for Their Use', in Nehal Bhuta and others (eds.), *Autonomous Weapons Systems: Law, Ethics, Policy* (Cambridge University Press, 2016)

Scharre, Paul, 'Robotics on the Battlefield, Part II: The Coming Swarm', Centre for a New American Security (2014), https://s3.amazonaws.com/files.cnas.org/documents/cnas_TheComingSwarm_Scharre.pdf

Scharre, Paul, 'Autonomous Weapons and Operational Risk', Centre for a New American Security (2016), https://www.files.ethz.ch/isn/196288/CNAS_Autonomous-weapons-operational-risk.pdf

Schifrin, Nick and others, 'How Drone Warfare Has Transformed the Battle Between Ukraine and Russia', PBS News Hour (13 December 2024), https://www.pbs.org/newshour/show/how-drone-warfare-has-transformed-the-battle-between-ukraine-and-russia

Schmitt, Michael and Jeffrey Thurnher,' Out of the Loop: Autonomous Weapon Systems and the Law of Armed Combat', *Harvard National Security Journal*, 4 (2012), 279-281, http://harvardnsj.org/wp-content/uploads/2013/01/Vol-4-Schmitt-Thurnher.pdf

Schreiner, Max, 'AI in War: How Artificial Intelligence Is Changing the Battlefield', The Decoder (21 January 2023), https://the-decoder.com/ai-in-war-how-artificial-intelligence-is-changing-the-battlefield/

Seba, Tony, 'Clean Disruption: Why Conventional Energy and Transportation Will Be Obsolete by 2030', Presentation to Swedbank (17 March 2016), http://www.swedbank.no/idc/groups/public/@i/@sc/@all/@lci/documents/presentation/cid_1987411.pdf

Sharkey, Noel, 'Saying No to Lethal Autonomous Targeting', *Journal of Military Ethics*, 9, 4 (2010)

Sharkey, Noel, 'The "Evitability" of Autonomous Robotic Warfare', *International Review of the Red Cross*, 94, 886 (Summer 2012)

Sharkey, Noel, 'Automating Warfare: Lessons Learned from the Drone', *Journal of Law, Information and Science* (2012), www.austlii.edu.au/au/journals/JlLawInfoSci/2012/8.html

Sharkey, Noel, *Staying in the Loop: Human Supervisory Control of Weapons'*, cit. Nehal Bhuta and others, *Autonomous Weapons Systems: Law, Ethics, Policy* (Cambridge University Press, 2016)

Shaughnessy, Ian, 'The Ethics of Robots in War', *NCO Journal* (21 February 2024), https://www.armyupress.army.mil/Journals/NCO-Journal/Archives/2024/February/The-Ethics-of-Robots-in-War

Singer, PW and Allan Friedman, *Cybersecurity and Cyberwar: What Everyone Needs to Know* (Oxford University Press, 2013)

Singer, Peter and August Cole, 'Humans Can't Escape Killer Robots but Humans Can Be Held Responsible for Them', *Vice News* (15 April 2016), https://news.vice.com/article/killer-robots-autonomous-weapons-systems-and-accountability

Singer, Paul, *Wired for War: The Robotics Revolution and Conflict in the Twenty First Century* (Penguin Publishing, 2011)

Slim, H, *Killing Civilians: Methods, Madness and Morality in War* (Columbia University, 2008)

Smallwood, David, 'Augustine's Law Revisited', *Sound and Vibration* (March 2012), http://www.sandv.com/downloads/1203smal.pdf

Sparrow, Robert, 'Robots and Respect: Assessing the Case Against Autonomous Weapon Systems', *Ethics and International Affairs*, 30, 1 (2016)

Strachan, Hew, *The Direction of War – Contemporary Strategy in Historical Perspective* (Cambridge University Press, 2013)

Strachan, Hew and Sibylle Scheipers (eds.), *The Changing Character of War* (Oxford University Press, 2011)

Stop Killer Robots Campaign (2024), https://www.stopkillerrobots.org/stop-killer-robots/facts-about-autonomous-weapons/

Suarez, Daniel, 'the Kill Decision Shouldn't Belong to a Robot', Ted.com (2013), https://www.ted.com/talks/daniel_suarez_the_kill_decision_shouldn_t_belong_to_a_robot

Suchman, Lucy and Jutta Weber, 'Human-Machine Autonomies', cit. Nehal Bhuta and others, *Autonomous Weapon Systems: Law, Ethics, Policy* (Cambridge University Press, 2016)

Swank, R, *Combat Neuroses: Development of Combat Exhaustion*, Vol. 55 (Archives of Neurology and Psychology, 1946)

Syed, Matthew, *Rebel Idea* (John Murray Publishing, 2021)

T

Tamburrini, Guglielmo, 'On Banning Autonomous Weapon Systems: From Deontological to Wide Consequentialist Reasons', cit. Nehal Bhuta and others, *Autonomous Weapons Systems: Law, Ethics, Policy* (Cambridge University Press, 2016)

Tonkens, Ryan, 'The Case against Robotic Warfare', *Journal of Military Ethics*, Vol 11, No 2 (August 2012)

U

UK Ministry of Defence, JSP 383, 'The Joint Service Manual of the Law of Armed Conflict', (28 August 2013, https://www.gov.uk/government/collections/jsp-383

UK Ministry of Defence, 'Strategic Trends Programme: Future Operating Environment 2035' (2015), https://assets.publishing.service.gov.uk/media/6286575de90e071f69f22600/FOE.pdf

United Nations, Office for Disarmament Affairs, 'Convention on Certain Conventional Weapons – Group of Governmental Experts on Lethal Autonomous Weapons' (2023), https://meetings.unoda.org/ccw-/convention-on-certain-conventional-weapons-group-of-governmental-experts-on-lethal-autonomous-weapons-systems-2023

United Nations Institute for Disarmament Research, 'Safety, Unintentional Risk and Accidents in the Weaponization of Increasingly Autonomous Technologies', 5 (2016), http://www.unidir.org/files/publications/pdfs/safety-unintentional-risk-and-accidents-en-668.pdf

US Air Force, 'Autonomous Horizons; System Autonomy in the Air Force – A Path to the Future – Human-Autonomy Teaming', Office of the Chief Scientist, AF/ST TR 15-01 (June 2015), https://www.af.mil/Portals/1/documents/SECAF/AutonomousHorizons.pdf

US Army, 'Light Cavalry Gunnery: Target Acquisition' (Field Manual Publications, 1999), http://www.globalsecurity.org/military/library/policy/army/fm/17-12-8/ch3.htm

US Army, 'The Human Dimension White Paper: A Framework for Optimizing Human Performance', Combined Arms Center (2014), http://usacac.army.mil/sites/default/files/documents/cact/HumanDimensionWhitePaper.pdf

US Department of Defense, 'Defense Study Board' Summer Study on Autonomy' (June 2016), (, https://www.hsdl.org/?abstract&did=794641
US Department of Defense, 'The Role of Autonomy in DoD Systems', Task Force Report (2012), https://fas.org/irp/agency/dod/dsb/autonomy.pdf
US Surgeon General's Office, 'Mental Health Advisory Team (MHAT) IV Operation Iraqi Freedom 05-07', Final Report (November 2006), http://www.combatreform.org/MHAT_IV_Report_17NOV06.pdf

V

van Creveld, Martin, *Technology and War: From 2000 BC to the Present Day* (Simon & Schuster, 2010)
Vanderelst, Dieter and Alan Winfield, 'An Architecture for Ethical Robots Inspired by the Simulation Theory of Cognition', *Cognitive Systems Research*, 48 (May 2018)
Venckunas, Valius, 'Loyal Wingmen: The Cyberpunk Future of Aerial Warfare', AeroTime Hub (30 March 2023), https://www.aerotime.aero/articles/25825-loyal-wingmen-the-cyberpunk-future-of-aerial-warfare
Vergouw, Bas and others, 'Drone Technology: Types, Payloads, Applications, Frequency Spectrum Issues and Future Developments', cit. B. Custers and others (eds.), *The Future of Drone Use*, Information Technology and Law Series, vol. 27 (TMC Asser Press, 2016), https://doi.org/10.1007/978-94-6265-132-6_2

W

Walker, Paddy and Peter Roberts, *War's Changed Landscape? A Primer on the Forms and Norms of Conflict* (Howgate Publishing, 2023)
Walker, Paddy, 'Killer Robots? The Role of Autonomous Weapons on the Modern battlefield', MA thesis, Buckingham University (2013), https://www.tandfonline.com/doi/epdf/10.1080/03071847.2021.1915702?needAccess=true
Wallach, W, 'Towards a Ban on Lethal Autonomous Weapons: Surmounting the Obstacles', *Communications of the ACM*, 60, 5 (2017)
Warwick, Kevin, *Artificial Intelligence: The Basics* (Routledge, 2012)
Wead, Reverend Sean, 'Ethics, Combat and a Soldier's Decision to Kill', *Military Review*, March-April 2015, https://www.armyupress.army.mil/Portals/7/military-review/Archives/English/MilitaryReview_20150430_art013.pdf
Weiss, L, 'Autonomous Weapons in the Fog of War', *IEEE Spectrum* (2012), http://spectrum.ieee.org/robotics/military-robots/autonomous-robots-in-the-fog-of-war
Werner, Pieter, 'Swarm Robotics Market Set for Explosive Growth by 2028, *RockingRobots* (9 October 2023), https://www.rockingrobots.com/swarm-robotics-market-set-for-explosive-growth-by-2028/
Wheeler, Winslow, 'How Much Does an F-35 Actually Cost?', www.warisboring.com (27 July 2014), https://warisboring.com/how-much-does-an-f-35-actually-cost/
White, Olivia and others, 'War in Ukraine: 12 Disruptions Changing the World', McKinsey and Partners (9 May 2022), https://www.mckinsey.com/

business-functions/strategy-and-corporate-finance/our-insights/war-in-ukraine-twelve-disruptions-changing-the-world

Wilke, Christiane, 'Civilians, Combatants and the History of International Law', Critical Will: Law and the Political (28 July 2014), http://criticallegalthinking.com/2014/07/28/civilians-combatants-histories-international-law/

Wolpert, David, 'The Lack of Distinction Between Learning Algorithms', *Neural Computations*, 8, 7 (1996), https://www.mitpressjournals.org/doi/abs/10.1162/neco.1996.8.7.1341

Work, Robert and Shawn Brimley, '20YY; Preparing for War in the Robotic Age', Center for a New American Security (January 2014), https://www.cnas.org/publications/reports/20yy-preparing-for-war-in-the-robotic-age

Worcester, Maxim, 'Autonomous Warfare: A Revolution in Military Affairs', *ISPSW Strategy Series: Focus on Defence and International Security*, Issue 340 (April 2015), https://www.files.ethz.ch/isn/190160/340_Worcester.pdf

Y

Yampolskiy, Roman and Joshua Fox, 'Artificial General Intelligence and the Human- Mental Model', cit. Amnon Eden and others, eds. *Singularity Hypotheses: A Scientific and Philosophical Assessment* (Springer, 2012), https://intelligence.org/files/AGI-HMM.pdf

Yashinski, Melisa, 'Teaching a Single-Arm Robot to Fold Towels', Science Robotics (15 November 2023), https://www.science.org/doi/10.1126/scirobotics.adm8151

Yudkowsky, Eliezer, '2015: What Do You Think About Machines That Think?', *The Edge* (2015), https://www.edge.org/response-detail/26198

Z

Zaffar, Hanan, 'UAE Unveils Hunter 2-S Swarming Drone in Abu Dhabi', *Defense Post* (1 March 2022), https://www.thedefensepost.com/2022/03/01/uae-hunter-swarm-drone/

INDEX

A

abstraction considerations:
accountability issues, 146, 162; and force multiplication, 110-111; and LOAC, 137, 141, 151; and technical debt, 217; Article 57, 149-150; AWS internal representation, 233-234, 239, 250; classification challenges, 148-149; coding errors, 245-246; cognition considerations, 227; Course of Action procedures, 109-110; in data management, 204-206, 251; learning considerations, 226; non-linearity in training, 206, 301; overview and challenges, 9, 99-100, 101, 105, 137; planning routines and abstraction, 105-106; proportionality and distinction, 143-150; sensor stimuli ambiguity, 268

accountability considerations, 146, 162; and meaningful human control, 280-289

action methodologies, 238-239; action selection conundrum, 257-259; machines' desired states, 257

active protection systems, 115-116

agency/principal-agent considerations, 228-229

algorithmic circularity, 183

ambiguity: accountability considerations, 146; and tactical advantage, 144; art versus science in isolating drivers, 106-107; Course of Action procedures, 109-110; management of instructions, 240; performance ramifications, 111; proportionality and distinction, 143-150; ramification overview, 9; types of ambiguity, 240

anchoring routines: introduction, 14, 252, 304; goal setting, 252-255

approximator considerations, 186

architectural considerations, 190-193

art versus science as an AWS driver, 106-107

Article 36: introductory framework, 3; and LOAC, 152-156, 280-281

artificial general intelligence: introduction, 23; timelines, 23-24

artificial intelligence: algorithmic bases, 184; coding errors and AWS deployment, 245-246; data collection, 6, 107-109; definition of terms, 4-5, 6, 16-17, 20; discontinuities, 6-7; incrementalism and AWS deployment, 13, 47-48,

194, 227, 293, information management, 230; introduction, 6, 186-189; proportionality and distinction, 143-146
Atkins' ethical governor argument, 75-77, 172, 241-242, 255
attention methodologies, 229-232; information drift considerations, 229-230; non-attended information, 229-230
attrition considerations, 164
Augustine's Law, 74-75
automaticity, 7, 19, 301
automatic target recognition, 105
autonomous weapon systems (AWS): 'degree of force' considerations, 80, 81; 'digital trail' considerations, 80-81; action selection, 100, 226; architectural considerations, 189-193; ethical advantages and AWS deployment, 78-79; heterogeneity considerations, 213-214, 236, 265; introduction, 47-48, 179-180; notion of artificial responsibility advisor, 79
autonomy definitions: 19-20; Abstract, ix-x; assumptions, 5; drivers, 4, 28-29, 95-100, 107-109; ethical considerations and AWS deployment, 29, 107-109; lack of abstraction, 19-20

B
battlefield unpredictability, 120, 141-142: behaviour volatility and AWS deployment, 225; confidence levels, 177; proportionality and distinction, 143-145

behaviour considerations: introduction, 259-261, and MHC, 299-300
biases: bias types, 120; coding biases, 250-251; introduction, 49-50; multiple fighting assets in collaboration, 119-120
budget overruns, *see* economic considerations, 271

C
calibration considerations, 271
CASE (Change Anything, Change Everything), 213, 218, 259
centaur model, 16
circularity in AWS routines, 121
civilian damage considerations, 80, 93, 150; accountability considerations and AWS deployment, 146, 150-152; Article 57, 149-150, and meaningful human control, 280-289
classification challenges, 148-149, 202-203
coding errors, 245-246
coding issues: and planning tools, 106-109; biases, 120; calibration considerations, 271; coding biases, 250-251; coding errors, 245-246; coding for ambiguity, 53; coding for AWS action, 239; coding methodologies, 236-245; introduction, 30-31; proportionality and distinction, 143-144
Cohort, *see* Delivery Cohort
context: as ballast, 39-40, 92-93; AWS' future-orientation, 37, 103-104; context in machine learning, 185-186; contextual definitions, 35-38

command considerations,
277-280; and doctrine, 277;
and leadership, 278-279;
irreducibility and AWS
deployment, 278; role of MHC in
command equation, 284
communications: connectivity
considerations, 121; spoofing,
121-122
conclusions around AWS
deployment, 290-307
configuration of AWS: calibration
considerations, 270-272;
confidence levels, 233;
introduction, 221; tuning
routines, 271
configuration conundrums and
AWS deployment; 221-222
control considerations: agency/
principal-agent considerations,
228-229; calibration
considerations, 271; coding
errors, 245-246; composite
agency, 229; control models,
122, 126-127; hybrid models,
258-259; introduction, 99;
meaningful human control,
280-289; methodologies, 211-214;
swarm characteristics, 124,
126-127; toggling considerations
(human/machine), 111-112
conflicts of interest, 238
continua (technical progress
lines): AWS type, 91, 97-100;
continuum of methods, 38-39;
degrees of autonomy, 111;
deployment model continuum,
104-105; ethical continuum,
29, 137; introduction, 16;
operational continua, 29-30,
85-86; targeting and control,
99-100; teaming models, 111-112

Convention for Conventional
Weapons (CCW): Article 36
and LOAC, 152-156, 189,
campaigning therein, 143;
constraints of the organ, 291;
introduction, 12; MHC debate,
286, 290
correction routines and AWS
deployment, 220
countermeasures, 168
crewed platform considerations,
70-71
crewing: and Ukraine, 63, 93;
coding biases, 250-251; Delivery
Cohort considerations,
193-197; constraints, 74-75,
77-78, 89-90; multiple fighting
assets in collaboration, 119;
ratio considerations, 111;
skills considerations, 89, 95;
workload prediction, 162-163;
maintenance considerations, 272

D

dangers of circularity, 37
data: and complexities under
LOAC, 141-142; and planning
tools, 106-109; calibration
considerations, 271; capturing
context, 237-238; classification
challenges, 148-149, 202-203;
coding biases, 250-251; coding
errors and AWS deployment,
245-246; conflicts of interest,
238; correction considerations,
220; data inputs and decision
space, 247; data noise, 205, 218,
227, 243; data quality, 242-243;
data scrubbing, 249; definition
of a fact, 236; dependencies,
219; establishing associations in
received data, 238; extraction,

transformation and loading, 249-250; feature extraction issues, 209; feedback loops and AWS deployment, 122, 218, 257-258; habituation and sensitisation in AWS, 231; incentive considerations, 252; issue of abstracts, 237; labelling, 144, 202-203, 210; long term priors and correlations, 217-218; memorisation versus memory-making, 231; new prevalence of data, 87-88, 112; overfitting considerations, 209; proportionality and distinction, 143-144; training set incompatibility, 205-206; undeclared consumer phenomenon, 218; weightings, 144-145, 210

data inputs and weapon's decision space, 247

decision aids, 120

defence planning: capacity planning, 163; course of action procedures, 109; introduction, 10-11; role in AWS deployment, 44-45; role of context, 44-45, decision aids, 120

defence and offence considerations, 101; and swarm characteristics, 127-128; MHC considerations, 282

degrees of autonomy, 111

Delivery Cohort: accountability considerations, 146, 150-152, 162, 174-175; action sequence considerations, 237; agency / principal-agent considerations, 228-229; and human dignity, 176; and LOAC, 139-143; Article 36 and LOAC, 152-156; attention methodologies, 229-232; behavioural considerations and AWS deployment, 162, 165-166; calibration considerations, 270-272; capacity planning, 163; coding biases, 250-251; complexities, 27, 39, 60, 99, 193-197; composition, 175, 193-194, 196; conclusions for Delivery Cohort, 296-308; configuration conundrums; 221-222, 225; correction considerations, 220; contextual ramifications, 296; credit assignment, 225; data inputs and decision space, 247; data quality considerations, 242-243; data training considerations, 209; duty to explain, 300; fault-free assumptions, 135, 221, 224, 242; goal setting, 252-253, 255; goals, 26-27, 193-197; incentive considerations, 252; information drift considerations, 229-230; intention and action considerations, 151, 210; introduction, 11, 193-197; lasting technical advantages, 160-161; leadership considerations, 278-280; long-term priors and correlations, 217-218; management of instructions, 240; meaningful human control, 280-289; norm development and constraints thereof, 42-43; pace of error correction and issues arising, 258; procurement inertia, 158; proportionality and distinction, 143-150; reasonableness test and AWS deployment, 144-145; responsibilities, 174-175,

194; spectrum considerations, 134-135; suspension of attacks, 150; swarm deployment, 128-129; team culture considerations, 195-197; threat inflation, 163; validation and testing, 272-275; value calibration considerations, 256; workload prediction and AWS deployment, 162-163
deterrence considerations, 125-126
doctrine in AWS deployment: introduction, 50
drivers in AWS deployment; deployment empirics for AWS, 89, 103-104; historical drivers, 56, 58, 290; introduction, 54-55; new data prevalence, 87-88; operational drivers and AWS deployment, 89, 103; procurement drivers, 55-56
drone developments: and Ukraine, 161, 167-168; current operational breadth, 67-68, 85-86, 103-104; development cycles, 2-3, 85-86; first person view developments, 62-63; geographic diaspora, 2, 66-67, 103-104; use ramifications, 170-171
dual use technology: complexities for definition and containment, 67, 160; hardware, 264; global procurement, 85-86, 92; introduction, 57-58, 64; smartphone developments and AWS deployment, 86-88; technology creep, 66-69, 92

E
economic considerations, 64: and Ukraine, 167-168, 169; attrition considerations, 164; budget overruns, 271; budgetary considerations, 70-73, 91, 93, 102; economic utility function, 255; economies of scale and AWS deployment, 113-114, 123-124; F-35 and AWS deployment, 72-73; lobbying considerations, 91-92; proliferation considerations, 166-172; removing human supervision, 70-72, 93; role of sunk costs, 130, 159; role of state programs, 92, 93
electronic warfare considerations, 64, 88-89
ethics: accountability considerations, 146, 174; and planning tools, 106-109; Article 36 and LOAC, 152-156; consequences of errors, 240; role of contextual drivers, 29; ethical drivers, 75-82, 167; ethical constraints to AWS deployment, 172-177; Ethical Governor argument, 75-77, 172, 241-242, 255; human shortcomings, 76; intention and action considerations in AWS deployment, 151, introduction, 29-30; complexity of multiple ethical frameworks, 173; suspension of attacks, 150
Extraction, transformation and loading of data in AWS, 249-250

F
F-35 and AWS deployment, 72-73
failure modes: Amazon example, 294; and machine learning models, 199-200, 210; attrition considerations, 164; calibration considerations, 271, capacity planning, 163; failure causes in AWS models, 130-135, 224;

coding biases, 250-251; coding errors, 245-246; data inputs and decision space, 247, 251, 259; error types, 220-221; fault-free assumptions by Cohort, 135, 221, 224, 242; legacy factors and AWS deployment, 163-164, 168; malfunction considerations and AWS deployment, 133; pace of error correction, 258; error ramifications, 130-131, 133; technical debt, 216-223; transient associations, 251; types of ambiguity, 240; validation and testing and AWS deployment, 272-275
fault-free assumption for AWS, 135, 221, 224, 242
feature extraction, 209
feedback loops, 122, 218, 257-258
firmware: introduction, 215
flexible autonomy, 118-122
force multiplication, 110, 119; swarm characteristics, 124
forgetting routines, 236, 260; catastrophic forgetting, 302
fragility and AWS deployment, 88-89, 190-193, 216-223, 272, 301
Future of Life Institute, 18

G
Geneva Conventions and LOAC, 138-143
global procurement considerations, 85-86, 92
goal setting in AWS: anchoring, 252-255; and threshold setting, 145-146, 254-255; in decision aids, 120; introduction, 117; satisficing approaches and AWS deployment, 254

Gospel and Lavender planning tools, 106-109
geopolitical considerations, 45, 123-124

H
habituation and sensitisation, 231
hacking, 133-134
Hague Conventions, 139-140
hardware for AWS: AWS architectural considerations, 190-193; degrees of freedom, 263-264, 265-266; effectors and actuators, 263-264; fragility and AWS deployment, 88-89, 190-193, 216-223, 272; immutable physical laws, 241; overview, 8-9, 262-265; power considerations, 263-264, 267; scaling issues and AWS deployment, 192; system complexity, 133; technical debt, 216-223; trade-off examples, 269-270
high regret outcome considerations, 247, 274
historical drivers in AWS deployment, 56, 58
historiography, 14
human dignity and AWS deployment, 176
human considerations: absent creativity and qualia, 208-209; accountability considerations, 146-148, 149, 162, 174; AWS architectural considerations, 190-193; behavioural considerations and AWS deployment, 162, 165-166, 244-245; capacity planning, 163; coding biases, 250-251; coding errors, 245-246; configuration conundrums;

221-222; ethical shortcomings, 76; goal setting consideration, 117, human aptitude, 9, 45-51, 102, 113-114; human dignity, 176; human norms, 41-42, 93, 102; human traits and role in AWS deployment, 48-49, 50-51; intention and action considerations, 151; leadership considerations, 278-280; meaningful human control, 280-289; primacy of human in chain, 291-292; whole brain emulation, 190-191
Human Rights Watch: 'Killer Robots' Research, 19, 148
hybrid deployment models, 258-259, 291

I
incentive considerations, 252
in-field challenges for AWS integration, 135
infrastructure profusion, 254
innovation: and Ukraine, 62-63; drivers towards innovation, 3, 62; innovation and autonomy, 3;
in-the-loop/out-of-the-loop: introduction, 11, 19; out-of-the-loop control problem, 121
intangibles in AWS deployment, 49-50, 61, 102; art versus science in isolating drivers, 106-107
integration issues: and procurement, 157-158; configuration conundrums; 221-222; frictions, 50, 69-70, 158; leadership considerations, 278-280
interim tools: planning tools, 105-110

international humanitarian law – international human rights law juxtaposition, 18, 138-139

L
labelling, 144, 202-203, 210
Lanchester's Square Law, 90
lasting technical advantages, 160-161
Laws of Armed Conflict (LOAC): 3000.09, 117-118; and Geneva Academy, 137-138; and planning tools, 106-109; Geneva Convention and LOAC, 138-143; Hague Conventions, 139-140; human Rights, 139; intention and action considerations, 151; introduction, 18, 30, 136-138; juxtaposition with international humanitarian law and international human rights law, 18, 138-139; Martens Clause, 152; meaningful human control, 280-289; proportionality and distinction, 143-150
learning and adaptation: introduction, 112'; learning architectures, 197-207
learning types, 223-224
legacy complications, 159, 163
lethality considerations, 64-65, 108, 171, 281; and meaningful human control, 280-289
logistics: in-field challenges, 135
loitering platforms, 117-118
long-term priors and correlations, 217-218

M

machine learning (ML): approximator considerations, 186; context and terms, 4; introduction, 184-186

machine learning neural training: calibration considerations, 271; coding biases, 250-251; coding errors, 245-246; conclusions, 300-302; counter-intuition, 198; gradient descent challenges, 203; introduction, 13, 198-199; learning architectures, 197-207; learning types, 223-224; local minima considerations, 218; long term priors, 217, non-linearity in training, 203-204; pace of error correction, 258; reasoning considerations, 197-198; smoothing routines, 198, 210, 234; training versus using modes, 199-200, 225; unlearning routines, 225, 302; value assignment in training, 203

machines' desired states, 257

manning: *see* crewing

Martens Clause, 152

meaningful human control (MHC): introduction, 16, 31; definitions, 32-33; role in AWS, 134; workload prediction, 162-163; control issues around MHC, 281-282

memory and processing considerations, 202

mismatching capability and tasking, 162, 168, 204, 303

missing pieces, 172, 207-210, 236

morals portfolio for AWS, 241; and MHC, 299

Moravec's Paradox, 181

Models of deployment: and machine learning models, 199-200; collaboration-based models, 91, 99; control models and AWS deployment, 211-214; flexible autonomy, 118-122; human teaming models, 110-116, 118; introduction, 95-100

maritime assets, 129

multiple fighting assets in collaboration, 119; partnering characteristics, 121

N

non-state actors, 103, 160, 170

norms of war: and planning tools, 106-109; and proportionality and distinction, 145-146; and swarming characteristics, 127-128; Article 57, 149-150; dangers of circularity, 37; drone developments, 1-2, 98, 113-114; norms in AWS deployment, 42, 100; procurement norms, 158-160

notion of nullification and AWS deployment, 160-162, 303

O

OODA (observe, orient, decide, act) loop considerations, 83, 100, 142

operational challenges to AWS deployment: and complexities under LOAC, 142; and Ukraine, 82, 84, 90-91, 93; attrition considerations, 164; control models, 211-214; flexible autonomy, 118-122; hardware considerations, 83-84; introduction, 10; lasting technical advantages and lack

thereof, 160-161; maintenance considerations, 272; multiple fighting assets in collaboration, 119; nullification, 160-162, OODA loop considerations, 83, 100, 142, operational drivers to AWS deployment, 82-94, 105-106, patching, 135, practical challenges and AWS deployment, 271, proportionality and distinction, 143-150; ramifications of failure, 130-131, 133, rear area transformation, 83, risk measurement in AWS, 130-131, 168; task fluidity in AWS, 121; urban considerations, 100; validation and testing, 272-275
overfitting considerations, 209
oversight considerations, 276-280: meaningful human control, 280-289

P
pace of technical development, 43, 48
pace of error correction, 258
patching, 135, 195, 220, 256
philosophical considerations in value selection, 257
planning routines, 105-106
plausible deniability, 175
polling frequency of AWS sensors, 249
power developments, 84, 267
power considerations, 263-264, 267
predictability considerations, 119
principal-agent considerations, 228
probabilistic determination in AWS: proportionality and distinction, 143-146; sensor data management, 235-236

procurement: and Geopolitics, 70, 85, 123; and Ukraine, 63, 93; Article 36 and LOAC, 152-156; cost curve considerations, 168; procurement drivers, 63, 69-75, 91, 157; procurement frictions, 62, 105-106, inappropriate incentives, 274-275; legacy complications, 159, 163, 168; nullification, 160-162
performance considerations, 105-106; procurement complexity, 133, 221-222; procurement denominators, 101-102; reducing complexity, 73-74, 121; risk measurement in AWS, 130-131, 212; role of heuristics, 62; role of state programs, 92, 106-107, 135; validation and testing, 272-275
proliferation considerations, 166-172
proportionality and distinction, 143-150
processing functions, 248-252
precursor systems, 113-115
Project Maven, 105-106
public debate issues, 286-287
psychological drivers, 63-64

R
ratio considerations, 111
reasoning considerations, 197-198
reasoning and cognition: definitional considerations, 227; introduction, 227-229
regulatory bodies: absence of oversight, 7
relevant commander, 1
remote warfare: Ukraine, 1, 21
representation (initial), 235-236, 239

required transformation, 276
Revolution in Military Affairs: introduction, 11, MHC, 32, smartphone developments and AWS deployment, 86-88; within continua of warfare, 38-39
risk: measurement in AWS, 130-131, 168; NASA precedents, 131; meaningful human control considerations in risk mitigation, 283-284
Rules of Engagement, 210

S
sensor profusion, 262-263, 265-270: management and handshake, 268; 'Object identification' conundrum, 268; sensor data management, 235-236; sensor stimuli ambiguity, 268
situational awareness: awareness and uncertainty, 51-53; introduction, 51; 'scope compliance' uncertainty, 243
skills considerations, 89, 95
specific technical developments for AWS, 84-85, 87-88, 91, 98, 126
state programs as driver, 92, 93
Stockholm International Peace Research Institute, 57, 96-100, 105, 113, 275
'Slaughterbots', 35-37, 288
smartphone developments and AWS deployment, 86-88
smoothing routines, 198, 210, 234
software for AWS: calibration considerations, 271; circularity in AWS routines, 121; firing sequence considerations, 242, 249; highly coupled nature, 121, 126, 130-131, 132; overview, 8-9; pace of error correction, 258; polling frequency of AWS sensors, 249; processing functions, 248-252; smartphone developments and AWS deployment, 86-88; validation and testing, 272-275
spectrum considerations, 134-135
supervision considerations, 291-293
swarming: and deterrence, 125-126; capabilities, 123-124, 127-128; constraints, 129-130; control issues, 126-127; domains other than aerial, 129; introduction, 123-124, 125
swarming model for AWS, 123-130
system complexity: anchoring and goal setting in AWS deployment, 253-255; and machine learning models, 199-202; AWS architectural considerations, 190-193; confidence thresholds, 234; configuration considerations (glue code); 221-222; control models and AWS deployment, 211-214, gradient descent challenges, 203, highly coupled nature, 121, 126, 130-131, 132; introduction, 26; pace of error correction, 258; refreshing and conditionalisation routines, 235; swarm constraints, 129-130; system predictability, 31; toggling considerations (human/machine), 111-112, 118, 122; Vincennes incident, 132-133

T

targeting: and planning tools, 106-109; complexities under LOAC, 141-142; imprecision considerations, 182

targeting and control, 99-100; and MHC, 283-284

task complexity: and planning tools, 106-109; and weapon actions, 96; causes in AWS models, 130-135; control models, 211-214; data inputs and decision space, 247, 251, 259; goal setting, 252-253; introduction, 96; mismatching capability and tasking, 162, 168, 204, 303; multi-jurisdiction, 274; multiple fighting assets in collaboration, 119; proportionality and distinction and AWS deployment, 143-144; 'scope compliance' uncertainty, 243; swarm constraints, 129-130; toggling considerations (human/machine), 111-112, 122; use of classifiers, 244; validation and testing, 272-275; Vincennes incident, 132-133; What You See Is All There Is (WYSIATI), 120

teaming: human machine teaming (HMT), 113; type introduction; 112-113

technological determinism, 3, 15, 293; global drivers, 59-60; incrementalism, 47-48, 194, 293; interim steps, introduction, 16; missing pieces, 172, 207-210, 236; non-linearity, 15, 131-132, 137, 301; persistence considerations, 58-59; portfolio effects, 60-61, 64, 98-99; smartphone developments and AWS deployment, 86-88; specific technical developments, 84-85, 87-88, 91, 98, 126

technical debt: calibration considerations, 270-272, 303-304; Course of Action procedures, 109-110; introduction; 32, 216-223; memory and processing considerations, 202; mismatching capability and tasking, 162, 168, 204, 303; smoothing routines, 198, 210, 234; validation and testing, 272-275

technical feasibility: introduction, 15; machine learning complexity, *see* failure modes, also 197-203

technology creep, 66-69, 92

temporal considerations, 31, 200-201, 238; delays, priorities, urgencies, 238, 251-252, 288-289

Terminator franchise, 46-47; whole weapon independence, 17

threat inflation, 164

threshold management, 146, 298

threshold setting considerations, 145-146, 254-255

timelines, 20-25; missing pieces, 207-210

timeline uncertainty, 24

toggling considerations (human/machine), 111-112, 118, 122

training versus using modes, 199-200, 225

training set incompatibility, 205-206

trust considerations, 300-301

U

uncrewed ground vehicles: introduction, 65; and Ukraine, 65
undeclared consumer phenomenon, 218
unlearning routines, 225, 302
urban considerations, 100, 147
use cases: importance, 14; geopolitical aspect, 21; munition types, 21-22; role of norms, 65-66, 68, 101-105; Ukraine, 68-69
utility function: data inputs and decision space, 247, 251, 259; high regret outcome considerations, 247, 274; introduction, 14, 247-248; optimality notion, 247; practical implementation in AWS, 212-213
use of force, 247

V

validation and testing, 272-275
value setting, 256-257; calibration considerations, 256
versos: absent creativity and qualia, 208-209; and machine learning, 183-186; Delivery Cohort complexities, 27, 39, 60, 99, 193-197; integration constraints, 171; introduction, 13-14, 15, 22, 40-41, 59-60, 79, missing pieces, 207-210, 236; swarming constraints, 129-130; Vincennes incident, 132-133

W

war's character, war's nature, 48, 297
weightings, 144-146, 210
Western constructs of war, 103, 160
wetware, 180: and complexity layering, 183
whole weapon independence, 17
whole brain emulation, 190-191
What You See Is All There Is (WYSIATI), 120
wingman platforms, 113-114
Wirth's law, 193, 262
workload prediction, 162-163

ABOUT THE AUTHOR

Formerly commissioned into the Fifth Royal Inniskilling Dragoon Guards, Dr Paddy Walker is Managing Director of the Leon Group. He is a Senior Research Fellow at the Humanities Research Institute at the University of Buckingham, an Associate Fellow at RUSI and an Associate at the Institute for the Public Understanding of War and Conflict at the Imperial War Museum. Previously London chair of NGO Human Rights Watch, Paddy is a Board Member of NGO Article 36 and co-authored *War's Changed Landscape: A Primer on Conflict's Forms and Norms*, also published by Howgate, with Professor Peter Roberts in 2023.

www.ingramcontent.com/pod-product-compliance
Lightning Source LLC
Chambersburg PA
CBHW041303110526
44590CB00028B/4236